The North American Waterfowl Management Plan

This book provides a compelling insider's account of the North American Waterfowl Management Plan (NAWMP), a groundbreaking initiative that has transformed avian conservation across the continent.

In the early 1980s, waterfowl conservation faced a dire crisis. By the mid-1980s, a bold and visionary idea took shape, culminating in the creation of the NAWMP. Now approaching its 40th anniversary, this extraordinary endeavor has conserved millions of habitat acres, mobilized billions of dollars in funding, and united thousands of partners from Latin America to the Arctic. Drawing on his experience as a founding member of the NAWMP science community, the author provides unparalleled insights into this remarkable conservation journey, honoring the visionaries who paved the way. Through a detailed exploration of NAWMP's origins, growth, achievements, and challenges, this book highlights the Plan's enduring legacy and its relevance to today's conservation efforts. It offers valuable lessons for those striving to replicate its success in protecting other migratory species. At a time of escalating ecological crises, NAWMP stands as a testament to what can be achieved when diverse stakeholders come together with determination and shared purpose. This book not only celebrates NAWMP's transformative impact but also inspires a new generation of large-scale, cooperative conservation initiatives.

This book is essential reading for conservation professionals, policymakers, natural resource students, and anyone passionate about protecting nature.

Michael G. Anderson is Emeritus Scientist at the Institute for Wetland and Waterfowl Research, Ducks Unlimited Canada, an Institute he helped found in 1991. He has been engaged in national and international waterfowl management working groups for 40 years, including the North American Waterfowl Management Plan's (NAWMP) Science Support Team (NSST) and the Prairie Habitat Joint Venture's science team, both of which he helped establish. He served on the NAWMP International Plan Committee and chaired its science sub-committee for many years. He helped conceive and organize the North American "Future of Waterfowl Management" forums in 2008 and 2017. He was appointed to the NAWMP Revision Steering Committee charged with revising the 25-year-old Plan in 2012, and most recently served on the 2024 Update Steering Committee. In 2007, he was honored with the International Canvasback Award for contributions to NAWMP, the highest award bestowed by the NAWMP community. In 2015, he was inducted into the North American Waterfowl Conservation Honour Roll for his lifetime contributions.

Routledge Studies in Conservation and the Environment

This series includes a wide range of inter-disciplinary approaches to conservation and the environment, integrating perspectives from both social and natural sciences. Topics include, but are not limited to, development, environmental policy and politics, ecosystem change, natural resources (including land, water, oceans and forests), security, wildlife, protected areas, tourism, human-wildlife conflict, agriculture, economics, law and climate change.

Rewilding and Ecological Justice
The Ethics and Politics of Wildlife Regeneration
Cristian Moyano-Fernández

Becoming Nature Positive
Transitioning to a Safe and Just Future
Marco Lambertini, Joseph W. Bull, Leroy Little Bear, Harvey Locke, Eva Zabey, Dorothy Maseke, and Carlos Manuel Rodríguez.

The Economics of the Wildlife Trade
Theory and Implications for Conservation
Frederick Chen and Michael 't Sas-Rolfes

Rewilding and Reintroductions in Britain
Human Interactions with Conservation
Virginia Thomas

Conservation Ethics
The Challenges of Preserving Nature
Pierfrancesco Biasetti

The North American Waterfowl Management Plan
Building a Continental Conservation Partnership
Michael G. Anderson

For more information about this series, please visit: www.routledge.com/Routledge-Studies-in-Conservation-and-the-Environment/book-series/RSICE

"Anderson's meticulous research has documented the development of a preeminent model of wildlife management that demonstrates how science and partnerships can have positive impact on a continental scale. This is a 'must-read' resource for anyone in conservation".

– **Paul Schmidt**, *Retired Co-chair of NAWMP, Assistant Director, USFWS, and Chief Conservation Officer, Ducks Unlimited, Inc.*

"*The North American Waterfowl Management Plan* is one of the greatest conservation success stories of our time. By documenting the evolution of the Plan, Anderson not only preserves its legacy, but also provides a potential roadmap to success for other fledging conservation initiatives".

– **James Ringelman**, *PhD, Ducks Unlimited (retired)*

"Drawing on years of research and management experiences, Anderson's historical treatise provides deep insights into the development and evolution of the North American Waterfowl Management Plan. Anderson's thorough yet accessible narrative unveils critical steps that guided one of the world's most successful wetland-wildlife conservation programs".

– **Robert Clark**, *PhD, Environment and Climate Change Canada*

"*The North American Waterfowl Management Plan* is a carefully crafted chronicle of arguably the most successful public-private, international conservation program ever undertaken. Its evolution is artfully documented from concept development to implementation and expansion and serves as a model for future conservation challenges".

– **Robert Streeter**, *PhD, Assistant Director USFWS (retired)*

"Only an elder statesman who lived it like Mike Anderson could tell the story of NAWMP from its origins as the largest agro-ecosystem restoration project ever attempted. There are reasons that waterfowl today fare better than other birds, and it's all here".

– **Tom Nudds**, *PhD, Emeritus Professor, University of Guelph*

"A seminal work on waterfowl and wetlands conservation, this book illuminates the enduring impact of the North American Waterfowl Management Plan. Comprehensive and compelling, it is an essential resource for conservationists, policymakers, and scholars dedicated to sustaining wetlands and wildlife".

– **David Gordon**, *Division of Bird Habitat Conservation, USFWS (retired)*

"*The North American Waterfowl Management Plan* was a bold and audacious undertaking, now spanning 40 years since its inception. Only one person with deep knowledge, insight, and involvement in every step of the process could have summarized the remarkable accomplishments of this groundbreaking effort. Mike Anderson has done a masterful job recording this journey, with an insider's perspective and calm, thoughtful reflection on both its successes and remaining challenges. This book is a *tour de force* that not only should be on the shelf of every waterfowl scientist and manager but also serves as a casebook study on how to truly make a difference for conservation at a continental scale. It should be required reading for every past, present, and aspiring conservation biologist, wildlife manager, and policy maker in the natural resources".

– **John M. Eadie**, *PhD, University Distinguished Professor Emeritus, University of California, Davis*

The North American Waterfowl Management Plan

Building a Continental Conservation Partnership

Michael G. Anderson

Routledge
Taylor & Francis Group
LONDON AND NEW YORK

earthscan
from Routledge

Designed cover image: Getty Images

First published 2026
by Routledge
4 Park Square, Milton Park, Abingdon, Oxon OX14 4RN

and by Routledge
605 Third Avenue, New York, NY 10158

Routledge is an imprint of the Taylor & Francis Group, an informa business

British Library Cataloguing-in-Publication Data
A catalogue record for this book is available from the British Library

ISBN: 978-1-041-22439-6 (hbk)
ISBN: 978-1-041-22438-9 (pbk)
ISBN: 978-1-003-73231-0 (ebk)

DOI: 10.4324/9781003732310

Typeset in Times New Roman
by codeMantra

Access the Support Material: www.routledge.com/9781041224389

To the visionary leaders who conceived the Plan and the thousands of dedicated managers, scientists, administrators and field staff who carried it through four decades of progress. May we continue to build better on their legacy.

Contents

Preface

In the early 1980s, waterfowl conservation was in crisis. The vital waterfowl breeding regions of the Central Prairies were locked in drought. Clouds of dust were blowing across bare fields. Wetlands were drying out, and populations of several duck species had plunged, some to historic lows. Widespread practices of intensive cultivation and summer fallowing led to spring dust storms reminiscent of the 1930s. And despite years of conservation work, it wasn't enough to sustain the birds.

New research in North Dakota demonstrated that mallards – the most important species on the continent for hunters – were not producing enough young to maintain their population. At the same time, federal and state officials were locked in arguments over the effects of hunting on duck populations, pitting Canada against the United States, and conservative-minded northern states against pro-harvest southern states.

It was a time of uncertainty, strife, and pessimism among waterfowl managers. Fortunately, a few leaders, such as Jim Patterson of the Canadian Wildlife Service, looked over this landscape and declared, "There has to be a better way!". He found like-minded leaders, including Harvey Nelson and Rollie Sparrowe of the US Fish and Wildlife Service, who agreed.

Since the 1916 Migratory Bird Treaty between the United States and Canada, and subsequent enabling legislation, the federal governments have had legal responsibility for managing waterfowl. In the late 1970s, separate teams in Canada and the United States began drafting and seeking consensus around national waterfowl plans in preparation for international negotiations that commenced in 1984. Mexico was consulted too but was not yet ready to fully engage; that would come later.

Signed in 1986 by the US Secretary of the Interior and Canada's Minister of the Environment, the landmark North American Waterfowl Management Plan (NAWMP or Plan) laid out a general strategy for cooperation between the two countries. It was bold, visionary, expansive, and unfunded!

By 2025, six Plan updates later, and with some 264 million acres conserved, waterfowl ranked among the healthiest bird populations in North America (Birds Canada 2024; NABCI 2025). Yet the work is far from finished. The challenge remains to secure gains, reverse recent setbacks in policy protection for wetlands, expand the conservation footprint, and engage more North Americans in this vital work.

But the path traveled by Plan partners over the last 40 years more closely resembled the erratic flight of a green-winged teal through flooded alders than a canvasback streaking straight from its night roost to a sago pondweed bed. This is a story of that journey, based mostly on the words of those who traveled it. I am fortunate to have been involved in NAWMP from the beginning in many different roles, but more importantly, I had the chance to know and work closely with nearly all the people you will meet in the pages that follow.

Marking 40 years of achievements in 2026, the Plan's innovative provisions include continental-level planning linked to local actions; regional partnerships (>7,000) of public agencies, nongovernment organizations, and private citizens; and an abiding commitment to learning and adapting as time passed. All these, coupled with a shared passion for birds and wetlands, have been keys to the success and durability of the Plan.

Beyond chronicling this unprecedented conservation journey and paying homage to those who laid the trails, I hope that the NAWMP experience will prove instructive for our successors and for others seeking to accomplish similar things with migratory wildlife (UNEP-WCMC 2024).

As Clay Routledge (2023) wrote, "membership in an enduring group connects us to others across time and space and unites us in shared ideas and goals". I hope this book will help accomplish those connections. I also write for the conservation students of today who need models of success to counter temptations to despair in the face of so many threats to the natural world. At 40 years and counting, NAWMP illustrates what can be accomplished when people of goodwill and determination commit to doing difficult things together.

Mike Anderson, November 2025

Box P.1 Dr. John Cornely – Historian (1946–2023)

John and I had hoped to tackle this project together, but illness intervened, and we lost him all too soon. John deserves credit for recognizing in the 1990s that an oral history collection of those involved in the early years of NAWMP would provide precious records for posterity.

John was born in Kansas, earned a degree in Biology, and then served four years as a communications officer in the Air Force. Afterward, having completed graduate degrees at Texas Tech University and Northern Arizona University, John began his USFWS career at Malheur National Wildlife Refuge in Oregon. Later he supervised the region's dusky Canada goose program and, in 1988, moved to Colorado to become the Regional Migratory Bird Coordinator, where he became increasingly involved in NAWMP.

John also became a leader of the USFWS Heritage Committee recording oral histories of dozens of retired employees and individuals instrumental in founding NAWMP. He continued his conservation work in retirement, serving in leadership roles in The Trumpeter Swan Society; as Chairperson of the FWS Oral History Committee (2014–2023); and as an emeritus member on the Board of Directors for the National Wildlife Refuge Association. This book is possible only because of John's foresight.

Note: Throughout this book, unless otherwise noted, quotations attributed to individuals are from the USFWS National Conservation Training Center (NCTC) oral history archives provided to me by John Cornely, who collected most of them, or directly from NCTC. These materials are in the public domain and readily available to scholars. Financial figures are in US dollars unless otherwise noted. Acronyms used in this book appear in Appendix A, and a glossary of people quoted in Appendix B. Scientific names of species are mentioned in Appendix C. Appendix D features 10 people, nominated by others, to represent the thousands across North America who helped make NAWMP successful. Other online appendices are found at (www.routledge.com/9781041224389) under support material.

References

Birds Canada and Environment and Climate Change Canada. 2024. *State of Canada's Birds*. (https://naturecounts.ca/nc/socb-epoc/main.jsp). Birds Canada. Port Rowan, ON.

North American Bird Conservation Initiative. 2025. *The State of the Birds, United States of America, 2025*. (www.StateoftheBirds.org). North American Bird Conservation Initiative. Washington, D.C.

Routledge, Clay. 2023. *Past Forward. How Nostalgia Can Help You Live a More Meaningful Life.* Sounds True. Boulder, CO. 118pp.

UNEP-WCMC. 2024. *State of the World's Migratory Species*. UNEP-WCMC. Cambridge. 88pp.

Acknowledgments

I cannot possibly acknowledge all the people in our three countries who have dedicated substantial parts of their careers to the Plan over the past 40 years. I hope that the voices of a few of those people in the following chapters and appendices reveal some of the passion, diversity of views, and commitments that made NAWMP uniquely successful.

I am indebted to the entire Plan community for helping me tell the NAWMP story. For general advice and counsel through the birth of this book, I thank especially John Cornely, Dave Case, Jim Ringelman, Bob Streeter, Seth Mott, Dave Paullin, Jim McCuaig, Tim Sopuck, Dean Smith, Stew Morrison, Dave Gordon, and Tony Roberts.

For contributing their voices to the project in the form of individual essays, I thank Bob Streeter, Dale Humburg, Dave Case, Dave Paullin, Deanna Dixon, Eduardo Carrera, Diane Eggeman, Jim Ringelman, John Eadie, Michael Brasher, Sett Mott, Tim Sopuck, Dean Smith, and Tom Moorman. Dean Smith and Dave Case also helped temper the Epilogue helping me focus on resilience instead of loss.

For 77 vital content reviews of the 24 chapters, I thank Andy Raedeke, Anne Bartuszevige, Tony Roberts, Barb Avers, Bob Ford, Bob Streeter, Dale Humburg, Dave Case, Dave Howerter, Dave Gordon, David Smith, Deanna Dixon, Fred Johnson, Gerald McKeating, Gray Anderson, Greg Soulliere, Jim McCuaig, Jane Fitzgerald, Jeff Nelson, Jim Ringelman, Joe Benedict, Mark Koneff, Karla Guyn, Kate Martin, Matt Connolly, Mike Carter, Mike Tome, Mike Brasher, Patrick Devers, Paul Schmidt, Rex Johnson, Bob Clark, Bob Davison, Rollie Sparrowe, Seth Mott, Mitch Weegman, Steve Wendt, Terry Rich, Tim Sopuck, Tom Moorman, Tom Nudds, Tom Rothe, and Trevor Swerdfager. I thank Wildlife Habitat Canada for financial support to offset publication costs and Ducks Unlimited Canada for logistical support.

Some people whose encouragement over the last 40 years was particularly important to my NAWMP journey include Harvey Nelson, Bruce Batt, Seth Mott, Gordon Kerr, Rod Fowler, Stew Morrison, Matt Connolly, Deanna Dixon, Bob McLean, Jeff Nelson, Jim Ringelman, Dale Humburg, Mike Tome, Fred Johnson, Dave Sharp, Paul Schmidt, John Eadie, Bob Clark, Ken Cox, Sandy Macauly, Dave Gordon, Gerry McKeating, Ken Williams, David A. Smith, Bob Streeter, Charles Baxter, Eduardo Carerra, Jim Patterson, Greg Siekanic, Llwellyn Armstrong, Bob Emery, Todd Arnold, Brian Joynt, Terry Neraasen, Dave Duncan, Jane Austin, Jay Rotella, Scott Stephens, David Pashley, Bob Ford, Michael Carter, Dave Paullin, Dave Howerter, Jim Devries, Lew Cowardin, Ron Reynolds, and Ray Greenwood. My employers, the North American Wildlife Foundation through the Delta Waterfowl and Wetlands Research Station supported my engagement in the Plan from 1985 to 1990. Ducks Unlimited Canada continued that support from 1990 until and far beyond my retirement in 2013.

For encouragement through the many months of work on the project, I thank especially my wife Diane and my three sons, Eric, Karl, and Hirst. I deeply appreciate their understanding and support in shouldering much of the routine work of sustaining life during my obsession. Hirst also kindly redrew Figure 18.2 for me.

Hannah Ferguson and Serena Watkins at Taylor & Francis, and Romaine Heldt at Codemantra were supportive and very helpful from first submission and throughout the publication process.

1 Why NAWMP Matters

The North American Waterfowl Management Plan (NAWMP) became a proven model for organizing continental conservation. While hunters, as the first stakeholders, brought unusual strength and traditions to the effort, the model should nonetheless be instructive for other conservation campaigns if other stakeholders can be aroused.

Partnerships have been essential for the Plan and NAWMP provides many examples of people working together to achieve something substantial. This too should be an actionable lesson for any other conservation cause. The state/federal/private NAWMP coalition fostered the essential attitude in 1986 of "We need to make this work. We can figure this out. We can do this!" It was the people, the people who cared about the resources and what the birds needed who lit the fire. Sure, there was a catalytic ecological crisis in the 1980s, but the natural world is awash with crises at least a serious as for waterfowl in that era.

Today, many people of my children's generation are deeply anxious. So many things are awry with the world my generation is handing down to them that I don't doubt it can feel overwhelming at times, unfixable, especially in the absence of concerted action to address global issues.

We need to celebrate NAWMP in part because here is a massive-scale, complex conservation program that is working. We aren't done by any means, not close, but the partners have accomplished a lot. The conservation world needs more winners and the NAWMP experience can help show the way.

Some of the lessons from NAWMP are clear and simple:

- People and organizations need to care about outcomes, not egos, before partnerships can work. Credit matters little; getting things done matters a lot.
- Mutual trust and the recognition that collaboration is the only way things can be achieved at large scales are essential.
- Individual integrity is the only way to build and maintain such trust. It may seem like an old-fashioned value, but it matters decisively.
- Shared passions were vital, as hunters brought to the Plan in large measure, but NAWMP found creativity and strength in its diversity of people, perspectives, geographies, and organizational cultures.
- Partnerships are vital but so are leaders, the people with an inner fire that won't let an idea die!

Many of our efforts were intellectually challenging, stimulating, and difficult, but for the most part eventually rewarding. We not only got things done but cemented respectful, long-term friendships across three countries. For many of us, personally, these were cherished outcomes.

DOI: 10.4324/9781003732310-1

Partnerships were built initially with the obvious stakeholders but recruiting others that did not realize at first that they were stakeholders added strength and diversity. The conservation campaign for broad ecological values is relevant to all.

Migratory birds with their flyway connections have special properties with the power to bring people together, but so do streams, grasslands, forests, deserts, and more. Seek a unifying concept. Watersheds are a natural one. At larger scale, maybe an archipelago, a mountain range, or an ecosystem can inspire a coalition? The oceans and atmosphere are obviously other things that we all have a stake in.

There are many possible rallying constructs to bring people together. Use them. Don't wait for governments to lead. Be the spark that starts the fire! Politics is often a seesaw of hopes and despair, but promising times come around, and you want to be ready for opportunities.

How can you get started? Seek people across affected sectors. Find kindred spirits. Listen to everyone. Follow the science with your plans. Create a "can-do" culture. Then, make your program work. Then make it better. Never stop improving. Never stop evolving. Celebrate your successes; share the credits. But don't be seduced by early success because things change. Remain vigilant and nimble. But start!

Believe that you and your fellow crusaders can make the world better, because you can!

NAMWP has made North America better for people and wildlife as the following story reveals.

2 The Birth of NAWMP

In 2026, we celebrate the 40th Anniversary of arguably the world's largest, most complex, and most successful terrestrial wildlife conservation initiative. With 26 established Joint Venture partnerships (>7,000 partners) covering much of North America, some $8.3 billion raised, 264 million acres of conserved waterfowl habitat, populations of most waterfowl species are more secure than virtually all other groups of avian species on the continent (Rosenberg 2019). Along the way, the Plan has been widely emulated and celebrated as an unprecedented success. The roots of the North American Waterfowl Management Plan, however, go back decades before its signing in 1986. Understanding the developments that made NAWMP possible is essential for assessing its value as a model for other large-scale conservation initiatives.

While the NAWMP partnership was indeed born in a crisis of prairie drought, low duck populations, new assessments of habitat loss, and resurgent angst about harvest regulations, the waterfowl management community was poised to respond effectively because of developments stretching back another half century.

Concerns about protecting migratory birds, including waterfowl, from indiscriminate exploitation led the governments of the United States and Canada to sign the landmark Migratory Bird Treaty in 1916, followed by enabling legislation in both countries over the next couple of years (Anderson et al. 2018). Mexico was added as a signatory party in 1936. Japan (1972) and the USSR (1976) followed.

One of the key provisions of enabling legislation for the 1916 Treaty was to vest responsibility for migratory bird protection with the federal governments of Canada and the United States, rationalized because migratory birds cannot be managed effectively by any single country, state, or province. The primacy of federal authority was challenged in the United States by a Missouri vs. Hollard lawsuit (U.S. Supreme Court 1920) but was upheld holding treaties above state law. States and provinces retained authorities to manage most other wildlife and soon developed agencies with responsibilities to set open and closed seasons for hunting, fishing, and trapping and, logically, sought ancillary roles in managing waterfowl too.

Crises of the 1930s

The 1930s brought crises beyond the economic debacle that everyone knows well. For most of us alive today, it was our parents or grandparents who knew the privation and stress associated with lost jobs and minimal social assistance. In the farming country of the US Midwest and Canadian Prairies, drought, coupled with farming practices not designed with soil and moisture conservation in mind, turned North America's breadbasket into an infamous Dust Bowl.

Waterfowl species breeding on the prairies suffered through year after year of drought. Without small ponds to nourish breeding pairs in spring, nor larger wetlands that could sustain

DOI: 10.4324/9781003732310-2

broods from hatch to fledging, populations plunged; though without the benefit of standardized bird surveys, we have only anecdotal evidence for the depth of the decline.

Governments, hamstrung with the economic crises of the decade, sounded alarm but had little capacity to respond. They could and did reduce bag limits and season lengths. Private citizens, keen duck hunters mainly, became alarmed and stepped up to help. The birth of both Ducks Unlimited and the Delta Duck Station in the 1930s proved to be developments of lasting importance to waterfowl conservation.

A group of wealthy waterfowlers formed the More Game Birds in America Foundation in 1930 and began exploring the "prairie duck problem" with reconnaissance efforts and reports issued in 1931 and 1933. In 1935, they undertook the first extensive survey of breeding waterfowl in Prairie Canada and the Northern United States (Leitch 1978). The International Wild Duck Survey, involving scores of observers, counted ducks and broods across the agricultural regions of Alberta, Saskatchewan, and Manitoba, as well as along sampling routes through Minnesota and the Dakotas. Others counted birds by air in Canada north to the Canadian Shield and as far as Great Slave Lake. Their observations led people understandably to focus on conserving large prairie wetlands as "brood salvage waters" in the face of the desiccation of smaller ponds. By 1937, the organization had consolidated its plans, organized a Board of Directors, and was reborn as "Ducks Unlimited" with the mission of raising funds in the United States and sending dollars to Canada to conserve what it believed were essential wetlands there. A sister corporation, Ducks Unlimited Canada, was hatched a year later for the purpose of delivering that conservation work. Eighty-nine years later with robust support from private donors and numerous agencies, their work continues.

Following suit in 1934–1936, the US Bureau of Biological Survey (a precursor of the USFWS) sent their own teams to Canada to learn more about waterfowl population levels and nesting conditions in the prairie and parkland regions. These were so revealing that they established the first four Flyway biologist positions to continue such work. That cadre, together again after WWII, began work in 1947 on the establishment of annual waterfowl surveys. Two important books, Waterfowl Tomorrow (Linduska et al. 1964) and Flyways (Hawkins et al. 1984) chronicle these early cooperative endeavors, mainly by the US and Canadian federal agencies, as they sought a better biological foundation for waterfowl conservation.

In 1937, Minneapolis businessman James Ford Bell, founder of the General Mills corporation, brought Aldo Leopold's graduate student, H. Albert Hochbaum, to his hunting property on Delta Marsh, Manitoba, and charged him with establishing a research agenda that would enhance the scientific basis for prairie waterfowl conservation. Hochbaum's classic study of canvasbacks and other breeding ducks (Hochbaum 1944) at Delta was the beginning of a continuing private sector investment in the scientific underpinnings of waterfowl and wetland management (Wait 2022).

Also born in the crisis of the 1930s was a program destined to provide more than $800 million in wetland conservation funding in the United States by 2024. A first tranche appeared via the 1929 Migratory Bird Conservation Act, which designated $7.9 million for waterfowl refuge acquisition. A more sustainable approach via the federal Migratory Bird Hunting and Conservation Stamp, soon to be known simply as the "Duck Stamp", emerged thereafter. Featuring Jay N. "Ding" Darling's iconic mallard pair the $1 stamp was first required of waterfowl hunters in the 1934–1935 season and raised $591,466 (Gabrielson 1958). It was followed the next year by Frank Benson's canvasback stamp and so began a 90-year and counting flow of hunter dollars into habitat conservation. Although intended initially to help fund land acquisition for the fledgling National Wildlife Refuge program, in the early years, most funding was used for operations and maintenance of existing properties. With most conservation work put on hold during WWII, it took until 1954 before meaningful funds could be applied to new land acquisition. By 2024, more than 5.7 million acres of waterfowl habitat had been protected with Duck Stamp revenue. It took another 50 years for a parallel program to appear in Canada with the birth of Wildlife Habitat Canada in 1985.

The US Federal Aid in Wildlife Restoration Act of 1937, usually referred to as the Pittman–Robertson Act for its sponsors, Nevada Senator Key Pittman and Virginia Congressman Absalom Willis Robertson, imposed a federal tax (11% in 2025) on firearms, ammunition, and archery equipment, with proceeds distributed to the states for conservation investments. This was another development in an action-packed decade of hunters stepping up to help support wildlife conservation. The Cooperative Research Units program was also established in 1935 to enhance graduate education in fisheries and wildlife sciences and to facilitate research on topics of mutual interest among federal and state agencies and universities. Today, there are 42 Cooperative Research Units in 40 states involving the US Geological Survey, a host university, one or more state agencies, and the Wildlife Management Institute. Several of these Units have a long history of pursuing applied research and educating students in waterfowl ecology.

Developments in the FWS during these years also led to the beginning of banding programs and initial attempts to census waterfowl populations with ground counts, aerial counts, photographic imagery, and more (Hawkins et al. 1984). Hawkins noted that the first bandings began in the late 1920s. By 1935, Frederick C. Lincoln, the man most responsible for growing the North American bird banding program, reported more than 2.5 million birds had been banded and released (Hawkins 1984). Lincoln soon recognized the potential of aircraft as a means of enumerating wild waterfowl, which led to numerous early attempts to design aerial waterfowl surveys, initially on concentrations of wintering birds (Lynch 1984).

1940s–1950s

Patterns of band returns soon led Lincoln and others to recognize that annual migrations of waterfowl were largely, though not solely, north-south journeys and that for management purposes it might make sense to organize human institutions in a similar pattern. Thus, the idea of management by administrative "flyways" was conceived and, based in part on the movement data and in part on the practicality of political boundaries, cooperative oversight of waterfowl management was laid out in four parallel regions – the Pacific, Central, Mississippi, and Atlantic Flyways.

The state and federal agencies soon responded to this logic. In 1946, several northeastern Atlantic states formed a joint black duck committee to share information on this prized species. By 1947, the North Mississippi Flyway Technicians had self-organized and, by the following year, so did the Central Flyway state and federal biologists and a Pacific Flyway Study Committee. The first year (1948) of differing waterfowl harvest regulations by Flyway followed with season lengths from west to east of 40, 35, 30, and 30 days, respectively.

A formal resolution by the International Association of Game, Fish and Conservation Commissioners in 1951 (presently AFWA) formally created the Flyway Councils and their Technical working groups, including a FWS representative for each Flyway. They also recommended the inclusion of a Flyway member on the FWS Service Regulations Committee and a provision of open public input on regulation matters via routine publication of government intentions in the federal register.

In Canada, federal/provincial dialog related to waterfowl and other game management was formalized in 1922 with the creation of regular federal/provincial wildlife directors' conferences.

Canadian officials slowly and unevenly became involved in Flyway Council business as well, though traditionally they abstained from issues related to US harvest policy.

Thus, by the mid-1950s, the Flyway Councils were in place, banding programs were expanding, breeding population surveys were beginning, research was underway, early habitat conservation efforts had begun, and, crucially, waterfowl managers and many hunters across the continent were engaged in cooperative efforts to manage the birds they cared about.

The 1950s proved to be generally good years for waterfowl populations and waterfowl hunters. Moisture was abundant in the mid-continent breeding areas, ducks were plentiful, and harvest regulations were liberal. Still, astute witnesses of rampant, often government-assisted drainage in the United States and Canada sounded the alarm about wetland loss. Ira Gabrielson (1958), the first Director of the FWS when it was formed in 1940, declared, "There is general agreement....that the preservation of marshlands is the key to the preservation of waterfowl populations" and that the continuing rapid disappearance of suitable wetland habitat for ducks and geese was "a major problem of national significance". Yet, governments did little to respond.

The early 1960s turned dry again leading to reduced hunting opportunities and growing contentiousness over the causes of the decline and appropriate responses. It was a decade featuring much tinkering with hunting regulations (e.g., special teal seasons, the point system wherein different species carried different weights in a bag, species' closures, the first split seasons designed to maximize hunter opportunities). If not a full-fledged crisis, the decade at least stirred considerable angst among waterfowl managers and the sense that researchers and managers had work to do to chart a less bumpy path forward. Toward that end, in 1961, an International Migratory Bird Committee was set up by the Secretaries of Interior and Agriculture from the United States and by the Canadian Ministers of Natural Resources, Northern Affairs, and Agriculture. Working at the conceptual/strategic level, they kicked around issues of conflicts between human land use and waterfowl production, habitat loss, coastal marsh conservation, and more. One concrete outcome was to support a seminal international meeting of waterfowl managers in 1964.

Patuxent, 1964

The blue-ribbon group was convened at the Patuxent Wildlife Research Center in 1964 by scientists from the Canadian Wildlife Service and the US Bureau of Sport Fisheries and Wildlife (Buell 1965). Six CWS scientists participated, 19 from the BSFW, and seven State biologists from the Flyways. Their charge was to develop a plan for future waterfowl research and management. Research received much of the attention over the four-day meeting, but they also tackled the vital question of population goals. A summary by Art Hawkins presented the group's conclusion that: "The primary goal [of waterfowl management] is to maintain habitat of sufficient acreage and quality to support a continental breeding population with a species composition similar to the 1950s". They rationalized that these waterfowl and habitat conditions were needed to meet anticipated hunting and other recreational demands in the foreseeable future. Specifically, they settled on the goal of average total duck numbers in the May Waterfowl Breeding Pair and Habitat Survey of 32 million birds, recognizing that this would fluctuate from year to year as it always had (25–40 million). They did not recommend species-specific targets.

Flyway documents following the Patuxent meeting (Grieb 1964; Douglas et al. 1970) reported that the target period finally adopted by the United States and Canada was the average total duck estimates for 1956–1962, which was characterized by Walt Crissey as likely more attainable than using only the wetter years of the 1950s. So, the target became 34.2 million total ducks, with a range of 25–42 million, and with projected fall flights of 77 million (1962) to 120 million (1957) total ducks from the surveyed area. They further suggested that achieving this would support an average of 3 million hunters (2.5 million in the United States and 500K in Canada and Mexico combined) with a seasonal bag of about six to eight birds each.

Those at Patuxent in 1964 struggled with estimating how much habitat the birds would require to sustain those numbers. Some thought that no more loss of wetlands and associated uplands on the mid-continent breeding grounds was the most critical thing, but admitted that it did not seem achievable, and research was needed to better prescribe a course of action. They expressed hope that the birth of the Northern Prairie Wildlife Research Center (NPWRC) in Jamestown, ND, and the promise of a similar institution being developed by CWS in Saskatoon, SK, might help better elucidate habitat requirements. They also recognized ongoing losses of wintering and migration habitat that needed attention, although it was generally thought then to be less urgent than stemming losses on breeding areas. With uncertainties acknowledged, the interim target became habitat like 1956 for about 40 million breeding ducks. Finally, they also recommended that, because about 80% of US harvested ducks were associated with the prairies of Canada and the United States, a great proportion of habitat conservation work should be focused there.

It's hard to find a road that hasn't been traveled before. This meeting happened 22 years before NAWMP, and many years before the separate US and Canadian National Waterfowl Plans were developed. It is striking today to realize that more than 60 years ago biologists had envisioned population targets not much different than current NAWMP objectives and that they envisioned a clear connection between those biological objectives and human harvest objectives and the primacy of mid-continent habitat conservation for achieving those population goals. Many of the same ideas and logic that were expressed in the 1986 NAWMP Plan are found in this faded 40-page photocopied report that I've kept in my "pre-NAWMP" folder for decades. For instance, in the 1964 meeting notes, Walt Crissey (BSFW) urged that habitat goals should be set for greater than average conditions to take advantage of abundant breeding populations whenever they might occur, an idea that was finally expressed 50 years later in the 2014 NAWMP Addendum and is still part of operating guidance to Joint Ventures today (NAWMP Committee 2024).

Following the consensus at Patuxent, Munro (1965) hinted at a hoped-for CWS habitat program of $50 million over ten years, to protect 2/3 of the approximately 6 million ponds in Prairie Canada. That basin-leasing program piloted in 1967 protected 176,500 acre years for $968,000/year but never grew to the scale hoped for (Tener and Loughrey 1970) and soon faded away (Burnett 2002, pp. 145–60). Meanwhile, the US permanent easement and wetland purchase programs were ongoing but at a rate and budget far less than needed to counter ongoing losses. Pittman–Robertson funds invested by states in waterfowl habitat, private duck club growth in the South, and Ducks Unlimited projects in Canada were all contributing to the habitat objectives too, but were much less, collectively, than the impact needed to attain the lofty goals discussed at Patuxent in 1964.

Perspectives of the 1970s

Directors of the Canadian and US wildlife services kicked off the decade with companion presentations at the North American Wildlife and Natural Resources Conference. John Tener and Alan Loughrey (1970) celebrated the creation of the first Canadian Migratory Game Bird Hunting permit in 1966, which finally provided data on the numbers of Canadian hunters and harvest estimates. They offered eight objectives for CWS waterfowl programs including assessing the aspirations of Canadians for both hunting and nonconsumptive recreation, conserving rare species, addressing the needs of indigenous peoples, ameliorating waterfowl damage to agricultural crops, reducing waste/loss of birds, and supporting basic research.

Gottschalk and Studholme (1970), anticipating growing ecological pressures from human population growth, offered six goals for US waterfowl management: (1) preserve and manage species, subspecies, and identified populations of migratory birds so that none will become endangered; (2) enhance public understanding, appreciation, and use of migratory birds within limits imposed to maintain desired population levels; (3) increase, where desirable and possible, the numbers of ducks, geese, and swans, to the carrying capacity of their habitats; (4) maintain a suitable, acceptable, and feasible, distribution of migratory birds, especially waterfowl, with due consideration for the birds' behavior, welfare and habitat, and public recreational opportunities; (5) control offending birds or bird populations to prevent harm to either man's health or society's welfare; and (6) preserve, develop, and manage habitat necessary for the well-being of all migratory birds and needs of people utilizing them.

At the same symposium, John Woodworth (ID) speaking for the states and H. Albert Hochbaum (Delta Waterfowl) expressed contrasting views about the responsibilities of governments to provide hunting opportunities vs. sustaining populations. Woodworth, however, endorsed the effort underway by the International Association to update a North American waterfowl policy.

The topic of improving international cooperation in waterfowl management took center stage at the North American conference six years later. A.G. Loughrey (Canada), Nathaniel Reed (the United States), and Jack Grieb (State perspective) reviewed progress following the establishment in 1961 of an International Migratory Bird Committee. Working at the strategy level, that group had probed issues of conflicts between human land use and waterfowl production, especially on the prairies, habitat loss, coastal marsh loss, and more. But little in the way of concrete actions followed.

By 1975, other committees had been struck to address more tactical or technical matters – one focused on developing common policies between the countries and resolving conflicts (chaired by Directors Greenwalt and Loughrey), and another committee of advisors to the Directors, charged with seeking solutions to migratory wildlife issues. Among other things, they were working on defining waterfowl population objectives. The general flavor of the dialog was of the United States wanting to maximize ducks to shoot, while Canada was seeking to optimize duck production and management of problems with people, principally grain farmers.

Reed (1976), speaking about migratory birds generally, urged that the three countries determine what information is lacking on a species-by-species basis, assign priorities for projects aimed at filling information gaps, identify critical habitats and monitor changes, and develop a systematic means of exchanging scientific information.

Jack Grieb (1976) from Colorado used mallard management in the Mississippi Flyway as a case study of the challenges. Canada and the United States set harvest regulations by different criteria. He noted that IAFWA established a migratory wildlife committee in 1966 to come up with a joint waterfowl management policy for both countries, but six years of work did not result in an agreement. He urged that the committee be reactivated to recommend a Canada/US harvest allocation agreement, and that a renewed International Migratory Bird Committee be established with a greater role for the Flyway Councils and including Mexico.

At the same conference, Lynn Greenwalt (Director, FWS) boldly offered a "management plan for waterfowl" (1976). He declared, "the primary purpose of waterfowl management is to maintain the waterfowl population base at the highest level possible within the constraints imposed by other overriding needs and demands of our society". While acknowledging that the quantity and quality of habitat were essential, most of his presentation focused on issues related to hunting. The Plan he offered would have:

1 Quantitative population objectives where sufficient information makes these useful.
2 Diligent effort to obtain such information where it is missing.
3 Identification of population management units, especially for ducks.
4 Refining the understanding and ability to measure the effects of hunting regulations and harvest on waterfowl populations.
5 What constitutes a proper distribution of waterfowl resources?

Remarkably, in all these 1976 visions of continental waterfowl management, there was hardly any mention of habitat conservation or enhancement, despite multiple acknowledgments of its importance. Important, but the conversations were dominantly about regulations, something the agencies were able and accustomed to managing.

Despite slow advancements in habitat conservation, duck populations gradually improved during the 1970s. With the exceptions of 1973 and 1977, the prairies were relatively wet during the decade. In Prairie Canada, on average, about a million more wetland basins held water in the 1970s (4.29 million) than during the 1960s (3.26 million) (USFWS 2024). The tinkering with regulations continued. A seminal paper by David Anderson and Ken Burnham (1976) posited a way in which the harvest of mallards might largely compensate for natural mortality, encouraging those who long-believed harvest mattered little to the size of duck populations, at least within the range of harvest experienced in recent decades. So, the long-standing debate about the effects of harvest on waterfowl was framed more precisely in arguments about the additive vs. compensatory nature of hunting mortality. But the debate raged on and the path to testing rigorously the effects of harvest on survival and breeding population size was fraught with challenges.

Although early analyses were mixed regarding the degree of compensation for mallard harvest, there was an emerging suspicion that regulations and harvest were not as influential as many historically thought, shifting concerns back to habitat conditions as having the strongest influence on duck populations. Nevertheless, debates about hunting regulations continued.

Stabilized Hunting Regulations, 1979–1985

With frequent changes in regulations in the past, the impact of harvest on populations was analytically confounded because the two essentially were "coupled". Years of high populations and wet conditions were often accompanied by liberal harvest regulations. Conversely, times of low populations and drier habitat conditions usually led to more restrictive harvest frameworks. If hunting regulations could be held constant for a time, encompassing both favorable and unfavorable years for duck breeding success, biologists reckoned they could perhaps evaluate simultaneously – without the influence of variation in regulations – nonhunting mortality, hunting mortality, and recruitment to breeding populations (Brace et al. 1987). Thus, the so-named "stabilized harvest regulations" concept emerged in 1979 as an agreed-upon experiment, initially by Canada and the Prairie Provinces but joined by the United States in 1980 and accompanied by additional resources for evaluation of the framework including augmented spring banding and assessments of breeding area habitat conditions.

Much of the coordination was accomplished among John Rogers, Jim Patterson, George Finney, Kent Brace, Dale Caswell, Dick Pospahala, Bob Blohm, and others. Pospahala and Brace were tagged as overseeing the associated field investigations. Brace et al. (1987) provide a list of the objectives agreed upon by the federal services and the Flyway Councils. With a focus on mallards, the augmented monitoring activities were undertaken to help interpret patterns observed in the study.

The results of the experiment were reported in a special session at the North American Wildlife and Natural Resources Conference in 1987. Frameworks were kept constant through the 1985 hunting season, with the last associated data captured in early 1986. The period was characterized by a drying trend through the prairie pothole region with slowly declining duck production rates and breeding populations.

Both waterfowl abundance and hunter effort were correlated with harvest per day and total harvest (Trost et al. 1987). Trost (1987) also detected a negative relationship between harvest rates and survival rates for mallards, especially for males, but some relationships were ambiguous, leading to a tentative conclusion that hunting mortality for males may be more additive than for females, which suffer higher natural mortality than males during the breeding season.

Regarding habitat conditions, Turner et al. (1987) found that over all years and three provinces, 59% of wetland basins and 79% of wetland margins were adversely affected by agricultural activities, and that these rates were somewhat higher in grassland ecoregions than parklands.

During the same years, Greenwood et al. (1987) uncovered surprisingly poor mallard nest survival in a geographically diverse study across AB, SK, and MB. With 1,572 mallard nests (4,138 total nests), they found average hatch rates of only 12% (6–17% by year) but noted that nest survival was higher in landscapes with a higher percentage of grass cover. These results, along with those of Turner et al. on widespread agricultural impacts on wetlands, provided a sense of urgency for developing an international habitat conservation initiative.

In summing up the stabilized regulations experiment, Rollie Sparrowe and Jim Patterson (1987) offered that while the effect of hunting on mallard survival remained uncertain, a cautious and conservative approach to regulations seemed warranted. Meanwhile, there was a clear signal that progress on habitat conservation was essential.

In hindsight, maybe the most valuable element of the study was that it largely set aside the incessant annual debates about hunting regulations long enough for the two countries to finally focus on the challenges of habitat conservation. As George Finney noted (NCTC interview), the critical thing was to find a way to separate work on a long-range continental plan from the all-consuming snarl of frequent debates around setting hunting regulations, both within and between countries. The Flyways were heavily focused on the allocation of harvest within the Flyways and between countries but also concerned that any continental plan might bring unwelcome "interference" in regulation setting between countries. Creation of the stabilized regulation experiment allowed most of the debate about annual adjustments to be set aside for five years.

Planning, Planning, Planning

Angst over harvest management dominated the time and energy of many waterfowl biologists in the 1970s, but others, especially field biologists on the US and Canadian Prairies, continued to sound the alarm about ongoing habitat loss. Parallel losses on major wintering areas sparked concerns there as well. Consequently, the decade of the 1970s became, in many ways, the decade of regional planning, led mainly by the Flyways in the US and Federal/Provincial authorities in Canada.

Harvey Nelson (NCTC Interview) recalled that

> [We went through] a great variety of planning exercises back then. The development of the Flyway Management Plans and a lot of species plans and a variety of habitat related planning exercises…evolved to a point in the mid '70's…that there was concern developing over how

> many plans do we have and how are they all related. Of course, the same activity was underway in Canada and when we had joint meetings with Canadian Wildlife Service people and the provincial folks, some of these same concerns started to develop. We have a whole array of plans and how do we relate them and bring them together at some point?

So, leaders in the CWS and USFWS started a Program Review Committee to help ensure communication between the countries. Director of Migratory Birds, John Rogers, and Harvey Nelson largely led this from the FWS with Hugh Boyd, Jim Patterson, and others as their CWS counterparts. Again, in Harvey's recollections,

> We began to recognize the need for some sort of umbrella plan to provide a more cohesive approach, especially as habitat for migratory birds continued to decline, and so we began to talk about something other than individual country plans. Thoughts began to crystallize about the need for a big umbrella plan, but it wasn't until probably in '82, '83, '84, during that period when people got serious about doing this, that we began to talk about something that, that would be like a North American Waterfowl Management Plan and that was the genesis of it.

Canadian Perspectives

While still voicing concerns about probable overharvest of some species, federal and provincial biologists recognized that habitat loss/degradation was the greatest problem facing waterfowl populations and that addressing it would require coordinated action of public and private organizations. Rising numbers of hunters in Canada through the late 1960s and 1970s also provided context for these discussions. Unquantified but thought to be increasing take by First Nations people added uncertainty. Conflicts with other users of the land, especially prairie grain farmers suffering crop losses to fall-staging waterfowl, were a dominant issue of the day.

The conservation challenges seemed daunting and likely would require both substantial new money and conservation-friendly public policy adjustments to succeed. To coordinate the work, planners envisioned a national waterfowl management plan, below which regional plans would focus on regional issues. Regions so described would be federal government regions, and linked below those, a third layer of Provincial and Territorial plans would direct "local" programs. This concept respected the jurisdictional complexity of Canada and its Provinces and Territories where land and much of water management is the responsibility of the provinces. Private sector support was imagined, but the planning at this stage involved government folks only.

Around this time, we saw the first active lobbying for an international plan by a non-government organization. DU Canada's Executive Vice-President Stew Morrison recalls (pers. comm. August, 2025) being apprised of this notion of an international plan by Manitoba's Gene Bossenmaier at a Federal/Provincial Wildlife Ministers conference in 1976. In March of 1977, DUC's longtime Chief Biologist Bill Leitch addressed the Annual Meeting of the DUC Board of Directors and made a compelling case that DU's early focus on conserving large permanent marshes, while valuable, needed to be augmented with different work in the Prairie Pothole Region – working with farmers and ranchers to secure small wetlands and associated nesting cover where the gains in duck production would be greater. Though internally controversial, this helped broaden the thinking of DU's Canadian leadership.

After gaining some intelligence from the DU leadership in the United States the following year, Morrison met with Environment Minister Romeo LeBlanc and Senator Gil Molgat, a known DU supporter, to talk about the need for a comprehensive international waterfowl plan and to seek help in resolving conflicts with prairie agriculture. Morrison followed up with a letter reiterating his recommendation. He specifically urged that the private sector join the provincial, state, and federal governments engaged in the planning process. He noted that money from the United States would need to be channeled to Canada to help. Morrison also shared intel that Cecil Andrus, US Secretary of Interior, was supportive of a high-level meeting to lay the groundwork for a North American plan and suggested that LeBlanc (Canadian Environment Minister) and Andrus have a first meeting to seize what seemed like an opportunity to overcome the waterfowl habitat problem. DU Canada pledged its support. LeBlanc was soon succeeded as Minister of State (Environment) by Hon. Len Marchand, who responded to Morrison on behalf of the government two months later. It was a vaguely positive response that this seemed like a good idea but lots of preplanning was needed so the bottom line was "not yet". But the wheels of government were slowly turning.

Canada did assign people to continue investigating this, led within the Canadian Wildlife Service by Hugh Boyd a renowned goose biologist from the United Kingdom, but progress was slow with internal attention focused first on the National Waterfowl Plan.

George Finney (CWS) recalls being sent by Hugh Boyd to Saskatoon to work with Jim Patterson to help crystallize a Canadian National Waterfowl Plan, a task that would evolve over several years. Finney recalls that, by the early 1980s, teams from both countries had concluded that fooling around with population plans was missing the mark, that they had to be bold enough to try to deal with habitat destruction, which, until that point, the wildlife community had considered just too big to tackle.

Tom Nudds (pers. com. September 2025), then a postdoctoral fellow working with Patterson in Saskatoon, recalls Jim believed that the status of waterfowl habitat and the agro-economy were causally connected. Patterson believed there could be no sustainable agro-economy without sustainable agro-ecosystems. Patterson noted then that market-distorting subsidies, without considering ecologically damaging externalities, were comprising half of the entire contribution from the prairies to the country's GDP! So, in his mind, much of the habitat "fix" lay not with Environment Canada but in the agriculture, finance, and revenue ministries. That sort of systems thinking was rare.

Morrison recalled vigorous discussions with Boyd and others at the 1979 North American Wildlife Conference in Toronto about Canada's readiness to do its part in a continental plan. Not long after, Patterson, then Director of the federal Prairie Migratory Bird Research Centre in Saskatoon, was transferred to Ottawa to head up work on the national plan. Patterson and his family stopped by Winnipeg on the way east where he met with Morrison and began a strong working relationship that would continue through the launch of NAWMP and beyond. In the mid-1980s, Morrison seconded his then Chief Biologist, A. J. "Sandy" Macaulay, to Ottawa for a couple of years to assist Patterson and others in advancing the Plan. Sandy later became the first Prairie Habitat Joint Venture (PHJV) Coordinator.

US Perspectives

More explicitly than in Canada, it was a US Federal Wildlife Service goal to "perpetuate and improve waterfowl populations for the benefit of people". Although planners recognized the need to "protect and manage the habitat needed to maintain and increase waterfowl numbers", there was comparatively more discussion about providing opportunity for people to "use and enjoy" waterfowl, "optimizing" population levels consistent with habitat availability, and harvest-related issues generally. The Flyways generally supported this approach.

The National Waterfowl Plans in the Early 1980s

Following guidance from the joint CWS/FWS review committee, which endorsed working toward a continental plan in 1977, both countries set about coalescing their thoughts in national waterfowl management plans in preparation for international negotiations. In Canada, the process involved several rounds of consultations between federal and provincial wildlife agencies and eventually a public consultation draft (6th draft) in June 1980. After a comment period and further federal/provincial negotiations, a final plan was issued in March 1983 (Canadian Wildlife Service 1983).

In the United States, the FWS worked with the four Flyway Councils to produce a parallel national plan, again through several iterations but culminating in 1982 (U.S. Fish and Wildlife Service et al. 1982).

Both plans acknowledged the primacy of habitat conservation for the long-term security of waterfowl populations and the need to work cooperatively among governments, and secondarily with the private sector, to achieve habitat goals. Both nations also recognized the migratory nature of waterfowl and the need to work between adjacent nations to manage the birds. Both plans also included discussions of treaties, laws, and jurisdictional issues among the various governments involved. Both national plans pledged to avoid driving any waterfowl populations to the status of threatened or endangered and by implications underscored the importance of monitoring systems to track population changes over time. Both plans also expressed concerns about funding habitat work and research, especially to acquire better data on habitat values to waterfowl and trends in habitat conditions.

There were, however, revealing differences in emphasis and priorities based on differing political systems, recent history of waterfowl management, and the interests of various stakeholders. This informed how each country moved toward continental negotiations. A comparison of the principles and objectives in the two National Plans is provided in Appendix E (www.routledge.com/9781041224389). For Canada, especially from the Prairie Provinces, concerns about negative interactions of waterfowl with grain farmers were strong. With contemporary cereal varieties and harvest practices like swathing, damage to crops by field-feeding ducks and geese caused anger, economic loss, and ill-will between farmers and wildlife managers. Planners recognized that this had to be dealt with to gain political support for conservation. Damage compensation programs were in place but were thought to be inadequate. Lure crops, devices to scare birds, and other tactics were applied in some chronic trouble areas near fall-staging marshes, but the sense was more needed. George Finney described the sentiment that

> ducks were flying rats, they were principally pests eating crops, and that if we were going to encourage pestilence on the prairies, that....we needed to build into the plan compensation mechanisms for the farmers in order to help them endure the plague that we were bringing upon them.

Some provinces insisted that the plan had to include crop damage compensation, so Canadian planners had to work through that.

Fortunately, today such problems have been reduced considerably due to partial shifts from small grains to oilseed crops, faster maturing cereal varieties, and straight combining without swathing and thus exposing grain to trampling and easy feeding by waterfowl. But the issue persists in some places.

Canadians also focused more on the nexus between waterfowl, agriculture, and public policy, especially in the Prairies, where it was thought that some government policies (e.g., the Western

Grain Transportation Act) might be adjusted in ways to become less detrimental to wetland and grassland retention. The US Plan focused more on pollution prevention and disease management, more often problems for denser concentrations of migrating and wintering waterfowl than for more dispersed breeding birds.

Finally, Canada also expressed greater concern for better understanding the effects of hunting and nonhunting mortality on waterfowl populations, given the country's historically cautious stance on waterfowl harvest.

While both countries embraced the need to develop waterfowl population targets, the US Plan offered some specifics including:

- Achieve a combined duck breeding population index based on ten key species in the surveyed areas that reaches or exceeds the most recent 20-year average (35.7 million birds through 1980). Same for mallards, canvasbacks, and redheads individually.
- Achieve an upward trend in black duck populations as measured by the winter survey.
- Achieve goose, brant, and swan population objectives identified in species or populations management plans established jointly by the USFWS and the Flyway Councils.

Without a rigorous quantitative method of linking population objectives to habitat objectives, planners still were unable to prescribe overall habitat objectives; however, they did include a prioritized list of top habitat acquisition targets for the United States (Appendix D in U.S. Fish and Wildlife Service et al. 1982), which presaged the habitat priority regions designated in the 1986 NAWMP document. Given the much greater federal control over lands in the United States than in Canada, it is not surprising that plans for public lands featured more prominently in the US National Plan than in Canada's plan.

The US plan, developed with the Flyway Councils, also not surprisingly focused more on hunting and the management of hunting opportunity. It contemplated several specific actions about hunting elements (crippling loss, lead poisoning, quality hunting standards) and paid much attention to the distribution of birds and hunting opportunity among states and Flyways. In short, the objective was to keep things status quo. Now, >40 years later, with changing climate, waterfowl distribution remains a contentious issue but under little immediate human control.

It is notable too that neither national plan included First Nations as a cooperating authority in managing waterfowl. The US draft briefly noted plans to legalize and define harvest parameters for native Alaskans, but that was all.

A Broader Social Context

While these waterfowl plans were unfolding, broader concerns about wetland loss had also emerged as part of the environmental movement in the 1960s–1970s (Dave Gordon, USFWS, pers. comm. April 2025). Social activism helped stimulate significant environmental legislation in the United States in those years, including the Clean Water Act, which still exists today in modified form. One response by the Service was the development of the Cowardin et al. (1979) wetlands classification system which was intended to be used for a national inventory of wetlands. By the mid-1980s, Congress enacted the Emergency Wetlands Resources Act that directed the FWS to map America's wetlands and provide decadal wetlands status and trends reports to Congress. The Service created the National Wetlands Inventory to carry out these responsibilities. Thus, US federal wetland policy was developing at the same time NAWMP was being crafted, helping support FWS objectives for waterfowl conservation.

1984–1985: More Crises; A Different Outcome

All this planning meant that when the next catalytic crisis arrived, more people were ready to push. Ideas + crisis = opportunity.

By the mid-1980s, the duck world seemed in crisis yet again, but this time elements of a continental plan began to fall in place. Duck populations and the prairie breeding grounds were again in grim shape. Spring pond numbers in both Prairie Canada and the Northern United States dropped to record lows in 1981. By the spring of 1985, estimates of total breeding ducks in the traditional survey area were the lowest since the drought of 1962 (USFWS 2025). So too for individual populations of canvasbacks, mallards, northern pintails, and American wigeon (1986).

An unusually warm spring in southern Canada with attendant strong winds lifted clouds of dust into the sky from bare fields causing some of the worst soil erosion since the 1930s, and, in 1984, much of the North Central United States was blowing too. Agriculture Secretary Earl Butz (1971–1976) had infamously called on farmers to "get big or get out" and to "plant fencerow to fencerow". By the early 1980s, there was precious little perennial grass left in place to hold the dry soil.

Duck nesting studies just wrapping up in North Dakota (Cowardin et al. 1985) and underway in Prairie Canada (Greenwood et al. 1987; Hochbaum et al. 1987) documented dismal breeding success – rates of reproduction insufficient to maintain populations at current levels.

Studies of intensive duck management, mainly for mallards, using artificial nest structures, islands, engineered peninsula cutoffs, predator exclusion fences, lethal predator control, and the small-unit management studies at NPWRC were all designed in response to the duck crisis of the early 1980s. The US Environmental Protection Agency even permitted the use of strychnine-laced eggs for ground squirrel control inside fenced nesting cover.

New estimates of wetland loss and agricultural impacts on the remaining basins in Prairie Canada (Turner et al. 1987) during 1981–1985 revealed extensive and increasing impacts on waterfowl habitat. Between 1964 and 1974, about 11% of the remaining basins in the Minnedosa, MB, region were partially drained (Adams and Gentle 1978). Rakowski and Chabot (1983) later reported a 56% decline in wetland area in the same general locality. Add to all this the completion of the stabilized regulation study in 1986, which opened the door to once again clamping down on harvest opportunity as breeding populations fell. The proposed restrictions were especially important motivators in the central and Southern United States where the firsthand sense of an immediate habitat crisis did not exist as it did on the prairies.

It seemed clear that the approach to managing waterfowl over the past half century was not sustainable. Moreover, endless disputes over arguably marginally important matters like annual bag limits seemed destined to keep managers from focusing on more critical long-term challenges like habitat loss. As Jim Patterson, then Director of Migratory Birds for the Canadian Wildlife Service famously opined, "There has to be a better way!"

So, building on the momentum begun with the Canadian and US national plans, leaders of the USFWS and CWS agreed that the time had come to fashion a truly continental plan. Mexico was still in the discussion at this stage, but the federal agency there was not ready to engage then to the same degree, so the work of crafting NAWMP became a two-country effort but always with the goal of adding Mexico when possible.

The Program Review Committee set up in 1978 between the FWS and CWS was still in place. John Rodgers who headed the migratory bird group and Harvey Nelson were the main US representatives while Jim Patterson and others represented the Canadian side. Havey recalled in an oral history deposition that

> In '83 or '84 we had a meeting with the Canadian Wildlife Service in Ottawa....I was up there with Lynn Greenwalt who was the (FWS) director at that point. and we talked about these things, how should the two Services begin to address these problems in a systematic manner. It was at one of those discussions at dinner I believe in some restaurant in Ottawa that we said, "What we need is a North American Waterfowl Management Plan" and that's sort of how it, it generated and progressed, and the name became attached.

A North American Waterfowl Management Plan

The next key step was scoping what an international plan might entail, legally. The Joint Program Review Committee and the Canadian and US International Affairs offices, through channels within USDI, Environment Canada, and the State Departments, concluded that the most expeditious path would not be a formal treaty but rather an "international agreement in principle" for the purposes of joint resource management. That would obviate State Department and Senate approval in the United States and parallel procedures in Canada. At this stage, agreement also coalesced around a 15-year time horizon with five-year interval updates. This also appears to be the first point at which partnership concepts were explored in some detail.

With this overarching approach set, next came the decision to appoint a formal "negotiating team" of US and Canadian officials to move the concept of a joint plan forward. Rollin Sparrowe had come from Missouri to lead the Cooperative Fish and Wildlife Research Unit program in Washington, D.C. He soon added responsibilities as Chief of the Division of Wildlife Research, but in the fall of 1984, he succeeded John Rogers as Chief of Migratory Bird Management. It was a challenging time with drought on the prairies, plummeting duck populations, and nearing the end of stabilized hunting regulations in the United States and Canada. Still new in the role, Rollie was nevertheless appointed to lead the US delegation. His counterpart was Jim Patterson, holder of the parallel position in the Canadian Wildlife Service. The US delegation included State leaders from the Flyway system like Richard (Dick) Yancy of Louisiana, and a little later, when his health allowed, Jack Grieb from Colorado who since the 1960s had been engaged in big-picture planning for waterfowl conservation. Don Minnich, a planner in the Department of the Interior, rounded out the US contingent. Minnich, a friend of Interior Secretary James Watt, was then Deputy Assistant Director for Refuges and Wildlife and kept an eye on process more than biology. In addition to Patterson, CWS scientists George Finney (ON) and Robert (Bob) Bailey (SK) were on the team along with Kevin Lloyd (NWT), Robert (Bob) Andrews (AB) and Doug Risebrough (ON) (Figure 2.1).

The negotiating team began work in 1984. Finney, who later helped set up the Eastern Habitat Joint Venture and the Sea Ducks Joint Venture, recalled many challenges in getting agreement among the array of government participants while earlier developing the two national plans. However, by August 1984, the Canadian and American negotiating teams found that they had much in common. "The processes which had been largely running on independent tracks had taken us to more or less the same place, but then there was the question of how to do it", recalled George. Sparrowe remembers the first meeting amidst renewed tensions about harvest regulations. State and Federal agencies were concerned that the negotiating team and the nascent Plan might "trespass" into matters of prescriptive harvest regulations for the two countries. So, early on, the team decided that they would largely leave those matters to existing processes in the two countries and focus instead on habitat matters. They agreed on the importance of

Figure 2.1 The negotiating team at Indigo Island, Louisiana 1986 (top L to R, D. Minnich, R. Yancy, R. Andrews, D. Risebrough, K. Lloyd, and J. Grieb; bottom J. Patterson, R. Sparrowe, G. Finney and R. Bailey).

setting specific quantitative goals for duck populations, which they correctly assumed would be critical to building support for the Plan.

Rollie reflected (pers. com. 4/26/2025) that spending multiple days together in retreat settings like Indigo Island, Louisiana, Delta Marsh, Manitoba, and elsewhere the team began to trust one another and focus on conservation matters largely free of politics. Policy matters got lots of attention but when they encountered biological uncertainties (e.g., estimating habitat needed to achieve population targets) they turned the work over to technical support members like Bob Bailey (CWS) and Tom Dwyer or Bob Blohm (both FWS) who returned to the team with proposed solutions at subsequent meetings. The negotiating team spent a lot of time debating how the Plan might be funded (Chapter 3) and how it might be delivered and eventually hit on the concept of voluntary Joint Venture partnerships (Chapter 4). There were, it seems, few contentious matters to resolve, rather mostly logistical challenges and finessing approval of the concept by the federal governments. The team members seem to have been well chosen as individuals committed to moving the concept of an international conservation plan forward. Through a rapid series of meetings from late 1984 to 1985, the team came to an agreement on most matters. Solutions for finding funds and how to deliver the habitat work were less apparent.

The idea of delivering regionally seemed already embedded in the US Plan national plan and in the presumed regional structure of the Canadian Plan, but the idea of willing partners coming together to do this without anyone being clearly "in charge" was novel. A few years ago, I quizzed Sparrowe about how the Joint Venture concept crystallized and I share his answer in Chapter 4, but, for now, I'll just note that the idea was perhaps the most prescient recommendation to emerge from the negotiating team.

Praising the negotiating team 18 years later, authors of the 2004 Update said, "They looked beyond what *could* be done, to focus on what *should* be done".

There was also a lot of work between meetings for the negotiating team. Gary Myers (TN) recalled many updates delivered to IAWFA and the Flyways by Dick Yancy so that in the end there was little hesitation in endorsing the Plan. Presumably, others did similar things with entities like the Flyway Councils and the Canadian Wildlife Directors.

My first personal connection with the negotiating team was at their meeting in the fall of 1984 at Delta, Manitoba. On their single morning off, which turned out to be a snowy, blustery October day, most of the team went on a dryland goose hunt with Bruce Batt. Dick Yancy knew about Delta's reputation as a canvasback staging marsh and wanted to bag one. I agreed to take him despite the weather. Long story short, Dick was a frozen but happy Cajun as we made our way back to the lodge with a brace of canvasbacks. In an improbable juxtaposition of events, 31 years later, I shared the privilege of being inducted into the North American Waterfowl Conservation Honour Roll along with Richard Yancy who by then had passed away.

The negotiating team released a draft plan for public consultation in November 1985, and I recall vividly taking pains to review and comment on the draft along with Bruce Batt and our Delta President, Peter Curry. Little did I know that would be the beginning of a personal journey with NAWMP that continues today, 41 years later.

The Plan was finalized in one last draft by the writing team led by Jack Grieb (CO) and Bob Bailey (CWS) in preparation for signing in May 1986 by Tom McMillan, Canada's Minister of the Environment, and Donald Hodel, US Secretary of the Interior. But that day arrived only after some last-minute horse trading.

Rollie Sparrowe vividly remembers getting a call from the DOI's International Affairs staff attorney, being summoned to a meeting, and faced with blunt questions about why the DOI would want to sign another agreement with Canada. Their concerns had nothing to do with waterfowl or biology but rather feeling like they had rushed into a bad deal on acid rain and were leery about other deals with Canada! Sparrowe recalls that Susan Lamson, an International Affairs attorney, was dispatched to Ottawa to investigate further. At the same time, Gary Myers, then President of IAWFA, remembers getting a call from Jack Berryman (Executive VP of the Association), reporting that the Department of Interior's Solicitor's Office had gutted the Plan and they needed to figure out what to do. Myers told him to withdraw state support for the plan which apparently caused quite a stir because they had already scheduled the Secretary and the Minister for the signing ceremony. Myers and Berryman correctly judged that folks at DOI would be embarrassed to sign a document that 50 states did not support! Myers was also told that the United States had "gotten the short end of the stick" on the acid rain deal and President Reagan was determined that we would not get the short end of the stick in this campaign. So, the lawyers had worked it over to where there wasn't much of a commitment for anything from the United States in the Plan.

Don Minnich, who had been a member of the negotiating team and who was an ex-employee of the Colorado Division of Wildlife where Myers also previously worked, was now a DOI employee. In Denver, Minnich had been a planner and Myers the big game supervisor, who reported to Jack Grieb who also served on the NAWMP negotiating team. Anyway, DOI sent Minnich down to Tennessee, according to Myers, "to work on me to get me to come around on this plan". Myers remembered, "We began to put pieces back into it, back and forth, and finally got to within I think it was eight days of signing, when I checked again with the crafters of the plan [about their views]". "Jack Grieb still felt like that we shouldn't sign it". Myers also noted that Jack was in hospital at the time with terminal leukemia, but there were other US team members who felt it was worth signing. So, Gary recollects, "I threw in the towel, and the states supported it". But he went to see the Assistant Secretary of Interior before that and left assured that they would develop an operations plan and that "a lot of that stuff that they had taken out would go back in the operations plan". So immediately after the signing ceremony, Myers recalled going back to the assistant secretary's office and was reassured again that this stuff was coming back into the plan, "but it never did".

Comparing the November 1985 draft from the negotiating team with the one signed by the two governments, two differences stand out. On page 14 of the 1986 Plan under "Recommendations for Future Actions", the declaration of government commitments to funding the work was greatly softened, and greater expectations of funding by NGOs and private citizens were added. The tone became much more cautious, almost pessimistic about incremental government support. Gone too were any specific recommendations that the United States and Canada develop a more detailed operational plan by the end of 1987 that would include more specific commitments to the habitat work envisioned in the Plan.

Still, in hindsight, it was enough to get creative wheels turning elsewhere. Much different than an operational plan, the 1986 NAWMP was subtitled "a strategy for cooperation", which indeed it proved to be (Environment Canada and US Department of Interior 1986). It featured several prescient provisions and a few shortcomings that will be explored further in the chapters that follow but to preview a few key elements here:

Clear quantitative population objectives were essential for folks to rally around, and their choices of objectives were wise. Mid-continent duck populations of the 1970s seemed like a stretch from 1985 but perhaps still attainable. Previous national plans that urged a return to the levels of the 1950s seemed fanciful by comparison. Although an arbitrary choice, in terms of a balance between aspirational, hopeless, or trivial, the Team hit the sweet spot. They further reasoned that, in human terms, duck hunting regulations in the 1970s were relatively liberal and people seemed largely content with the opportunities available during that decade. Although they suggested that average habitat conditions in the 1970s were similar to the long-term average, in fact for the mid-continent, conditions were somewhat wetter than average. This choice was not reached through a great deal of scientific analysis but came from a conversation among Harvey Nelson, Jim Patterson, Jack Grieb, and Bob Bailey. These goals had the advantage of simplicity and would be easy to communicate to the public.

The 1986 vision was focused squarely on habitat as the critical, fundamental factor most likely to affect waterfowl populations. This fact had been recognized by waterfowl leaders since the 1930s, but never squarely addressed at the continental scale. That focus made sense to almost every experienced waterfowl biologist. They identified the most urgent places to make progress as the work began. The 1986 map and the priority areas it identified had long-lasting and far-reaching implications (see Chapter 4).

The Plan was big, bold, and long-term, just like the problems it was designed to address. A 15-year program, millions of acres of habitat affected, and a $1.5 billion price tag were startlingly ambitious targets.

The concept of the Joint Ventures as regional coalitions of willing partners was genius and fundamentally important to the success of the Plan and we will treat them at length below.

As important as any of the Plan's provisions was the commitment to updating it at 5-year intervals. This implied a commitment to learning and adapting as experience was gained, to renewal as circumstances changed in the biological or social environment in which the Plan was operating.

The Plan also included a call for continuing stabilized hunting regulations and, pending a full review of the data from the 1980 to 1984 experiment, it also declared interim prescriptive changes to mallard and northern pintail hunting regulations if populations fell below certain threshold levels. The Plan also endorsed continuation of current black duck harvest restrictions but deferred to the normal national processes for enabling these restrictions. This was the only time the Plan entered issues of harvest regulations. These provisions disappeared in the first (1994) Update of the Plan and prescriptive elements of harvest regulations have been absent from the Plan ever since.

Absent at the onset was any specific commitment to funding the work. George Finney quipped that the newly signed Plan became known as the "No Wampum Plan" because there was no firm strategy about how to get money. So, how to get started?

The First International Plan Committee

The new Plan Committee[1] met for the first time in August 1986 in Chestertown, MD, with Walt Stieglitz of the FWS in the Chair. The first meeting was devoted to (1) identifying the roles of the Committee and operating procedures, (2) coming to a common understanding of what was in the Plan, and (3) a discussion of three of the Joint Ventures identified in the Plan. Much of their time dealt with early progress in the Canadian PHJV. The Black Duck Joint Venture was next and a steering committee, chaired by Patterson, was appointed to work on an initial plan, ahead of the next PC meeting. Finally, a task group of FWS and CWS staff was designated to review monitoring priorities for the Arctic Goose Joint Venture (AGJV).

In December 1986, the Committee met again in Montebello, Quebec, and focused on reports from the AGJV scoping committee and the black duck group. The AGJV planners recommended focusing, in descending order of priority, on the lesser snow/Ross' goose complex, mid-continent white-fronted geese, small Canada geese, and brant. The black duck group reported progress in developing banding and survey priorities with the intention of producing a strategic plan in 1987. Rollie Sparrowe led a discussion of a first draft communications plan prepared by the Service Public Affairs Office.

By the PC's third meeting in March 1987 in Willows, CA, much of the conversation revolved around IAFWA's "First Step" initiative (Chapter 3). They also reviewed a comprehensive report on AGJV needs and projected costs. Black duck planning progress, a stabilized regulations report, and early thoughts about a potential special plan for canvasback and redheads completed the agenda.

These minutes demonstrate that the PC wasted no time plunging into the details of operational planning. Evidently, they had enough of grand strategic planning over the previous decade – they were ready to get projects rolling. But by early 1987, thoughts were turning again to longer term actions, funding and marketing which presaged a higher level leadership focus for the Committee.

Box 2.1 Dr. James H. Patterson – Visionary (1942–2002)

Jim Patterson's belief in the early 1980s that waterfowl conservation needed to find a different path was fundamentally important to the creation of NAWMP. The Plan is his legacy as much as anyone's. Unfortunately, unlike most other early leaders, Jim's reflections were not captured in the oral history records from NCTC and thus few of his own words are featured here.

Clearly though, his vision involved a focus on habitat conservation over harvest management, and recognition that agency land acquisition alone could not achieve sufficient change without the support of favorable public policies, sustainable yet profitable agricultural practices, and mitigation for crop damage by the birds. Though not fully developed, his view of multiple benefits of soil and water conservation as a path toward habitat conservation was evident. The citation at his induction in the North American Conservation Honour Roll noted his key role in "forging the partnerships that brought about fundamental changes in continental waterfowl management through the establishment of NAWMP in 1986".

While his achievements were many, the Plan, which Patterson helped form as Director of Migratory Birds for the Canadian Wildlife Service, remains strong to this day. Jim was also instrumental in the creation of Wildlife Habitat Canada in 1984 and served as president for the International Waterfowl Research Bureau and the International Association of Fish and Wildlife Agencies during his career. His last job in Ottawa was as the first Director of International and Government Relations for Ducks Unlimited, crafting a new niche for DUC that continues to this day. Reflecting on his friend, Stew Morrison described Patterson as a tireless "doer", someone who got things done. Jim seemed to be everywhere in those early years. He was astute politically and networked a lot. One example was engaging the help of the Canadian Embassy in Washington early on. His "can do" attitude was infectious even though it ruffled a few feathers along the way! (Figure 2.1).

Box 2.2 Dr. Harvey Nelson – Visionary and Organizer Par Excellence (1925–2010)

Harvey Nelson was the undisputed leading force in the creation of NAWMP in the United States from the earliest planning through his role as Executive Director of the US NAMWP office from 1986 until his retirement in 1992. Harvey was known and respected throughout the continent. People admired him and trusted him and his character and intelligence made him a force for constructive change from farmlands of the Midwest to the Hall of Congress.

Harvey grew up in Minnesota near the famous hunting grounds of Pelican Lake and Lake Christina where his love of waterfowl was born. After serving with the US Navy in WWII, he earned a B.S. degree in wildlife management from the University of Minnesota and a Masters from Montana State. He joined the USFWS in 1950 serving in numerous roles regionally and in Washington, D.C., notably creating and serving as the first Director of the influential NPWRC in Jamestown, ND. Later, he was offered the reins of the FWS's North Central Region from 1980 to 1987 before delaying retirement to become the first Executive Director of NAWMP.

After retirement he remained a legendary advocate for waterfowl conservation helping support the Minnesota Waterfowl Association among several others. Harvey was a man with an unmatched conservation legacy. Carrol Henderson, then of the Minnesota DNR, recalled that Harvy attended a trumpeter swan management meeting the day before he unexpectedly passed away at age 85. Happily, we have many recollections from Harvey of his time with NAWMP in the NCTC oral history archives as this book will attest.

Figure 2.2 The early US NAWMP office staff (L to R, Dave Sharp, Bob Streeter, Carl Madsen, and Harvey Nelson). Photograph by John Cornelly, USFWS files.

Note

1 US Members: W. Stieglitz, Chairman, H. Nelson, R. Sparrowe (all USFWS), K. Wich (Atlantic Flyway Council); L. Soileau (Mississippi Flyway Council); R. Hopper (Central Flyway Council), and E. Hunt (Pacific Flyway Council). Canada Members: J. Patterson, Vice-Chairman, and B. Bailey (Canadian Wildlife Service); R. Andrews (Alberta); K. Lloyd (Northwest Territories); D. Simkin (Ontario); B. Meadows (New Brunswick), and D. Sherrat (Saskatchewan).

References

Adams, Glenn D. and Gary C. Gentle. 1978. "Spatial changes in waterfowl habitat, 1964–1974, in the Manitoba Newdale plain." *Canadian Wildlife Service Occasional Paper* 39: 29pp.

Anderson, Michael G., Ray T. Alisauskas, Bruce D. J. Batt, Robert J. Blohm, Kenneth F. Higgins, Matthew C. Perry, James K. Ringelman, James S. Sedinger, Jerome R. Serie, David E. Sharp, David K. Trauger, and Christopher K. Williams. 2018. "The migratory bird treaty and a century of waterfowl conservation." *Journal of Wildlife Management* 82: 247–59.

Brace, R. Kent, Richard S. Pospahala, and Robert L. Jessen. 1987. "Background and objectives on stabilized duck hunting regulations: Canadian and U.S. perspectives." *North American Wildlife Natural Resources Conference* 52: 177–85.

Buell, Noble E. (Editor). 1965. *Report on Waterfowl Research Planning Workshop. Patuxent Wildlife Research Center. November 16–19, 1964*. Unpublished Report. U.S. Bureau of Sport Fisheries and Wildlife. Washington, DC. 40pp.

Burnett, J. Alexander. 2002. *A Passion for Wildlife: The History of the Canadian Wildlife Service*. University British Columbia Press. Vancouver, BC. 346pp.

Canadian Wildlife Service, Environment Canada, 1983. *Waterfowl Management Plan for Canada: An Overview*. Ottawa, ON. 39pp.

Cowardin, Lewis M., Virginia Carter, Francis C. Golet, and Edward T. LaRoe. 1979. *Classification of Wetlands and Deepwater Habitats of the United States.* U.S. Fish and Wildlife Service. FWS/OBS-79/31. Washington, DC. 142pp.

Cowardin, Lewis. M., David S. Gilmer, and Charles W. Shaiffer. 1985. "Mallard Recruitment in the Agricultural Environment of North Dakota." *Wildlife Monograph* 92: 35pp.

Douglass, Donald W., Roy B. Anderson, Eugene F. Bossenmaier, Billy Joe Cross, Larry R. Gale, John M. Keener, and Arthur S. Hawkins. 1970. *Lessons from the Sixties and Challenges of the Seventies in the Mississippi Flyway*. Mississippi Flyway Council. Lansing, MI. 63pp.

Environment Canada and U.S. Department of the Interior. 1986. *The North American Waterfowl Management Plan: A Strategy for Cooperation*. 19pp.

Environment Canada, U.S. Department of the Interior, and Secretario de Desarrollo Social Mexico. 1994. *1994 Update to the North American Waterfowl Management Plan: Expanding the Commitment.* 47pp. (https://nawmp.org/sites/default/files/2018-01/1994%20NAWMP%20Update.pdf).

Gabrielson, Ira N. 1958. "Waterfowl – a problem of national significance." *Naturalist* 9: 8–12.

Gottschalk, John. S. and A. T. Studholme. 1970. "Waterfowl management in the Seventies." *Transactions North American Wildlife Natural Resources Conference* 35: 297–304.

Greenwalt, Lynn A. 1976. "A management plan for waterfowl." *Transactions North American Wildlife Natural Resources Conference* 41: 194–203.

Greenwood, Ray J., Alan B. Sargeant Douglas H. Johnson, Lewis M. Cowardin, and Terry L. Shaffer. 1987. "Mallard nest success and recruitment in Prairie Canada." *Transactions North American Wildlife Natural Resources Conference* 52: 298–309.

Grieb, Jack R. 1964. *Report on the Patuxent Waterfowl Research Workshop*. Unpublished Letter to the Central Flyway Technical Committee. 10pp.

Grieb, Jack R. 1976. "Improving international cooperative waterfowl management efforts: Accomplishments and needs." *Transactions North American Wildlife Natural Resources Conference* 41: 227–32.

Hawkins, Arthur S., Ross C. Hanson, Harvey K. Nelson, and H. Milt Reeves (Editors). 1984. *Flyways: Pioneering Waterfowl Management in North America*. U.S. Department of the Interior. Washington, DC. 517pp.

Hawkins, Arthur S. 1984. "The U.S. response." Pages 2–9 in Arthur S. Hawkins, Ross C. Hanson, Harvey K. Nelson and H. Milt Reeves (Editors). *Flyways: Pioneering Waterfowl Management in North America*. U.S. Department of the Interior, Washington, DC. 517pp.

Hochbaum, H. Albert. 1944. *Canvasback on a Prairie Marsh*. Stackpole, Washington, DC. 207pp.

Hochbaum, George S. F. Dale Caswell, Bruce C. Turner and Daniel J. Nieman. 1987. "Relationships among social components of duck breeding populations, production and habitat conditions in Prairie Canada." *Transaction North American Wildlife Natural Resources Conference* 52: 310–19.

Leitch, William G. 1978. *Ducks and Men*. Ducks Unlimited Canada. Winnipeg, MB. 278pp.

Linduska, Joseph P. (Editor). 1964. *Waterfowl Tomorrow*. USDI Washington, DC. 770pp.

Loughrey, Alan G. 1976. "Improving international cooperative efforts: Accomplishments and needs." *Transactions North American Wildlife Natural Resources Conference* 41: 215–222.

Lynch, Jerry J. 1984. "A field biologist." Pages 35–40 in Arthur S. Hawkins, Ross C. Hanson, Harvey K. Nelson and H. Milt Reeves (Editors). *Flyways: Pioneering Waterfowl Management in North America.* U.S. Department of the Interior. Washington, DC. pp. 35–40.

Munro, David. A. 1965. "Waterfowl Management in Canada." *Transactions North American Wildlife Natural Resources Conference* 30: 212–22.

NAWMP Committee. 2024. *North American Waterfowl Management Plan Update: Expanding the Partnership*. USFWS, Washington, DC. 50pp.

Rakowski, Patrick W. and B. P. Chabot. 1983. *Changes in land use in the Minnedosa district of southwestern Manitoba: an update on the Kiel-Hawkins transects*. Unpublished Canadian Wildlife Service Report. Winnipeg, MB. 10pp.

Reed, Nathaniel P. 1976. "International cooperation in migratory bird management: United States Efforts." *Transactions North American Wildlife Natural Resources Conference* 41: 223–26.

Rosenberg, Kenneth V, Adriaan M Dokter, Peter J Blancher, Adam C Smith, Paul A Smith, Jessica C Stanton, Arvind Panjabi, Laura Helft, Michael Parr, and Peter P Marra. 2019. "Decline of the North American Avifauna." *Science* 366, no. 6461: 120–24. (https://doi.org/DOI:%2010.1126/science.aaw1313).

Sparrowe, Roland D. and James H. Patterson. 1987. "Conclusions and recommendations from studies under stabilized duck hunting regulations: Management implications and future directions." *North American Wildlife Natural Resources Conference* 52: 320–26.

Tener, John S. and Alan G. Loughrey. 1970. "The migratory waterfowl program in Canada." *Transactions North American Wildlife and Natural Resources Conference* 35: 287–95.

Trost, Robert E. 1987. "Mallard survival and harvest rates: A re-examination of relationships." *North American Wildlife Natural Resources Conference* 52: 264–84.

Trost, Robert E., David E. Sharp, Sean T. Kelly, and F. Dale Caswell. 1987. "Duck harvests and proximate factors influencing hunting activity and success during the period of stabilized regulations." *North American Wildlife Natural Resources Conference* 52: 216–32.

Turner, Bruce C., George S. Hochbaum, F. Dale Caswell, and Daniel J. Nieman. 1987. "Agricultural impacts on wetland habitats on the Canadian Prairies, 1981–85." *North American Wildlife Natural Resources Conference* 52: 206–15.

U.S. Fish and Wildlife Service, Pacific Flyway Council, Central Flyway Council, Mississippi Flyway Council and Atlantic Flyway Council. 1982. *A National Waterfowl Plan for the United States.* Washington, DC. Unpublished Report. 52pp.

U.S. Fish and Wildlife Service. 2025. *Waterfowl Population Status, 2025.* U.S. Department of the Interior, Washington, DC. USA. 67pp.

U.S. Supreme Court. 1920. State of Missouri v. Holland, 252 U.S. 416 (1920). (https://supreme.justia.com/cases/federal/us/252/416/).

Wait, Paul. 2022. *Delta Waterfowl.* Delta Waterfowl Foundation. Bismarck, ND. 219pp.

3 Who Will Pay for All This?

Wildlife Habitat Canada

While work on the Plan was underway, partners in Canada took a valuable step forward with the creation of Wildlife Habitat Canada and the Canadian habitat conservation stamp. The United States had created a duck stamp program in 1935 by which waterfowl hunters contributed funds for wetland conservation. Canada had established a migratory game bird permit in 1966, but the funds from that program supported CWS data collection and administrative work, not habitat. As Canada's national waterfowl plan was developing in the early 1980s, some of the same leaders of that effort, like Jim Patterson and George Finney of CWS and provincial leaders like Richard Goulden of Manitoba, began to build support for a program to help fund wetland habitat work. Private-sector leaders like Stew Morrison of Ducks Unlimited Canada weighed in too. The idea was to issue an annual conservation stamp, selected through a Canadian art competition, that would be an added requirement for waterfowl hunters, and the proceeds would be directed toward habitat conservation. The contest and proceeds would be administered by a new organization, Wildlife Habitat Canada. The proposal was approved in 1984 and renowned Canadian artist Robert Bateman was commissioned to create the first-of-country print for 1985.

Getting the stamp program approved was challenging. Earmarking money from the stamps to an NGO and for a specific purpose was unprecedented and vehemently opposed by Finance Canada. Their policy was simple: all funds raised by government automatically go into general revenue and any expenditure had to be approved by Cabinet and ultimately Parliament each year. The duck stamp crossed all those ages-old control processes, and it took perseverance by CWS staff and outside pressure to overcome the objections (Jim McCuaig, pers. comm. April 2025).

WHC was created as a Government-Organized NGO, not a typical, fully independent NGO. Thus, WHC, like the Wetlands Conservation Council that came later, remained closely allied with CWS. When I served on the WHC Board, we made our annual funding decisions without government involvement, but CWS was a close partner and always involved in an *ex officio* capacity in our Board meetings. WHC was thus able to act expeditiously and do things that a government agency could not do rapidly nor free of political considerations. The international umbrella of NAWMP helped enable this independence for WHC.

In 1984, a diverse Board was appointed to oversee the new organization, chaired by D. Stewart Morrison, then the Executive Vice-President of DU Canada. Other members were Richard Goulden (MB), artist Robert Bateman, educator David Suzuki, Art Smith (PEI), Louise Beaubien Lepage (QC), and Bob Deloury. David Neave, a provincial wildlife administrator from Alberta, was chosen as the first Executive Director. More about WHC's founding and current programs are available at https://whc.org/.

DOI: 10.4324/9781003732310-3

Stew Morrison (pers. comm. June 2023) recounted that Bateman's famous mallard painting was originally titled "Mallards on Thin Ice" to convey concerns about their current population status, but after much debate about whether that created a pessimistic or hopeful message, it was changed to the neutral "Mallard Pair Early Winter" and sold well in both Canada and the United States.

First Steps and Flagships

By 1986, the need to get funding jump-started in both countries was center stage. It seemed obvious to leaders like Nelson and Patterson that, with no promise of federal funding in the signed agreement, a coalition of willing partners would be essential to get things rolling. There was precious little time to get money and the associated legal processes in place to implement the Plan. The cobbling together of resources by partners to launch the Plan set a pattern for the future.

George Finney (CWS) recalled that he and Jim Patterson went to the 1986 IAFWA annual meeting to discuss start-up funding and found several states willing to help. "The so-called First Step Projects were tremendously important in terms of proving the concept and helping justify federal cabinet submissions in Canada", Finney recalled. Two key actors at this stage in the United States were Gary Myers (TN), then President of IAFWA, and Chip Collins, Executive Director of the National Fish and Wildlife Foundation (NFWF; Box 3.1). Myers remembers getting a call from Walt Stieglitz, then FWS Assistant Director, Refuges and Wildlife, shortly after the Plan was signed. Stieglitz would soon be appointed the first US Chair of the International Plan Committee, and he asked Myers if he would serve on a scoping committee to "figure out how to get some money and get this thing done". Also recruited were Larry Gale, Director of Missouri Department of Conservation, Hazard Campbell from NY, then president of DU Inc., and Chip Collins. The federal governments had signed the Plan but promised no money, so as Gary said, "How do we get a billion and a half dollars to implement a plan over 15 years…We didn't have a clue!" So, at the fall meeting of IAFWA, just months after the Plan was signed, Myers invited a bunch "duck guys" up to his suite along with his new committee, "to talk about how we're going to find money for this thing". Myers suggested that maybe he could get a million dollars collectively from the states. Hazard Campbell offered that maybe DU could match that, and Chip Collins offered that NFWF would try to match that total. Jim Patterson (CWS) suggested that they would seek a similar match from the Canadian Provinces and the federal government.

Box 3.1 NFWF

The NFWF was created by Congress as a charitable 501(C)(3) organization in 1984. As a private enterprise, it was designed to have greater flexibility than government agencies to make deals, arrange contracts, and facilitate novel conservation enterprises. The Foundation's main source of funding was revenue from fines, mitigation money, and other legal penalties assessed for violating federal environmental laws. It could also receive annual appropriations from Congress. Its mission was to work with public and private sectors to protect and restore habitats for fish, wildlife, and plants. With federal, corporate, and foundation clients, its business model was collaborative. Based in Washington, D.C., NFWF is overseen by a Board of up to 30 Directors comprising private citizens from diverse sectors and approved by the Secretary of Interior. It is among the largest US conservation foundations, in FY 23, allocating $1.3 billion, matched with another $410 million to assist 797 projects. https://www.nfwf.org/sites/default/files/2024-03/nfwf-conservation-investments-2023.pdfNFWF.

Later at the IAFWA business meeting, incoming President Bob Brantly from Florida designated Myers to take the lead on this First Step funding initiative for the Association. One of the first challenges was how to get money to Canada. The best advice Myers got from the Department of the Interior was just do it and figure out what's legal as you go along! Myers laid this simple matching money scheme out in a meeting with the IAFWA Executive as "the only thing we can think of". Herb Doing said NY would put $100K on the table. Next, Myers called Larry Gale (MO) and asked for another $100K. Gale said he didn't know but he would try. Encouraged, Myers continued to call state directors. Jim McCuaig echoed this attitude in Canada too, "There were no roadmaps, so we just did what seemed best….and rarely sought permission. We preferred to ask forgiveness if it really became a problem; but it rarely did, although many responses were, 'You can't do that, can you?'".

Where to Turn?

It soon occurred to Myers that if they were going to ask states for money, they needed to know what they were going to do with it! "We needed to find some place in Canada that produces a lot of ducks that fly all over the place and will benefit a lot of states". Rollie Sparrowe's staff dug around and came back with Quill Lakes, Saskatchewan. Rollie sent Myers a map that showed those banded birds went all over the place. An added attraction of the Quill Lakes project was habitat for endangered species including piping plovers, peregrine falcons, and whooping cranes. So, Myers traveled to Saskatchewan to meet with Dennis Sherratt, then Director of the Provincial wildlife agency and some of his staff and asked, "Could you figure out some kind of needed project that would spend about 4 million dollars at the Quills?" with only a vague plan about how the funding might come together. Sherratt, evidently intrigued, assigned biologist Ross Melinchuk to work on the ideas along with Jeff Nelson, a biologist who at that time worked for Ducks Unlimited Canada. Soon enough Ross and Nelson put together a game plan of how they could usefully invest money in the Quills. "Then, Myers noted, we needed other places [in Canada] where it would be useful and geographically appropriate to invest dollars from other states".

Melinchuk had recently led a related Saskatchewan Prairie Pothole Project near Redvers (Melinchuk and MacKay 1986). Funded in 1985 by just-created Wildlife Habitat Canada and Saskatchewan Parks and Renewable Resources, the project's primary purpose was to assess landowner receptiveness to a variety of options to secure long-term protection of small, privately owned wetlands and associated upland nesting cover – the habitat that NAWMP would soon target in the Prairie Pothole Region.

Enabling NFWF Expansion

All of this sounded promising. IAFWA was well on the way to the $1 million state pledge, but two roadblocks appeared: (1) the match capacity and legal authority for NFWF to transfer funds out of the United States and (2) unexpected reticence to match by DU Inc. The NFWF had a cap – they could not spend more than a million dollars on any single thing, so those interested in facilitating NAWMP matching funds had to get that cap raised. Also, it was unclear if they could do business in Canada. NFWF needed authority to do that too. As Myers noted, "it's pretty hard to get money appropriated to do something you lack clear authority to do!"

Fortunately, Bob Davison entered the scene. Davison was formerly of the National Wildlife Federation where he had worked on wetland issues with Alan Wentz, and now served as a senior staff member for Democratic Senator George Mitchell on the Senate Committee of

Environment and Public Works. That committee handled all the programs of the US Fish and Wildlife Service as well as laws such as the Clean Water Act. Bob originally had been hired in 1985 by then Subcommittee Chairman Republican John Chafee, which empowered Davison considerably when the Democrats took control of the Senate in 1986.

After the NAWMP was signed, Davison began attending IAFWA and Plan Implementation Committee meetings and began crafting legislation to amend the act that established NFWF to enable it to accept donations and appropriations of federal dollars to support conservation projects in the United States and other nations. Davison found this an intriguing challenge as there was no substantial source of federal funding for the habitat work called for in the Plan and a lot of creativity was being employed to get funding started. It "just blew me away", Davison said to see how states were willing to come together and use state money to match other dollars to fund waterfowl habitat work in Canada!

> I could envision the federal government maybe doing it, but it was hard to imagine the states could politically get away with using their dollars to fund habitat conservation in another country, even though there were good biological reasons for doing that … but politically it was counterintuitive, and I thought rather daring actually and it intrigued me.

Soon, the Foundation's establishment act was changed to increase the NFWF cap limit from $1 million to $30 million and to permit the use of a combination of federal and private money to support projects in the United States and Canada. Myers recalls testifying to the committee about the need to do this. He remembers a conflicted Frank Dunkle (then FWS Director) who agreed that it was good to grant the authority, but his administration was NOT at that point in favor of spending money in Canada, and it took several more months and a good deal of state pressure for the federal administration to finally acquiesce. NFWF also had to augment its administrative capacity to take on this larger role.

Ducks Unlimited Commits

In the fall of 1986, DU Inc. seemed ready to commit to matching state dollars, but as time passed in 1987 signals from DU Headquarters became less clear. Dale Whitesell, their long-serving staff leader, had cooled on the government program. DU Canada, the sister organization in Canada, had remained four-square behind the Plan. Myers recalls getting a call from Rollie Sparrowe telling him to, "cool it with trying to move forward with this thing because of the DU situation". A week later, Myers recounted,

> it was on a Friday afternoon [when] I got a call from Jim Patterson, Jim had a layover in Chicago and had dropped by DU headquarters… they were having a big meeting there, and Matt Connolly from Massachusetts, then working with the North American Wildlife Foundation, had just been recruited to head the staff of DU and it was now clear they were going to match the [state] money for NAWMP.

Through hard work and creativity by many people, First-Step funding became a reality in 1988. Myers and IAWFA had pledges of $1 million in total from 12 States. Ducks Unlimited matched that $1 million, and, using newly appropriated dollars, NFWF added another $1.75 million and that grand total was more than matched again ($5.1 million) by federal and provincial governments and nongovernment players in Canada. NAWMP in Canada was truly launched.

That was a critical development. In reviewing scores of various meeting notes from 1987 to 1988, I noticed there remained a great deal of angst about sending US funds to Canada. This persisted despite the clear biological need to do so if the Plan was to succeed. The idea seemed to be nonstarter with some in Congress and among some FWS staff. The alternative in their minds was to focus first on top-priority work in the United States and rely on state and private money (mainly DU) for Canada. Then perhaps in the longer term, they might direct some federal money outside the United States. Fortunately, State leaders and continental thinkers prevailed.

Although the 1986 Plan called for a cost-sharing formula of 75% US and 25% Canada based on the historical proportions of ducks harvested by each country, for the First-Step projects, the target became a 50:50 cost split for the projects in Canada. It ended up about 54% Canadian. This exception was pushed to send a strong signal about commitments of both countries to the work and hopefully dull somewhat the concerns about sending US funds to Canada. It was enough money from US sources flying across the border that Canadians were compelled to seize the opportunity.

Jim McCuaig, soon to be Harvey Nelson's counterpart as the NAWMP Coordinator in Canada, recalls that after signing the plan it was obvious that the Canadian federal government needed to ante up its part, but there was no way that either the Canadian Wildlife Service or Environment Canada would find that much money in existing budgets. One of Jim's first tasks was to design the Cabinet Document that sought such incremental funding. "A strong point, of course, was having huge support from the United States to fund projects in Canada". McCuaig recounts, "we succeeded in Cabinet, in getting that financed, but it was not easy". Before the Cabinet submission, McCuaig described CWS as begging, borrowing, and appropriating money from all kinds of places but, with the First-Step and Flagship Projects up and running, "We were then able to…go back to Cabinet or Congress and declare 'See, it works. This is real'".

Jim also recalled,

> It was dicey … right up until the end, when a Briefing Note came from our Washington Embassy on the day we went to Cabinet. That was the same day that President Bush spoke to everybody and supported the Plan. And we were able to use that to get it through Cabinet.

Sheila Tooze of the Canadian Embassy staff was a great asset in Washington. "Her networks were amazing", McCuaig recalls.

> That briefing note was written by Sheila and me for Ambassador Derek Burney who attended Bush's speech to DU in Washington (Box 3.4). The Ambassador grabbed us both on the way out of the room and ensured that the briefing note would soon be in the Environment Minister's hands.

The Canadian First Steps

Early planning for conservation actions had already been underway in the Prairie Habitat Joint Venture (PHJV) and the Eastern Habitat Joint Venture (EHJV), so identifying additional First-Step projects was not difficult. Criteria included addressing the goals of the Plan, and projects that could show results quickly. CWS also needed Provinces who were ready to commit a share of the costs. McCuaig recalled that each Province that wanted to participate brought a candidate project to the table at a scale that fit the amount of money likely to be available. For Canada, seven were chosen: two in the PHJV and five in the politically more complex EHJV.

Each of these projects was described in a 1991 report of the North American Wetlands Conservation Council (Canada). By then, a total of 17,735 acres of wetland and upland habitat had been secured and another 8,371 acres developed, with work still underway at five of the seven projects. They were:

1 Buffalo Lake Moraine, Alberta
The area was typical "knob and kettle" glacial topography with a high density of small wetlands. Parts of this 528 square-mile agricultural region in central Alberta were being managed to enhance waterfowl nesting success by planting cover and improving conservation practices on farmed land.
2 Quill Lakes, Saskatchewan
The first and largest of the First-Step projects, sprawling over 7,000 square miles, this landscape included staging, molting, and migration habitat for waterfowl and shorebirds as well as extensive pothole-dotted nesting areas. Program elements included establishing dense nesting cover, minimizing agricultural crop damage by field-feeding ducks, installing safe nesting baskets, promoting conservation framing practices, and demonstrating managed grazing systems to improve nesting habitat.
3 Matchedash Bay, Ontario
Located on Georgian Bay, this 5,381-acre area was at high risk of continuing wetland loss. An area of value for breeding and migrating waterfowl, it was also a region of broad biodiversity values. Investments were made in both securing and restoring habitat.
4 Baie-du-Febvre/Nicolet, Quebec
This complex of floodplain habitats on the South shore of Lac Saint-Pierre along the St. Lawrence River was an important spring staging area for waterfowl. Dike construction for exclusive agricultural use threatened to reduce waterfowl and fish habitat. Habitat securement and impoundment developments for waterfowl, fish, and agriculture were featured.
5 Grove Pine/Big Brook, Prince Edward Island
Prime black duck habitat, this area of coniferous forest, stream-side meadows, and brackish marshes in the drainage area of the Fortune River was in private hands and at risk of development. The project consolidated a 1900-acre block of wildlife lands including 500 acres of wetlands.
6 Grand Lake Meadows, New Brunswick
Grand Lake Meadows was a complex of diverse floodplain habitats including lakes, sedge meadows, flooded hardwood swamps, and oxbow ponds, and some agricultural use of higher elevation lands. Believed to be one of the most valuable waterfowl production areas in New Brunswick, it also supported the only colony of Yellow Rails in Atlantic Canada. This First-Step project secured 3,100 acres, the beginning of a longer term effort to secure the whole area.
7 Salt Marsh Protection, Nova Scotia.
In the last 300 years, 58% of Nova Scotia's salt marshes had been converted to other land uses. The remaining low salt marsh areas are important to the marine food chain as well as migrating and wintering black ducks and Canada geese. The higher salt marsh areas are important for brood rearing black ducks and willets. The project secured 300 acres and tested the feasibility for large-scale securement of salt marshes.

The US Flagships

The first joint ventures in the United States (Chapter 5) were targeting March 1989 for drafts of their implementation plans. To kick-start the effort and provide evidence for the feasibility of such cooperative projects, NAWMP partners identified ten initial "Flagship" projects. In each of these, partners were challenged to demonstrate the full array of wetland habitat protection and enhancement options, including acquisition, technical assistance to private landowners, innovative wetland management techniques, and coalitions of contributors. Like a flagship in an armada, these projects were to set the stage for other projects to follow. The projects were on fast-track implementation schedules to demonstrate their potential for participants, funding sources, and the public.

A first related funding request to the USFWS was for \$82.8 million in FY 1990, of which \$50.4 million was for acquisition, \$21.1 million for habitat development, and \$4.5 million for technical assistance to landowners. State and Private sources matched this budget with another \$30.8 million.

The 10 US flagship projects comprised:

1 Lake Thompson Watershed, South Dakota
This 506 square-mile watershed in East Central SD had been declared a National Disaster area in 1984 and 1986 due to flooding. A task force following the floods recommended that wetland restoration higher in the watershed might be an effective preventive measure. Flood control, water quality, and wildlife habitat were all foci for the work. Acquisition, development, and technical assistance for private landowners were part of the \$6.1 million project.

2 Chase Lake Prairie Project, North Dakota
This ambitious project encompassed part of five counties in central ND. The overarching goal was to improve the balance between economic needs of private landowners and the needs of wildlife. It included economic incentives for landowners to improve wildlife habitat on private land, enhancing habitat on existing State and Federal lands, and acquiring additional protected acres via easements or purchase. In addition to breeding waterfowl, this pothole landscape included the largest White Pelican breeding colony in the United States, vital habitat for other colonial birds, marsh birds, shorebirds, and extensive native grasslands. With a menu of programs much like at Lake Thompson, the initial plan totaled \$6.7 million. Getting the State of ND onside for this project was challenging. The Governor wanted to shut down FWS acquisitions in the state after clashing over the controversial Garrison Diversion Project. Dale Henegar (NDGF Commissioner) was also reluctant to support Prairie Pothole Joint Venture (PPJV) acquisitions in this environment. A North Dakota Action Group came together to promote PPJV in ND and re-profile it as not mainly an FWS acquisition program (Box 3.2).

3 Upper Sacramento Valley, California
This Flagship project, one of the first to take shape in the United States, had a complex genesis. The Central Valley is one of the most significant waterfowl wintering areas in North America. Of the original 4 million acres of wetlands in the Valley, only about 280,000 acres remained in 1988. Nowhere else on the continent were so many waterfowl dependent on so few acres of habitat. The Upper Sacramento Valley contained the best remaining waterfowl habitat in the northern portion of the Central Valley. The tract comprises four National Wildlife Refuges (Sacramento, Colusa, Delevan, and Sutter), two large FWS easement areas (Butte Sink and Lurline Willow Creek), two state Wildlife Management Areas (Gray Lodge and Upper Butte Sink), and one privately owned Conservation Area (Audubon's

Bobelaine Sanctuary). The area had the highest potential for wetland habitat restoration, enhancement of public and private wetland areas, and expanded acquisition. The initial $10 million proposal included acquisition, habitat development, and an aggressive outreach program for private landowners.

4 Chenier Plains, Louisiana
The coastal zone of southwest Louisiana, curving around to southeast Texas and on to the Laguna Madre, includes vital wintering habitat for many species of waterfowl, waders, and shorebirds and breeding habitat for resident species like mottled ducks. Habitats are diverse, some managed, but many not. This Flagship Project proposed, presciently, that the $8 million include $275K to plan and evaluate the effectiveness of the projects being delivered.

5 Texas Mid-Coast
This proposal was nearly identical to the Chenier Plains project, but with the addition of $2.4 million in nonfederal acquisition money. Including nine National Wildlife Refuges and several state areas, the proposal also featured support to study trends in wetland losses to help plan future strategies for preserving coastal marshes. Proponents also intended to evaluate private land enhancement practices for wintering waterfowl.

6 Cache and Lower White Rivers, Arkansas
Throughout the lower Mississippi Valley, <20% of presettlement bottomland hardwood forests remained by the late 1980s, greatly diminishing the capacity of the region to support wintering waterfowl and other wildlife. The area then supported some 30% of wintering mallards in Arkansas, and the project goal was to secure some 145K acres in the 360K-acre floodplain at a price tag of about $8.5 million.

7 ACE River Basin, South Carolina
The ACE Basin comprises the Ashepoo, Combahee, and South Edisto River basins downstream to the point where they enter St. Helena Sound. The area extends inland from the Atlantic Ocean for more than 20 miles and encompasses approximately 200K acres. Nearly half of the area is a complex of fresh, brackish, and saline wetlands. The ACE Basin is one of the most productive and complex estuarine systems on the Atlantic Coast. Working with DU and the Nature Conservancy, ready to match $3.7 million in federal dollars by more than 2:1, the goal was to secure in perpetuity key tracts of this unique habitat.

8 Cape May, New Jersey
The Cape May Project area was a 15,500-acre peninsular tract on the Delaware Bay in New Jersey and included some of the best black duck habitat on the mid-Atlantic Coast. Until recently, it was a rural, agricultural area. The project proposed $10 million in federal acquisition, augmented with $6 million in state and private funds.

9 Lower Great Lakes/St. Lawrence Basin
This project focused on federal, state, and private wetlands along the Ohio coast of Lake Erie. Infrastructure development and technical assistance predominated in this $12.8 million project.

10 Montezuma Wetlands Complex, New York
The Montezuma Wetlands Complex Project was designed to unify the management of the federally administered Montezuma National Wildlife Refuge, the state-owned Howland Island, Crusoe Lake and Cayuga Lake Wildlife Management Areas, new state and federal acquisitions, and two privately held wetlands. This project would result in wetlands in a single complex of some 35K acres. Mainly an acquisition and infrastructure project, it totaled $7.4 million.

Box 3.2 You Need to Talk with Harvey

By David Case, D.J. Case, and Associates

In early 1987, Mike Johson said to me, "You need to talk with Harvey". Mike was the Waterfowl Biologist for the ND Game and Fish Department. Harvey Nelson was a regional director for the USFWS and would soon be the first US NAWMP Executive Director. Mike said the NAWMP really needed help with communications and encouraged me to meet with Harvey at the upcoming North American Wildlife and Natural Resources Conferences in Quebec City. The NAWMP was a big deal in the waterfowl conservation world. The possibility of being involved in the communications efforts seemed too good to be true.

The Communications Challenge

The 1986 NAWMP document didn't explicitly call for a major public engagement or communications effort. But, to achieve the level of landscape impact envisioned in the Plan was going to require an audacious continental-scale behavior change effort. And the funding to underwrite that effort was going to have to come from new sources. Fortunately, NAWMP authors, leaders, and early implementers understood the challenge. They recognized that lots of different people in lots of different places needed to be persuaded to do lots of different things to achieve the Plan goals.

Diving In

Following the meeting with Harvey, we began developing a scope of work for our first communications contract. Initially, the scope was very product focused – slideshows, brochures, videos, and fact sheets. Through discussions with Harvey and USFWS External Affairs staff Phil Million and Megan Durham, we agreed to broaden the scope to include organizational development for the emerging NAWMP enterprise and, especially, to take a more strategic approach to communications – an approach centered first on identifying and understanding the various audiences and less on jumping right to "communications products".

Our timeline was tight. We started in July 1987 and had to prepare a report and recommendations by early October. It was exhilarating and nerve-wracking at the same time, but it was every conservationist's dream…to be at the spearpoint of such an important endeavor. We began immediately conducting interviews – many of them hours-long deep dives with conservation leaders across the United States and Canada. We not only conducted the interviews by phone but also made multiple trips to Washington, D.C., and to conservation-related meetings in the United States and Canada.

The 1987 report for Harvey/USFWS would lead to other opportunities for us to work on NAWMP communications. Two early projects turned out to be particularly important and would inform our approach to conservation communications for years to come. First, in 1988, the ND Game and Fish Department hired us to facilitate the North Dakota Action Group in developing a NAWMP implementation plan for the State. Communications was a fundamentally important part of the project. The ND Action Group wanted to draw attention to what could be accomplished through NAWMP in ND and throughout the PPJV. This was our first chance to engage deeply with people close to the ground on NAWMP communications...local farmers, ranchers, bankers, hunters, and agency personnel. We gained valuable insights into integrated delivery of communications and conservation actions; in particular, how a continental conservation plan gets expressed through actions by people in communities on the landscape.

Implementation of NAWMP in ND was kicked off on September 26, 1989, with the Chase Lake Prairie Project Dedication. Writer Michael Furtman (2017) witnessing the event noted that "tour buses lined the road, and someone's pasture had been turned into a parking lot". As dignitaries spoke, flags of the United States, Canada, Mexico, and ND "snapped in the wind". "Then local landowners and dignitaries alike poured little vials of ND soil into a drainage ditch dug long ago to begin the wetland restoration".

Box 3.3 The Cosumnes River Preserve

By David Paullin, Central Valley JV Coordinator, Retired

The Cosumnes River Preserve south of Sacramento was our flagship project. The Preserve was a work in progress before the NAWMP and Central Valley Joint Venture (CVJV) came on the scene. This was originally an Audubon Society project, which was bought to protect one of the last old-growth riparian forests in the Central Valley. It was purchased to provide habitat for threatened and endangered species like yellow-billed cuckoos, Swainson's hawks, and valley elderberry longhorn beetles. Waterfowl and NAWMP were not part of the justification for the initial acquisition. The CVJV came on the scene about the same time that Audubon was wrapping up its acquisitions. Adjacent to the gallery forest were some agricultural fields within the river floodplain that could easily be restored to wetlands. So, Ducks Unlimited partnered with Audubon and our flagship project was "born". The important element here is that the CVJV hooked our "caboose" to the Audubon train which had been moving along for quite some time. Over my nine years (as coordinator) that model was repeated, over and over again, as agencies hooked their "cars" to the Consumnes train. The Preserve grew as more people wanted to join the project. It was like building something out of LEGO®s. A new program would pop up and an agency representative would bring a sack of money for some good purpose and their logo would be added to the Cosumnes River Preserve sign! It was a great example of partnerships that worked for people and conservation. The Consumnes Preserve website holds more of this interesting story (https://www.cosumnes.org/about-the-preserve/).

A Glance Back at the First Steps and Flagships Today

From the vantage point of 2025, a look back at the 17 projects is encouraging. Many of them were targeted strategically as keystone endeavors in systems where additional work was needed. For instance, the Chenier Plains and the mid-Texas Coast stretch over 6.5 million acres of coastal marsh, rice fields, and conservation lands that are incredibly important to wintering waterfowl and resident breeding mottled ducks, a myriad of other birds, fish, and shellfish, and as buffers against storms and rising seas. Work in this vital system continues with many partners engaged.

The ACE Basin in South Carolina has become a crown jewel of wetland conservation along the South Atlantic Coast. Ducks Unlimited, The Nature Conservancy, the Lowcountry Open Land Trust, and private landowners have protected over 160,000 acres of wildlife habitat from commercial and residential development since 1988, quickly exceeding their original goals. Notably, more than 50,000 acres are available for public use through two Wildlife Management Areas, two State Parks, a National Wildlife Refuge, and a National Estuarine Research Reserve. Over 70,000 acres are protected through conservation easements and work with private landowners.

The Upper Sacramento Valley has seen annual progress for waterfowl habitat since 1989 and also provides habitat for endangered Chinook Salmon that rely on the watershed.

Revisiting the Chase Lake Project in ND 30 years after its dedication, outdoor writer Michael Furtman (2017) described progress there as "tangible, measurable, and marvelous". The Project was sited within the Chase Lake Wetland Management District, which now protects some 140,000 acres of wetlands and grasslands. Craig Bihrle (2019) noted that since the first North Dakota NAWMP project got underway 30 years ago, the state had hosted 85 projects that conserved about 1.7 million acres. NAWMP projects "benefit waterfowl and other wetland-obligate birds", said Terry Steinwand, then Director of the ND Game and Fish Department quoted by Bihrle. But there's more to it. "The vast majority of those funds", Steinwand added, "go to landowners who voluntarily enroll land in NAWCA projects…it really is a win-win for places like North Dakota".

Focused on restoring natural flows and habitats associated with Arkansas's Cache River, the Cache and Lower White River project included a substantial portion of the Cache River National Wildlife Refuge. Partners like the United States Army Corps of Engineers, Audubon, and the Nature Conservancy have been instrumental in restoring the natural meanders and flooding regimes of this extensive system.

The Friends of the Montezuma Wetlands Complex celebrated the acquisition and management of extensive protected areas in New York and the work there continues as recently highlighted as a case study of successful cooperative partnership conservation (Wagner et al. 2022).

In Canada, Matchedash Bay became the first step in Canada's EHJV that has now invested >$600 million to secure >1.8 million acres and improve some 600K acres across the joint venture. The Grand Lake Meadows in NB covers >26,000 acres of diverse habitats in 21 parcels and is the largest wetland area in the Province, designated as a Protected Natural Area. Grove Pine/Big Brook became a Provincial Wildlife Management Area in PEI, protecting some 2,700 acres along the tidal reach of the Fortune River.

Quebec's Baie-du-Febvre is now a federal Migratory Bird Sanctuary along the St. Lawrence River featuring a diversity of habitats of importance to birds and fish, particularly Greater Snow Geese. The site is part of UNESCO's Lake Saint-Pierre World Biosphere Reserve and is recognized as a wetland of international importance by RAMSAR.

The Quill Lakes system has been challenged by continuing drainage in the upper basin leading to water management problems in the lower basin that have affected both farmers and breeding shorebirds. These issues aren't entirely resolved but conservation in higher reaches of the watershed continues.

Conservation projects in Alberta's Buffalo Lake Moraine provided important lessons for successfully expanding work there and helping to inform the growth of similar pothole habitat conservation projects across Alberta, Saskatchewan, and Manitoba.

Individually, these projects proved useful, but more importantly, as the first-steps founders envisioned, these initial projects provide "proof of concept" that encouraged the explosive expansion of work in the NAWMP Joint Ventures in the decades ahead.

Getting Organized for the Longer Term – a US Implementation Board

In August 1987, Harvey Nelson and Jim Patterson invited a group of conservation leaders for an initial NAWMP development meeting at the Interior Building in Washington, D.C. Invitations were issued to other agency and NGO leaders[1] from Canada and the United States for the purpose of offering advice to the first Plan Committee about fundraising, support services, and implementation.

Eight months later as the First-Step projects got underway, Nelson, with the support of the US section of the Plan Committee, again brought together several US private-sector leaders to help advance the Plan. They agreed to form an Implementation Committee (later renamed a Board) to assist the government-populated Plan Committee with actions needed to grow the Plan. Their principal role would be to provide guidance and assistance to the Executive Director (Nelson) and the US section of the Plan Committee. They were to nurture relationships and help maximize public outreach and support for implementation, at both national and regional levels. Fundraising and political influence were seen as other vital roles.

Organizations invited included the Wildlife Management Institute, International Association of Fish and Wildlife Agencies, Ducks Unlimited, The Nature Conservancy, Izaak Walton League, Berry Brooks Foundation, NFWF, National Wildlife Federation, North American Wildlife Foundation, National Association of Conservation Districts, National Audubon Society, National Rifle Association, The Wildlife Society, and American Forest Foundation. Two committees were struck initially: a habitat committee comprising DUI, NFWF, IAFWA, TNC, and others to be named and an information and services group including WMI, NWF, NAWF (Delta), Izaak Walton, Audubon, NACD, NRA, Berry Brooks, American Forest Foundation, TWS, and others to be named. Larry Jahn (WMI) was appointed Interim Chairman and called the second meeting in June where Matt Connolly (DU) was elected Chair and the Board's Terms of Reference were approved. The Terms highlighted actions the Board could take to expedite implementation of the Plan including guidance, raising awareness, encouraging funding and evaluations, and building partnerships.

At the Implementation Board's inaugural meeting, FWS Director, Frank Dunkel, challenged the group to help provide "big picture" advice and assistance. Harvey Nelson shared recent hires of JV coordinators. Chip Collins at NFWF was looking for funds. Top priorities were Quill Lakes SK, and private matching money. The first of that comprised $200K from the California Waterfowl Association and $80K from the Iowa Heritage Foundation. Mike Berger reported that DU was ready to help deliver and help raise money for First-Step projects in Canada. Their internal goal was to raise $150 million above their regular program over the next 15 years for NAWMP projects. Jack Berryman (IAFWA) expressed concern about softening federal support after the Plan was signed and pledged the Association's help. Larry Jahn (WMI) reported that a scoping committee of himself, Gary Myers, and Hazard Campbell was doing active outreach for matching money. Jahn presciently noted that benefits for other wildlife and people must be recognized in promoting the Plan too, more than ducks. Jahn's vision not only included waterfowl, certainly, but also a broader means for engaging and building support for wildlife, wetlands, and associated habitats.

Jahn urged that the Implementation Board could be a bridge between concepts, principles, and application. Three things should be at the top of the Board's list he offered: review of plans, advice, and action! He saw value in keeping government and nongovernment committees separate, using each to do what they are best at, though others like Bob Davison saw value in keeping them more connected.

The first focus of the Implementation Board, however, became fundraising and communications. Chip Collins felt there was a serious dearth of information about the Plan "out there".

Sportsmen, Congress, most anyone needed to know more. Congress was not hearing from constituents about funding the Plan. Collins seemed undeterred, viewing NAWMP as an opportunity as long as we had a demonstrated need. Give people the feeling that the Plan was an umbrella under which they could achieve their goals, and the matches should come together. The group thus focused on communicating about the Plan across agencies, Congress, non-government associations, and beyond. The Plan had no agreed-upon strategy at that point, no compelling articulated case for support, nor a designated "voice" to assist with generating support. As Dave Case noted at the time, leaders had to get clear about what they wanted people to do. Give money? Write to legislators? Fortunately, Dave Case and Harvey Nelson connected, and D.J. Case and Associates landed a first communication contract with NAWMP in 1987 (Box 3.2).

Not all the environmental NGOs chose to join the Implementation Board. The Nature Conservancy, for example, needed to become convinced of broad benefits of the initiative, but most of those invited to the first meeting were willing to help in some manner. Two outside the fold, however, the Wilderness Society and Defenders of Wildlife, seemed set to oppose NAWMP in Congress if the Plan didn't broaden its biological goals.

Meeting again in May, the NAWMP staff reported the addition of Carl Madsen with his private-lands experience in Minnesota. For communications, Region 6 had added Carol Lively and Region 4, Dan O'Neil. Bob Streeter indicated intentions to build communications capacity in each of the US joint ventures. The FWS regional directors (or their assistants) remained largely "in charge" of the developing joint ventures. Streeter noted that regions 2, 3, and 6 were progressing though there was some resistance from FWS Refuge personnel who were not accustomed to planning at large scales and with the broader conservation community.

Addressing the meeting, Peter Meyers, Deputy Secretary of Agriculture, called 1985 Farm Bill the most important conservation program in the last 50 years, declaring that anyone farming highly erodible land must have a conservation plan and must be implementing it by 1990 to retain Farm Bill benefits. Swampbuster, Sodbuster, and more, up to $1 billion annually, would be spent to offset damage to natural resources by agriculture and that sounded promising. Synergies with NAWMP seemed likely.

Matt Connolly recalled (pers. comm. March 2025) that another key player in the genesis of NAWMP funding was Jim Range. Jim had been Senator Howard Baker's Chief of Staff when Connolly recruited him to become DUI's National Secretary, a volunteer board position. Range's employer, Waste Management Inc., donated Washington office space for DU and it was staffed with a focus on NAWMP funding. Range and Connolly had previously worked on the passage of the Magnuson-Stevens Act in 1976 to foster long-term sustainability of marine fisheries. It had elements that seemed relevant to funding NAWMP. It too had its origins at the grassroots level. Later, people like Representatives Gerry Studds and John Dingell and Senator Ted Stevens also were recruited to play strong roles in supporting NAWMP.

Meanwhile, by the summer of 1988, a new wave of drought hit Prairie Canada. Dawn Redmond, Canada's NAWMP communications lead, emphasized the seriousness of the situation, especially in Saskatchewan. Much publicity ensued including that with the drought and conclusion of the period of stabilized hunting regulations, drastic restrictions on duck harvests on both sides of the border might be in the offing. The early teal season was canceled in the United States while authorities waited for the breeding population survey results before addressing the main seasons. As troubling as that was, it seemed to be good timing to promote the need for NAWMP. Yet surprisingly, Redmond reported that, at an August meeting of Federal and Provincial Wildlife Ministers in Saskatoon, only one in ten ministers listed NAWMP as a high-priority topic! This despite an associated news conference announcing the kickoff of the Quill Lakes project.

Meanwhile, Canadian wildlife leaders like Jim Patterson, David Neave, Bob Bailey, and Gordon Kerr encouraged their US counterparts to think beyond First Steps. The need for habitat conservation was massive and long term. Fifteen years set an ambitious tone, but needs would stretch far beyond that time horizon. DU companies on both sides of the border were reiterating their commitment. David Wesley announced the opening of a DU office in Sacramento, and the hope to duplicate that in each of the major US priority habitat regions. Stew Morrison reiterated that DU Canada was well positioned to help deliver many of the new conservation programs in Canada.

In hindsight, the Implementation Board was vital at the beginning to help advance and focus discussions inside and outside the US federal government, but it faded away after the passage of NAWCA and creation of the Wetlands Councils in the United States and Canada. I wonder if such a body as the Implementation Board might still serve a useful purpose for the enterprise today, helping to renew commitments to the important work of the Plan 40 years on? I will return to that thought in Chapter 24.

The North American Wetlands Conservation Act – A Giant Boost for NAWMP

The First-Step and Flagship projects came to pass, but everyone knew that a larger, repeatable, and dependable source of funding would be needed to grow and sustain a continental conservation plan. Enter Bob Davison again. As previously noted, Davison had worked for Senator Chafee, the ranking Republican from Rhode Island on the Senate's Environment and Public Works Subcommittee. When the Senate changed hands in 1986, Democratic Senator George Mitchell became the Subcommittee chair but Mitchell, impressed with Davison, asked him to work for him doing essentially the same things he had done with Chafee. So, for about nine years, Davison was the only senior staffer working on fisheries and wildlife issues for the majority, and because of his history with Chafee, also for the Republican minority.

Bob had participated in the 1987 effort to amend the law to enable NFWF to accept donations and appropriations of federal dollars and support conservation projects in other nations. With his knowledge of what the NAWMP Implementation Board was trying to achieve, Davison began to think creatively about similar mechanisms that might enable funding international habitat work.

"The first-step projects moved ahead. It really happened", he recalled in an interview with John Cornely for the FWS Heritage Committee. That seminal experience got the wheels turning for Davison. Reflecting in 2012, Bob allowed that

> …the North American Wetlands Conservation Act is the thing I'm most proud of in my career….in large part, because no one asked me to do it. I just sort of saw that there was a need and tried to figure out how to [emulate] the success that people had figured out [with the First Steps]. I was really taken by that and [thought] if they can do this then maybe we actually could pass legislation that would make this possible more formally, because if the states could get away with it politically then it seemed like we ought to be able to get away with it politically at the federal level and it ought to be doable and you've got DU behind you and the Foundation [NFWF] and other support… then maybe it's really doable.

Davison and his counterparts in the House of Representatives were also concerned that if NFWF became the implementing mechanism for the Plan, that could dominate it's work and detract from NFWF's broader conservation mandate. With the support of Chip Collins and Amos Eno, they began thinking of other long-term administrative solutions. By November 1988, Davison had outlined what he called the North American Waterfowl Management Plan Implementation Act, which led by December to draft legislation. It was heavily waterfowl-focused, tried to track closely the way Plan implementation was envisioned, and so it included legislatively establishing joint ventures and a Plan Committee, although these were already coming together organically.

As early as 1988, Larry Jahn and others on the Implementation Committee had cautioned presciently that other wildlife must be recognized in promoting funding for the Plan. The visions for NAWMP held by people like Chip Collins, Jim Range, Larry Jahn, and other politically savvy leaders of the Implementation Committee focused on waterfowl but were more expansive…a means to engage and build support for wildlife, wetlands, and habitat conservation broadly. They were some of the people who made sure that the NAWMP funding act eventually reflected a broader vision and helped ensure its passage.

In February 1989, Davison was ready to present his draft to Senator Mitchell, who was by then also the Senate Majority Leader. It was profiled as legislation to protect wetlands, waterfowl, and other migratory birds through NAWMP. He cautioned that some environmental groups may not be supportive because they may think it is focused too narrowly on protecting wetlands for duck hunters, but Mitchell endorsed circulating it in his name. Given the fraught politics between the State and the FWS over wetland issues in ND, Senator Quentin Burdick of ND, the full committee chairman, demurred on cosponsoring Mitchell's bill.

Knowing what was developing, Larry Jahn at WMI astutely asked Bob if Senator Mitchell might be willing to give the keynote address at the North American Wildlife and Natural Resources Conference that is held in Washington every four years. To Bob's "surprise and delight", Senator Mitchell agreed and took that opportunity to announce that he would be introducing the "Wetlands Act". The draft bill of March 15 had morphed into the North American Wetlands Conservation Act. The change reflected a shift in emphasis talking about wetland habitat for waterfowl and other migratory birds rather than Plan implementation. In addition, Davison had become convinced in talking with people like Gary Myers and others at IAFWA that his first draft was perhaps overly prescriptive and might reduce implementation flexibility, which had contributed to some early successes with the Plan. "We needed to provide funding but leave room for creativity and evolution without losing accountability", Davison recalled. Despite these changes, the connection between the Act and NAWMP was not lost. On both the Senate and House sides as the bill was introduced by members, their transmittal statements still described NAWCA as a bill created to implement the habitat provisions of the North American Waterfowl Management Plan.

Having watched how federal matching funds for state endangered species conservation got zeroed out in some years, a principal concern for Davison was to provide some sources of funds for the Act that were not entirely dependent on annual appropriations. By early 1989, Davison had come up with a solution mimicking earlier fisheries legislation. It earmarked the annual interest acquired from the Pittman-Robertson arms and ammunition excise taxes collected and held by the federal government before transfers to the States. To that, he added fines collected for violations of migratory bird regulations, to provide a small pool of guaranteed funds that could be augmented by appropriations. A key challenge was to ensure that the Senate Appropriations Committee would allow these funding sources to be automatically appropriated each year rather than simply being made available to be spent if Congress chose to do so. That problem was solved when the Senate Appropriations Committee agreed to look the other way on a proposal by Majority Leader Mitchell that intruded into their jurisdiction. Another challenge was how to stimulate allocation of considerable funds to Canada (and later, Mexico). Various formulae were floated, based initially on the proportion of funds needed for Canada estimated in the 1986 Plan. In the end, the NAWCA prescription was 50–70% to Canada and the balance to the United States.

Matching fund requirements from US nonfederal sources were critical leverage, as were expectations of additional substantial matching funds from Canada. There was, of course, pressure to back off from that formula. Jim Patterson (CWS) and others from Canada were key to reassuring the United States of Canadian support. The Canadian Embassy was engaged too, facilitating multiple meetings between US and Canadian officials.

In June, Representative Gerry Studds, on the House side, introduced companion legislation to the Senate version of the North American Wetlands Conservation Act but with one new wrinkle. Their version included Mexico, because of Studd's staff person, Gina DeFerrari's interest, and knowledge about conservation needs in that country. A final important change in the mark-up process, stimulated by Congressmen, John Dingell and Silvio Conte, a prominent Republican and a Democrat, inserted the existing Migratory Bird Commission (MBCC) in the project approval process between the new NAWCA Council and the Secretary of Interior. With astute stickhandling, the final bill was passed in November 1989 (https://www.govinfo.gov/content/pkg/USCODE-2017-title16/pdf/USCODE-2017-title16-chap64-sec4401.pdf), less than a year after Davison's first outline, an accomplishment lightning fast by federal standards!

Maintaining NAWCA – The First 36 Years

From the beginning, NAWCA mandated biennial reports to Congress, and the Act (Sections 10 and 16) specified the content of these reports. The first was filed in 1991, covering 1989–1991. Thereafter, reports were made in 1992–1993 through 2022–2023 (https://www.fws.gov/sites/default/files/documents/2024-06/north-american-wetlands-conservation-act-progress-report-2022-2023.pdf.). Reports from 1998 to date are available online. The reports are comprehensive, typically 40–50 pp. in length.

A Partial History of Changes Over Time

With the geographic priorities expressed by NAWMP as a guide, the original Act set the allowable distribution of funds at 30–50% annually in the United States and 50–70% to Canada and Mexico combined. Up to 4% was authorized to administer the grants. When the Act was reauthorized by Congress in 2003, as with the original act, the allowable portion for administration remained low, capped at 4% of the total. The proportion of funds allocated to projects in the United States was set at 40–70%, with 30–60% targeted to Canada and Mexico. Offsetting this change of potentially reduced funds going north or south was a trend for increasing appropriated dollars for the Act, as happened in both 1990 and 1994. Also, experience had shown that the US nonfederal match necessary to reach an allocation of 70% federal funds to Canada/Mexico had never been achieved. In most recent years, the traditional split has been 50% US, 45% Canada, and 5% Mexico.

The Act has gone through reauthorizations at approximately five-year intervals with few material changes other than reapportionment of funds among countries. One benefit has been the gradual though erratic increase in the allowable limit of the annually appropriated funds from Congress. An additional US small grant program was introduced in FY1996 with a ceiling of $50K for individual projects.

With hard work by advocating organizations like Ducks Unlimited, AFWA, and others, reauthorizations for NAWCA have always been passed with strong bipartisan support and large majorities in both the House and Senate. But there have been moments of brinkmanship and uncertainty. Scott Sutherland who led DU's effective Washington office from 1990 to 2012 recalled (pers. comm. May 2025):

> I think it was in late 2002, NAWCA was the very last bill to pass before Congress adjourned. The bill had passed in the Senate earlier in the evening but as the House shut down, we thought we had failed to get the House to agree to the Senate's version. Majority Leader Dick Armey was retiring from Congress and had made his final farewell speech when a staffer went to him on the House floor and convinced him to take up NAWCA reauthorization. He did and it passed unanimously at about 2:30 AM. Dr. Richard Armey was an economist and a Conservative Congressman from Texas but he had grown up in North Dakota and was a supportive of economically sound conservation.

In October 2020, the Act was reauthorized as part of the American Conservation Enhancement Act with recommended funding at $60 million/year. At that point, standard grants were still capped (though exceptions were possible) at $1 million each and small grants at $100K each. Small grants were available to US applicants only and preference was given to new applicants to spread the funding around and attract new partners. By FY2024, US standard grants were considered eligible between $250K and $3 million each; small grants were pegged at $10K–$250K each. Canadian programs were eligible between $200K and $10 million, and Mexican projects from $10K to $1 million. In all cases, a minimum of 50% US nonfederal match was still required.

For the most recent funding window (2024), the MBCC authorized $46.2 million to conserve 91.4K acres in 17 states with matching funds of $99.1 million, plus $11.7 million in Migratory Bird Conservation Funds to US refuges. Twenty-two small grants received $4.7 million (matched by $5.5 million). Finally, late in 2024, the Act was reauthorized yet again for five years (FY2006–FY2030) with a recommended annual appropriation still pegged at $60 million.

One provision of the Act has proved limiting for meaningful evaluations of the Act's effectiveness. The restriction of "not more than 4% of funds" could be spent on managing the program, was widely interpreted as administration and evaluation combined, but more on that in Chapter 14.

In a related effort in the early 1990s, Davison worked with Representative John Breaux's (D- LA) staff on what was dubbed the Coastal Wetlands Planning and Protection Act. It earmarked the tax on sales of gasoline for off-road usage to support the protection of coastal Louisiana wetlands. Davison and others hoped to capture some of that money for NAWCA; however, in the House, Congressman Gerry Studds, and his staff, Will Stelle, insisted that the money be targeted for wetland conservation projects in the United States. So, that money became the one "pot" managed by the NAWCC but designated for use in the United States only.

In reflecting on the whirlwind of activity that brought NAWCA to fruition, Davison credited much of the inspiration to his relationship with Gary Myers. "Gary is the one who inspired me to do it", said Davison. "The matching money scheme he had led for First-Step funding seemed brilliant and his enthusiasm for NAWMP was infectious". Bob's satisfaction came from seeing his idea coalesce in legislation. Having Senator Mitchell transition from the subcommittee chairman to the Majority Leader at the same time certainly made the path easier, and Mitchell's willingness to invest time in the North American Wildlife Conference presentation was pivotal to building support for the Act.

Less well known were some other interesting steps along the path to approval. Jim McCuaig tells a story about Jim Patterson and the journey of the NAWCA bill through the

Senate. Patterson, in addition to being the CWS leader of NAWMP, had also ascended to the presidency of the International Association of Fish and Wildlife Agencies. IAFWA had been a major player in the First-Step projects and in advocating for NAWCA, so naturally the Senate Committee asked the President of IAFWA to testify about the bill. Jim journeyed to Washington to comply, but as he was readying to testify, he received an urgent "cease and desist" message from the Canadian Embassy, reminding him that he could not do that. No Canadian federal employee is allowed to go under oath to a foreign power. So, what to do? Well, Jim, Bob Davison, and the committee chair came up with a brilliant workaround. Jim appeared at the hearing, the Committee immediately adjourned for the day, ordered in coffee, and just sat and chatted with Jim about NAWMP and NAWCA with whoever happened to be in the room! No oath, no "testimony", no problem! Though McCuaig recalls that Embassy officials weren't thrilled.

At another early juncture in the NAWCA journey, senior officials at DOI had expressed reticence about the bill, leading people to speculate that it may largely have been about seeing control of a major funding mechanism elude their oversight. Davison recalled (pers. comm. March 2025) that Interior Secretary Lujan wrote to the Committee in July 1989 that the Administration was opposed to a NAWCA Commission or Council. Rather, he thought, the Secretary should only be required to consult with interested parties prior to making recommendations to the Migratory Bird Conservation Commission. At the same time, the Administration also opposed requiring any set percentage of funds to go to Canada or Mexico.

Jim McCuaig shared another recollection of the debate. He remembers being cornered on a porch at the Audubon Refuge in ND by Secretary Lujan's Chief of Staff. Lujan was not a fan of money flowing out of the country. McCuaig had previously landed in hot water for publishing a newsletter article upon Lujan's appointment titled, "A Cold Front from the South". The Chief of Staff asked him what Canada's response would be to removing the 50% clause from NAWCA and just trusting the United States to send some money? "I swallowed hard, said Jim, and told her that Canada would withdraw its support for the Bill". That was the last time Canada officially heard about that question, and NAWCA passed. "It still scared me to have had to speak for Canada that strongly and bluntly to power, but luckily it worked", reflected Jim.

Box 3.4 President Bush Calls for No Net Loss of Wetlands

Ducks Unlimited had scheduled a Waterfowl Symposium for Washington, D.C., in February 1989. Matt Connoly recalled to me (pers. comm. Feb 2025) that DU President Hazard Campbell, who had been working with Myers and IAFWA on funding the First-Step projects, had a brother who attended the same boarding school in Maine as newly elected President George H.W. Bush. Connolly recalls "dreaming" what if the new President came and spoke to us about wetlands? Connolly's prior experience with politicians was that they often depend heavily on getting input about messaging from their host, seeking things that would please an audience. In about 20 minutes time, Hazard called his brother and he in turn called Bush and got a thumbs up! "It was remarkable, Matt reflected, they did in fact route the speech to us and we inserted enthusiastic language expressing the president's support for no-net loss of wetlands and for Senator Mitchell's bill". Mitchell was a Democrat but from Maine, so Bush and Mitchell had a lengthy personal relationship. Bush had also long been a modest DU financial sponsor. As it unfolded, the President spoke to a cheering crowd. I was in that happy audience but had no sense of the goings on behind the scenes. Matt recalls Secretary Lujuan and FWS Director Dunkle looking anything but happy on the sidelines.

So, NAWCA was launched with parallel Wetlands Councils created in the United States and Canada. Bob Streeter (FWS) was reassigned in February 1990 to implement the Act and serve as the coordinator for the yet-to-be-populated North American Wetlands Conservation Council. Jim McCuaig was appointed to a similar role in Canada. In the next six months, the US Council was appointed, a Council staff was assembled, Council procedures were developed and approved, and wetland conservation projects were solicited from Canadian and US Joint Venture partners. Connolly, who had succeeded Larry Jahn as Chair of the NAWMP Implementation Board in 1988, was appointed to the inaugural US NAWCA Council and elected as its first Chair in 1990.

Streeter, in his NCTC recollections, described a "can-do attitude" permeating the system in the early days of NAWCA. Lots of procedures were hastened to conclusions and timelines accelerated, in part because of leadership and partner pressures to implement the Act as quickly as possible.

From the beginning, the approaches and the roles of the two national Wetlands Councils differed. The United States set up a competitive process wherein projects were submitted, scored by staff attached to Council members, and then ranked for distribution of the available funds. But, as Jim McCuaig noted, the sovereignty issue came up again as NAWCA evolved.

> Canada was not in a position of allowing decisions of what was to happen to land and things in Canada [to] be taken in Washington. So, we approached it differently … finding out from Washington, how much money would be available for Canada in any granting cycle [based on a formula applied to anticipated funds from various sources].
>
> Then Canada used what was at first a 77-16-7% formula for distributing those funds among the Canadian Joint Ventures and told the Joint Ventures to plan for that much money. It was then in their hands to develop and vet the proposals, reach consensus within the JVs [on what to invest], run those recommendations through the Canadian Wetlands Council and once approved at those levels in Canada to send them down to Washington.

McCuaig was Canada's courier and voice in this process for many years. He remembers that

> I always had two things in my back pocket: I knew what I'd do if there was less money available than we expected – and that happened at the very first meeting, and so we had to cut a project. But in other cases, we were ready with other projects in our back pocket if there was more money available….and it became a very effective system.

In 2025, NAWCA remains the only US federal grant program dedicated to the conservation of wetland habitats for migratory birds. The accomplishments of NAWCA have been extraordinary (Chapter 24). Federal grants have typically leveraged twice the legally required dollar-for-dollar match from nonfederal sources. Partners have included private landowners, state, tribal and local governments, conservation organizations, sporting groups, land trusts, and corporations. By any standards, this has been a remarkable set of accomplishments. Information on current NAWCA programs can be found at (https://www.fws.gov/service/north-american-wetlands-conservation-act-nawca-us-standard-grants).

Box 3.5 Gary T. Myers, Visionary, and Orchestrator

When Gary Myers retired in 2009 at age 73, he was the longest serving State Wildlife Chief in the US capping a long and impactful career as a leader in wildlife management in TN and North America. NAWMP was fortunate that in 1986 Gary Myers was already in a position of influence in the International Association of Fish and Wildlife Agencies, from where he led the first efforts to assemble funding for waterfowl conservation in Canada under the Plan by building a coalition of state agencies and NGOs to fund the First-Step projects. Indefatigable, astute, optimistic and clever, for Gary failure simply wasn't an option. His "can do" attitude was infectious and brought others along.

Gary also helped facilitate the earliest US habitat joint ventures in 1987 and 1988 including the LMVJV and the GCJV. When JVs later broadened their missions to include all bird groups, Gary was there to champion that expansion too. After the First-Step projects, Gary worked for the enactment of the North American Wetlands Conservation Act and particularly rallying the required nonfederal matching grants from state agencies. Later, he also supported the broader North American Bird Conservation Initiative, including the IAFWA Southern Wings" program that encouraged states to help fund projects in the Caribbean and Latin America.

Recognizing his catalytic role in beginning the flow of state and other matching funds for waterfowl conservation to Canada, DUC chose Gary as the first person inducted in the Waterfowl Conservation Honour Roll. Among Myers' other awards was the George Bird Grinnell Memorial Award for Distinguished Service to Natural Resource Conservation.

Upon Myers retirement, his successor Greg Wathen, Chief of Wildlife, Tennessee Wildlife Resources Agency, remarked, "Gary Myers, who just happens to hail from Tennessee, really considers the entire North American continent to be his conservation playground"....adding "Gary Myers' most enduring quality is his ability to get people to believe that they can achieve the unachievable". Indeed, that is Gary.

Box 3.6 Dr. Robert P. Davison, Architect of NAWCA

Bob Davison grew up in New Jersey, received a B.S. Degree in political science at Penn State, an M.S. in wildlife at New Hampshire, and a Ph.D. at Utah State studying population dynamics of coyotes. In 1980, he accepted a faculty position at South Dakota State University and then followed Dr. Alan Wentz from there to work for the National Wildlife Federation in Washington, D.C. With NWF, he became deeply involved in wetland conservation and endangered species issues. In 1985, he accepted a staff position with the Senate Committee on Environment and Public Works where he intersected with NAWMP.

Later, he served as Deputy Assistant Secretary for Fish and Wildlife and Parks during the first Clinton administration. After that, he joined the Wildlife Management Institute as the Northwest field representative, worked for Defenders of Wildlife as a senior advisor, and taught part-time at Oregon State University. Davison's vision, creativity, and finesse were key to the passage of the single most important source of funds sustaining NAWMP since 1989.

In Bob's oral history transcript, he said, "If you want to be a leader, find a parade and get in the front of it. And that's all I really did with the North American Wetlands Conservation Act. There was a great parade that already had started". Others would contest that it is overly modest! His legacy to NAWMP and NAWCA is enormous.

Note

1 Warren Bantz (KS), Hazard Campbell (DUI), David Case (D.J. Case), Chip Collins and Amos Eno (NFWF), Matt Connolly (DUI), Larry Jahn (WMI), Stew Morrison (DUC), Gary Myers (TN), Dave Neave (WHC), Dale Nelson (IA), Marvin Plenert (FWS), Jim Range (Waste Management), Dennis Sherratt (SK), and Rollie Sparrowe (FWS).

References

Bihrle, Craig. 2019. "NAWCA: 30 Years of Advancing Habitat." *North Dakota Outdoors* 82, no. 4: 16–19. North Dakota Game and Fish. Department, Bismarck.

Furtman, Michael. 2017. *On the Wings of a North Wind.* Reprinted, Birch Portage Press. Duluth, MN. 184pp.

Melinchuk, Ross and Ron MacKay. 1986. *Prairie pothole project Phase I. Final Report.* Wildlife Branch, Saskatchewan Parks and Renewable Resources, Regina, SK. 32pp.

North American Wetlands Conservation Council (Canada). 1991. *Report of Canadian First Step Projects for the 1988–1990 Period.* NAWCC (Canada) Report No. 91–1. (https://nawcc.wetlandnetwork.ca/Report%20On%20Can%20First%20Step.pdf).

Wagner, Matthew. R., Michael L. Schummer, Geoff Cripe, Matthew Halliday, John M. Coluccy, and K. Sarah Fleming. 2022. "Land protection partnership in the Montezuma Wetlands Complex, New York, USA." *Case Studies in the Environment* 6, no. 1. (https://doi.org/10.1525/cse.2022.1711654).

4 Creating the Joint Ventures and a Start in Canada

Joint Ventures can be difficult. Normally, every nongovernment organization competes with others for attention and money. Often, states, provinces, municipalities, and federal governments squabble over turf. Working together doesn't come easy. Yet, successful partnerships have been the fundamental key to NAWMP success.

Dave Paullin, the long-serving coordinator for the Central Valley Joint Venture (CVJV), speaks about parking egos at the door. "It's surprising, he said, how much you can get done if you don't insist on credit for it". The desire to be something more than your own organization is key; the recognition that you can accomplish more together than any one of you can alone.

The creation of every JV was unique. Some, like the CVJV, included several NGOs but only a single state. Others, like the Atlantic Coast JV, brought together 16 states plus Puerto Rico. Some focused mainly on a single ecosystem, like the Gulf Coast or Playa Lakes, while others like the Eastern Habitat JV (EHJV) in Canada encompassed a wide diversity of ecosystems.

In this chapter and the following one, I will explore the creation of the original seven habitat JVs, with a focus on a few that I know best from personal engagement. Two species JVs, the Black Duck and the Arctic Goose, were launched immediately too and are covered in Chapter 7. The others, now totaling 22 habitat and 3 species' JVs, I will describe more briefly in Chapters 6 and 10, including links to their current plans and profiles available online.

The Germ of an Idea

As George Finney (CWS) noted, in an NCTC interview, a major challenge for architects of both the US and Canadian national plans was how such a far-flung, ambitious continental-scale habitat program could be delivered. No single entity – federal, provincial, state, or nongovernment organization – had capacity nor jurisdiction to function everywhere within either country, let alone internationally. As the US/Canada negotiating team got to work, the problem was nowhere near resolution.

After more than one session brainstorming about this issue, Rollie Sparrowe (pers. comm. 2025) recalled sitting at lunch with Bob Andrews from Alberta, musing about this, going through again the mental checklist of assets and limitations of each organization as a delivery agent, and Bob opining that "maybe we need a different model, something like businesses might use to approach a common problem, maybe something like a joint venture?". Later that day, the negotiating team kicked that around and decided it was an idea worth floating in the draft of the Plan. "It seemed logical to think about forming these at regional scales that matched the team's present thinking about the half-dozen continental priority landscapes, but otherwise the idea, frankly, was still rather vague", Sparrowe admitted.

DOI: 10.4324/9781003732310-4

As the 1985 draft of the Plan was circulated for comment, the JV delivery model was only sparsely sketched out. Initially positing as a principle that "Joint Ventures will be considered as an approach to financing high-priority research and management projects of international concern that can only be addressed through a pooling of resources". Further, the draft noted, "Implementing this plan depends on responsible expertise existing in each country, province, state and region to formulate reasonable action plans for habitat preservation and management within their respective jurisdictions". Reviewing this recently, Jim McCuaig pointed out that a shift away from financing to implementing was a vital development. JVs were to design the programs and spend the money. Others, often in the same agency or NGO, would seek the money but this allowed the JV structure to concentrate on what it could do best – deliver conservation projects on the land.

By the time the Plan was signed in May 1986, ideas were a little crisper:

> Joint venture projects should be implemented through facilitating agreements negotiated and agreed to by all those wishing to participate.... An action group should be established for each joint venture. The planning should be set out as a proposal and forwarded to the North American Waterfowl Management Plan Committee for its review and recommendation.

Thus, from the onset, JVs were seen as self-formed organizations of interested partners willing to help share the costs of the enterprise and with wide latitude to approach the work as they saw fit. Most of the emphasis was on an initial "proposal" for review by the PC; operating procedures, scale, evaluations, revisions, and much more would be determined later.

Because each JV jurisdiction differed politically, ecologically, in land ownership patterns, waterfowl species of greatest interest, and more, each JV was destined to be unique. There were, however, some common challenges and common features of successful JVs.

Bob Streeter (USFWS) was the Director of the Plan Office in the United States during the formation of several JVs. A decade after leaving that post, he reflected in an NCTC interview that NAWMP

> has kept going because of the strength of the partnerships. And that each of the partners feels they are getting enough out of it to continue with it.... In my professional career, there's never been a program more fun, or more satisfying, to be associated with.

Jim McCuaig was Bob's counterpart in Ottawa during the days of JV formation in Canada. Jim believed that NAWMP had been a success mainly because of the people involved.

> It's people that make this work. It's not the thing. It's not the money. It's not the politics. It's the people in it. And.... what it is about the people is a common objective. Something outside themselves. This isn't a personal objective, you know, I want a get a promotion. I want to get money. I want power. No. None of that. It's – In this case, I care about waterfowl. I care about the land. I care about the landowners. I really want to make a difference. And that kind of shared objective, when it's set out in front of everybody, takes their eyes from their personal issues, and their personal problems, their jurisdictional issues, their internal service problems, whatever they are, and it focuses on that.

In his book *Getting to Yes*, Roger Fisher et al. (2011) advised readers not to talk about what "you" need, but to talk about what "we" need. The other vital thing about partnerships, McCuaig noted is that "everybody comes to the table and nobody is in charge…and the fact that nobody's in charge of it …is a key to this North American Plan functioning". McCuaig reflected, "If I'd have gone in there and said 'we're the feds, …and told people what to do, it never would have happened. Never".

While the JVs were finding their unique paths to partnerships, the national NAWMP Offices in Canada and the United States provided support. Some of that was financial, but mostly it was a coordination of efforts within each nation, a communication link to parallel efforts elsewhere, and a link to the international Plan Committee that was also trying to find its way to providing meaningful oversight for NAWMP. Although the Plan Committee's structure and function were described in more detail than for JVs in the 1986 Plan, it too needed to be populated, operating procedures defined, and relationships built with key subordinate parts, especially the JVs.

Regional Priorities – Where to Begin

The map of top priority conservation areas offered in the 1986 Plan gave guidance to where the accelerated habitat work should begin (Figure 4.1).

Because of their perceived importance to waterfowl and recent habitat losses, the top areas listed in the US National Waterfowl Plan (1982) were the Prairie Pothole Region, California's Central Valley, the Atlantic Coast from North Carolina to Massachusetts, coastal California, the Mississippi Delta, the Gulf Coast of Texas and Louisiana, and the Great Basin. The Canadian plan recognized that much of that country was valuable waterfowl habitat but that the greatest need for protection and enhancement was in the southern settled zone where agriculture and urban/industrial expansion had most diminished habitat abundance and quality. The 1988–1990 First-Step and Flagship projects had provided some initial foci for developing partnerships in each of these regions.

For practical purposes, some adjustments were made to these five shaded areas (Figure 4.1) at the outset of JV planning. The Prairie Pothole Region was divided into two JVs at the international border. Although the ecological challenges were similar in Canada and the United States, the agencies and public policies involved changed abruptly at the 49th parallel so it made sense to split it there administratively but with a commitment to sharing ideas and experience across borders.

For similar reasons, the US side of the Great Lakes/St. Lawrence region needed its own governance and by 1993 had morphed into the Upper Mississippi/Great Lakes JV, complementary to the EHJV in Canada that also grew beyond the St. Lawrence Basin.

The Lower Mississippi Valley and Gulf Coast region was also described as a single area in the 1986 Plan. However, the ecosystems of greatest interest differed between the Mississippi Alluvial Valley and the Coast. Administratively, there were differences too. Arkansas and Missouri, for example, had no direct interest in the coast; conversely, Texas had little interest in the Mississippi River. So, the decision was made early on to split the planning and delivery work for the bottomland hardwoods and the Gulf Coast marshes into separate initiatives. Recently though, the two JVs have been doing some common contingency planning around the anticipated impacts of climate change and sea-level rise. Also, in 2006, the Eastern Gulf Coast states were added to the JV family with an enlarged focus on the entire coastal plain.

Thus, the original five geographic areas of greatest concern in the 1986 Plan quickly evolved into the original seven habitat JVs. Links to extensive information on each JV, including historical documents, accomplishments, current strategic plans, current staff contacts, and

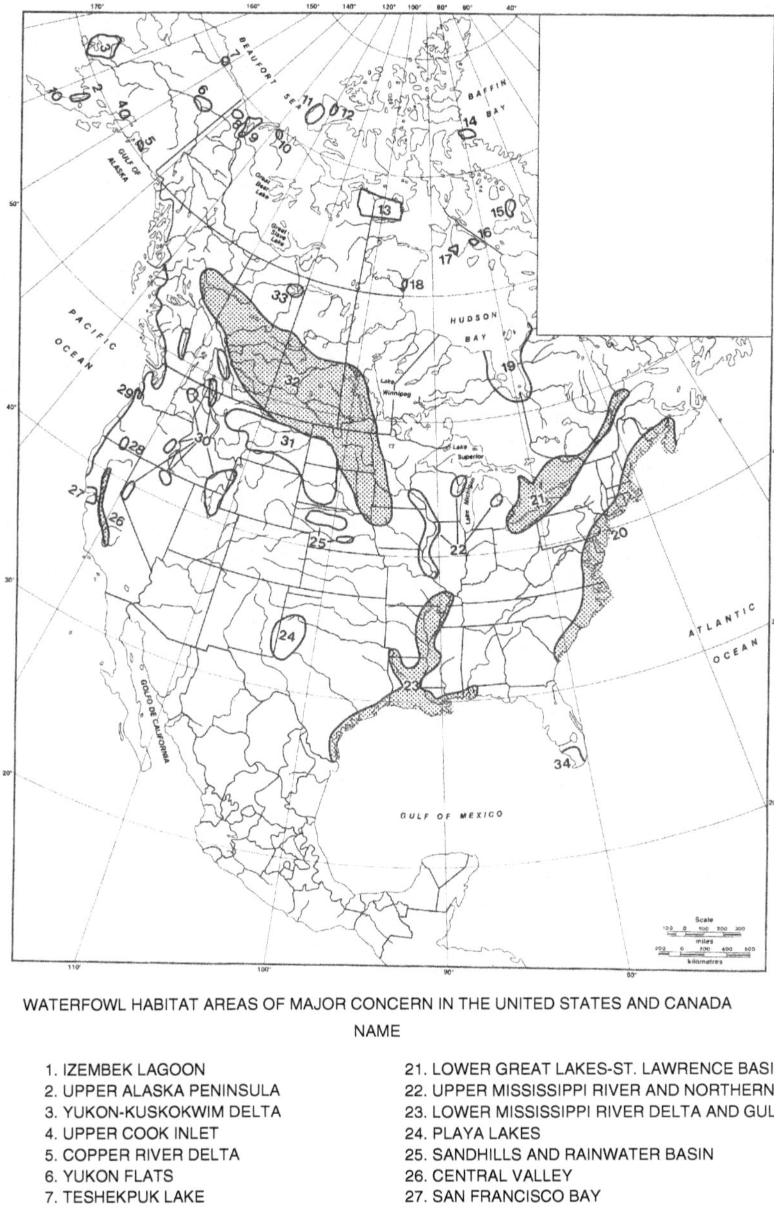

WATERFOWL HABITAT AREAS OF MAJOR CONCERN IN THE UNITED STATES AND CANADA

NAME

1. IZEMBEK LAGOON
2. UPPER ALASKA PENINSULA
3. YUKON-KUSKOKWIM DELTA
4. UPPER COOK INLET
5. COPPER RIVER DELTA
6. YUKON FLATS
7. TESHEKPUK LAKE
8. OLD CROW FLATS
9. MACKENZIE RIVER DELTA
10. ANDERSON RIVER DELTA (BS)
11. BANKS ISLAND No. 1 (BS)
12. BANKS ISLAND No. 2 (BS)
13. QUEEN MAUD GULF (BS)
14. BYLOT ISLAND (BS)
15. DEWEY SOPER (BS)
16. EAST BAY (BS)
17. HARRY GIBBONS (BS)
18. McCONNELL RIVER (BS)
19. JAMES BAY LOWLANDS
20. MIDDLE-UPPER ATLANTIC COAST
21. LOWER GREAT LAKES-ST. LAWRENCE BASIN
22. UPPER MISSISSIPPI RIVER AND NORTHERN LAKES
23. LOWER MISSISSIPPI RIVER DELTA AND GULF COAST
24. PLAYA LAKES
25. SANDHILLS AND RAINWATER BASIN
26. CENTRAL VALLEY
27. SAN FRANCISCO BAY
28. KLAMATH BASIN
29. MIDDLE-UPPER PACIFIC COAST
30. INTERMOUNTAIN WEST
31. NORTHERN GREAT PLAINS
32. PRAIRIE POTHOLES AND PARKLANDS
33. PEACE-ATHABASCA DELTA
34. SOUTHWESTERN FLORIDA

(BS) = MIGRATORY BIRD SANCTUARY
*MAP DELINEATIONS ARE GENERAL IN NATURE: SEE TEXT FOR DETAILS

SHADED AREAS ARE PRIORITY HABITAT RANGES

Figure 4.1 Priority habitat focus areas identified in the 1986 North American Waterfowl Management Plan.

individual JV websites, can be found at https://www.fws.gov/partner/migratory-bird-joint-ventures or via https://mbjv.org/who-we-are/directory/.

In this chapter and the next, I will sketch the development of these first habitat JVs beginning with the two in Canada and, in the next chapter, the first five in the United States.

Canada Gets Organized

There was a brief window between the completion of the National Waterfowl Plans (the United States in 1982, Canada in 1983) and the crystallization of NAWMP when some partners eagerly got to work. The Province of Manitoba was one jurisdiction that took up the challenge quickly. Director Richard Goulden and Chief of Habitat, Lorne Colpitts, provided the impetus, employing the existing structure of a multiagency Manitoba Waterfowl Technical Committee to assist. Barry Verbiwski (Provincial Waterfowl Biologist), Pat Rakowski (CWS Habitat Biologist), and wetland scientist Roger Pederson (Delta Waterfowl and Wetlands Research Station) led development of a Manitoba Waterfowl Management Action Plan in 1984–1985 (Verbiwski et al. 1984).

Beginning with the early draft objectives of NAWMP, and the proposed 15-year time horizon, the team set about estimating what needed to be done in Manitoba to achieve those goals. Parallel efforts occurred elsewhere but I focus on this example because Verbiwski, the leader of the writing team, graciously shared his files and recollections about the effort, which was lauded at the time by Jim Patterson as being among the best provincial start-ups.

The Manitoba Plan estimated how many ducks and geese would be required in Manitoba to meet continental objectives, given historical distributions of the birds, and how much habitat might be required to sustain these. It also reviewed data on crop depredation trends and the economic value of waterfowl hunting in the province. Planners believed that waterfowl populations had been declining in southwestern Manitoba due to loss of habitat, low reproductive rates, and increased harvest of adult hens during migration.

The plan declared five goals including population levels, preservation and management of habitat, research on population dynamics, lessening of crop damage, and encouragement of recreational use by Manitobans. Actions recommended included both population and wetland monitoring, law enforcement, education programs, and habitat enhancements on crown lands (including "Heritage Marshes" Box 4.1). Ambitious private land programs would take wing as the HELP program (Habitat Enhancement Land Use Program) after the provincial plan was written. Informal international communication was already evident at this stage. Verbiwski recalls, "In March of 1984, Bob Oetting dropped by my home to see where we were with the plan…and then called Harvey Nelson and the three of us added some sections to the Manitoba outline that evening". Oetting had previously worked with Nelson in ND.

When assembled, the Manitoba programs were projected to cost about $38 million annually to year 2000, including about 36 staff-years of added capacity. Of that $38 million, some $32 million would be needed for wetland enhancement and monitoring. That $38 million compared to a business-as-usual investment of $4 million, so this represented a real stretch for the provincial partners. In hindsight, this was an important start for Manitoba for at least two reasons: First, it drew the conservation partners in Manitoba together for joint planning ahead of the signing of NAWMP and, second, it got everyone thinking about the scale and multiple dimensions of such an ambitious project. Although this first effort lacked the data-driven modeling that the final plan in 1988 would have, it was a helpful start.

Box 4.1 The Manitoba Heritage Marsh Program

Planned as a ten-year cooperative conservation endeavor, the Heritage Marsh Program was launched in 1982. Focused on Manitoba's endowment of large marshes, the idea was to unite agencies to protect and enhance these wetlands. Delta Marsh, Oak Hammock Marsh, Whitewater Lake, Saskram Marsh, Oak and Plum Marshes, Netley-Libau Marsh, and Big Grass Marsh were among 11 candidate wetlands. Conservation elements would vary from place to place, but might include crown-land designations, crop damage prevention, land purchase, water management, and enhancing public use. A subset of these projects was achieved before NAWMP commanded most of the partners' resources elsewhere.

The PHJV

In the late 1980s, conservation planning action in Canada was at the Provincial/JV levels. Initially, NAWMP specified just two habitat JVs in Canada: the PHJV and the EHJV. The Prairie Habitat JV encompassed most of Manitoba, Saskatchewan, and Alberta from the US border northwards to the southern edge of the boreal forest – an area of some 200,000 square miles (Doherty et al. 2018). In 1986, the EHJV was expected to be smaller, imagined as a swath of land from the Great Lakes to the terminus of the St. Lawrence River. Though smaller in area than the PHJV, it encompassed parts of six provinces and was thus politically more complex than the western JV. Today, both JVs have expanded greatly to include additional landscapes and jurisdictions.

Provinces involved in First-Step projects (Chapter 3) were focused first on planning, funding, and delivering those products. Perhaps because they lacked a designated First-Step project, it was NAWMP partners in Manitoba who first kicked off a fulsome planning process for habitat delivery. I will treat the early evolution of the PHJV in more detail because I was personally engaged in it, and it comprised Canada's largest portion of NAWMP responsibilities.

The first PHJV technical planning meeting convened in Armstrong Lodge at Delta, Manitoba, on June 17, 1986, just a few days after the ink began to dry on the Minister's Signatures on the North American Plan. Eleven of us from Delta, DU Canada, the Canadian Wildlife Service, and the Manitoba government began debating step-down goals for duck numbers along with habitat programs, acres, and locations. Saskatchewan had already signaled their intention to focus on high-duck and high-mallard density areas in the parklands. Through the afternoon, we poured over Manitoba duck numbers by strata, kicked around indices of recruitment, hunter numbers, pond data from the federal May surveys, assumptions about production on DU managed areas, guesses about what we thought conservation tillage practices might achieve for nesting ducks, and the economics of purchasing vs. leasing land. In the end, I recall a sense of overwhelming complexity and deep uncertainty about what was needed on the land to materially affect the ducks.

Following that technical meeting, on August 15, a smaller group of six senior people from the same agencies convened in Winnipeg. Rich Goulden, then Director of the Manitoba Wildlife Branch, reported on the first two meetings of an *ad hoc* group of JV planners – the three Provincial Wildlife Directors, Jim Patterson (CWS), Dave Neave (WHC), and Stew Morrison and Sandy Macaulay (both DUC). Each province would be asked to draft an implementation plan, and the regional group would establish principles for the overall approach in prairie Canada.

Importantly, Goulden declared that it "would not be good enough to say that we've secured x number of acres of habitat. We must translate these acres into ducks flying south". That was an ambitious challenge given the knowledge of the day. McCuaig reflected on this as a vital mind flip. The primary goal was ducks, not acres or dollars, and the funders would need to know that the Plan was making progress. That attitude helped stimulate investment in rigorous biological evaluations, one of the important hallmarks of the PHJV.

Some key initial decisions were to engage agriculture in the planning work and to select a chair of the technical committee. A key technical challenge was how to organize all the information available on duck distributions, habitat preferences, breeding success, habitat trends, and presumptive effects of various habitat modifications – and how to combine these in a way that might allow scenario planning. Rick Wishart of DU Canada was appointed to chair the technical committee. After the challenging June meeting at Delta, Rick had done some preliminary work with DU staff on a computer planning tool to begin organizing some of this disparate information. Wishart's penchant for detail made him perfect for this seminal effort. Jeff Nelson, another biologist with DUC, was added to the team. Jeff's technical savvy and strategic outlook added strength to the group. A critical early step was to consult the scientists at FWS's Northern Prairie Wildlife Research Center in Jamestown, ND. Lew Cowardin, Doug Johnson, and others had been building a mallard production model based on data collected in ND and the Manitoba team thought that it might provide the basis for a decision support model for the PHJV if we could link it to a cost tool like DUC had begun to build.

Jeff and Rick's trip to ND, along with staff from Delta, set Manitoba and eventually the whole PHJV on a productive course to develop computerized models to help direct program planning and delivery. In time, with extensive commitments to field evaluations, the JV has developed world-class conservation planning tools (PHJV 2021a; Devries et al. 2023).

With a copy of the early mallard model in hand, DUC staff began integrating it with the DUC computer planning tool. Wayne Cowan (DUC) was engaged to help develop a conservation-agriculture component (zero tillage, delayed hay, rotational grazing, chemical fallow, etc.) which could be costed, but with little evidence yet for waterfowl responses to these treatments. Pat Rakowski (CWS habitat biologist), Barry Verbewski (MB Wildlife Branch), and George Hochbaum (CWS with modeling expertise) dove in as well. Greg Koeln (DU Inc., Chicago) added expertise in remote sensing and wetland mapping, with Brian Kazmerik and Mike Conchatre providing the programming expertise at DUC.

Jeff Nelson (pers. comm September 2025) recalls that this approach meant dramatic change. Internally, DUC engineers had been simply expanding literature values of x nest/acre to estimate cover requirements. They reasoned that if we find 10 nests/100 acres of cover type X, if we planted 1,000 acres of X, then we would have 100 nests. But this ignored the limitation of waterfowl behavior and pair densities. The FWS model would predict that if you have, say, 50 pairs in a square mile with 200 acres of cover, who select and nest in various covers, even if you planted 600 acres of cover, you will still have only 50 nests, not 150, at least not quickly, though their hatching success might improve resulting in more pairs settling there over time. Thus, the mallard model/CPT projections forced a major change in thinking about planning for upland habitat treatments.

A preliminary description of the CPT was shared with the Manitoba NAWMP Technical Committee in early September 1986, at which point the committee endorsed moving ahead with the DU planning tool as the primary basis for designing habitat programs in Manitoba. By October, the modeling work began in earnest with collecting and incorporating the best available data to parametrize the tool and, by November, DU staff had begun to broaden application of the model for planning in Saskatchewan and Alberta.

With a working tool in hand, the NAWMP technical groups in each province could begin scenario planning, linking the duck production estimates with assumptions about the social/political/financial realities of what might be possible on the dominantly private agricultural lands across the prairies. Engagement of agricultural agencies and experts on each of these committees was essential for conditioning the biologists' plans with thoughts about acceptability by rural communities and governments.

A hopeful though never fully realized vision for prairie habitat improvement was articulated early on by Jim Patterson (CWS) who imagined that conservation gains would need to be "kick started" via intensive waterfowl conservation programs (e.g., predator-fenced dense nesting cover) that would gradually diminish in importance and expenditures through the 15-year horizon of the Plan as conservation gains from improvements in public policies with positive ecological impacts would grow in importance. Thus the visionary PHJV Prospectus (PHJV 1987), completed in advance of the detailed provincial implementation plans, called for four components: (1) private and public lands in areas of high waterfowl production capability would be secured and managed intensively by wildlife agencies; (2) large wetland complexes important to waterfowl would be protected and, where necessary, enhanced; (3) habitat programs would be developed which would be complementary to and supportive of agricultural soil and water conservation initiatives; and (4) government policies that affect land use would be reviewed with the intent of achieving both agricultural and wildlife conservation objectives, resulting over time in broad changes to landscapes across the prairies.

This hoped-for resolution of policy conditions that had prevented sound soil, water, and wildlife conservation was achieved only to a limited extent in the early years of the Plan, though more progress has been made in the last decade (Chapter 24). A Federal Policy on Land Use included a clause that required wetlands to be "considered" in the execution of federal programs. This did somewhat ameliorate damage from federally funded road construction. A Federal Wetlands Policy followed, but in Canada the federal government has far less jurisdiction over land use than do the provinces, so progress with policy protection has been slow.

Provincial implementation planning took longer than expected as complications of program design and delivery mounted, but by July 1988 Manitoba had an implementation plan approved by the PHJV partners. It proposed major investments in fenced and unfenced dense nesting cover, inter-pothole grass seeding, and nesting baskets, complemented by agricultural demonstration programs around rotational grazing, chemical fallow, stubble mulch fallow, and winter wheat/minimum-till farming practices. All of these were designed to achieve specific habitat acreages and duck production objectives. We also proposed seven policy-change efforts targeted at municipal, provincial, and federal policies. Also included were a large-marsh conservation program, waterfowl crop damage prevention, and a program evaluation strategy.

Plans followed for Saskatchewan (June 1989) and Alberta (July 1989) that were similar but with slightly different mixes of programs. Alberta developed plans for their short-grass prairie biome in the south, for the parklands (spreading out from their first-step efforts in the Buffalo Lake Moraine), and for the Peace Parklands along with border with British Columbia, northwest of Edmonton. Saskatchewan proposed a mix of programs with greater emphasis on modifying agricultural practices through demonstrations and more land leasing programs vs. purchase and intensive management. Target areas chosen there were also mostly in the parklands but included the grassland region of the Missouri Coteau.

Thus, with a great deal of work by the partners, habitat conservation plans for the PHJV had come together during 1987–1989, in the absence of NAWCA or any other appropriations to guarantee a flow of funds to pay for these ambitious programs. At the time, people recognized the vital role of the First-Step projects in providing a proof-of-concept model for funding, but, in hindsight, I think the existence of the First-Step projects also gave heart to the planners who dove in and devoted many months of creative effort to developing the planning tools and action plans with the assumption that somehow, some coalition of funding partners would find a way to support these grand plans. In hindsight, that was truly an act of hope in the promise of NAWMP.

An important component of the PHJV plans, better developed than in any other JV at the time, was a companion evaluation plan (Delta 1987). Given the scope of the proposed work, its probable cost, and the vital nature of success on the prairies for the success of the continental Plan, partners realized that assessments of program effectiveness and subsequent adaptation of delivery strategies would be absolutely vital.

Delivery of the habitat work in each province was an additional matter for debate. The partners had agreed in principle to minimize new bureaucracy for delivery of NAWMP programs. But without knowing where the money would come from, from what source and by what route, most PHJV partner organizations imagined capturing a "piece of the pie".

Ducks Unlimited, with the advantage of staff and infrastructure distributed in all three provinces, and business connections to their sister organization in the United States seemed to be a natural delivery agent. Changing gears for NAWMP, however, would be a huge adjustment for Ducks Unlimited Canada. Previously staffed mainly with engineers and construction people focused on large managed wetlands, DUC had been slowly expanding its biological staff and, with people like Wayne Cowan designing farmer-friendly land use initiatives like Prairie CARE, the company was undergoing a major retooling including difficult staffing adjustments. DUC was also long accustomed to operating with private or provincial landowners as their only clients, so embracing the more complex world of planning through JV partnerships was a seismic change. DU Inc. also needed to be assuaged as their fundraising apparatus in the United States had been hugely successful selling acres of wetlands and miles of shoreline for big projects. NAWMP presented a new game for DU fundraising too.

The Canadian Wildlife Service with its national mandate logically had the lead coordinating function with the JV, but not the field staff in position to deliver programs. The provincial wildlife agencies also had a strong interest in program delivery. With some staff in place in target landscapes, and their ties with provincial agriculture departments, they had assets to offer too. It would take several more years for these logistical details to sort out. As the plans were developing, each province created their own delivery coordination agency.

The Manitoba government set up the Manitoba Habitat Heritage Corporation as a Provincial Crown Corporation in 1986 with its own Board of Directors comprising government, non-government, wildlife, and agricultural members with academic and community backgrounds. Headquartered in Winnipeg, in 1990, it was mandated to coordinate NAWMP delivery in the Province and Executive Director Rich Goulden added staff in high-priority delivery areas like Shoal Lake. In 2021, MHHC moved out of government and became a private, charitable organization with expanded responsibilities (see Sopuck Box 23.2).

In Saskatchewan, the government created a similar body in 1989, the Saskatchewan Wetlands Conservation Corporation (SWCC) with links to both the Wildlife Branch and water resources. It too created a Board to oversee the program with links to the provincial government and hired Bob Carles as the first staff leader. Unlike in Manitoba, this body was set up only after the Saskatchewan Implementation Plan was completed in June 1989 by a technical committee from CWS, DUC, SK Parks, Recreation and Culture, SK Rural Development, and the federal Prairie Farm Rehabilitation Administration that long had a presence in Saskatchewan. As in Manitoba, SWCC envisioned a substantial program delivery role for funds coming to Canada from the United States.

Alberta took a different approach. Much like in SK and MB, the implementation plan was drafted by a multiagency team comprising DUC, Alberta Agriculture, Forestry Lands and Wildlife, and CWS. Led by the provincial government, the partners created an Alberta Implementation Office for NAWMP and hired an experienced provincial manager, Morley Barrett, as Executive Director. By December of 1990, however, Barrett and company had settled on a different model for program delivery (Barrett letter to the PHJV Board, Dec 14, 1990). After extensive negotiations, Ducks Unlimited Canada and the Alberta Fish and Wildlife Division agreed to deliver jointly NAWMP land management programs in Alberta, using common planning, programming, tracking, promotion, and communication systems and utilizing as far as possible existing staff and offices to deliver the program cost-effectively. Alberta thus opted to minimize new structure in the delivery of NAWMP programs.

The directors of all three provincial delivery bodies were added as members of the governing PHJV Advisory Board that provided policy guidance and oversight for the entire PHJV enterprise, joining CWS, WHC, DU Canada, Agriculture Canada (PFRA), Delta Waterfowl and Wetlands Research Station, and the Directors of each of the provincial wildlife agencies. Chaired initially by CWS Regional Director, Gordon Kerr, with the support of A.J. "Sandy" Macaulay who had been seconded by DU to CWS for the new role of JV Coordinator, the JV got underway.

On May 4, 1989, the formal agreement creating the PHJV partnership was finalized. Gerald McKeating soon succeeded Kerr as Chair and led the JV through its first formative decade. Deanna Dixon replaced Macaulay in 1994 bringing extraordinary administrative and people skills to the increasingly demanding JV coordination role. Standing Committees of the PHJV at the onset included Evaluation (which I chaired) and Policy (chaired by David Neave, WHC). These committees continue to function under different leadership today.

Box 4.2 Dr. Alexander James "Sandy" Macaulay (1945–2003)

"Sandy" grew up in Edmonton, earned degrees in zoology and anthropology at the University of Alberta before moving to Manitoba where he studied the ecology of bulrushes under Dr. Jennifer Walker Shay. He went to work for Ducks Unlimited Canada serving initially as the Provincial Biologist for Saskatchewan, and from 1976 to 1987 as Chief Biologist in Winnipeg where he was instrumental in moving DU Canada from a company focused largely on construction and engineering to one based more on biological principles and an array of wetland types supported by objective benefit/cost analyses. He was seconded to Ottawa to assist Jim Patterson with implementing the North American Waterfowl Management Plan. With the Plan launched, in 1989, Sandy was hired as the first Coordinator for the PHJV. Paired with Gordon Kerr, the CWS Chair, he served in that role until 1994, supporting the JV's biological planning and evaluation processes and contributing to building the JV's institutional partnerships. An optimist by nature, he was a good fit for the initial coordination role. Sandy became an early proponent of using GIS technology for land management purposes. He founded a private company, Resources for Tomorrow, in 1993 and soon turned his full attention to it for the next decade.

Box 4.3 Richard Goulden – Seizing Opportunities (1941–1997)

Goulden describing NAWMP ca. 1988 said:

> We have a chance to play a part in a landscape drama... unfolding across the world's most richly blessed continent... the opportunity to recreate the setting for the return of great flocks of wild waterfowl, of songbirds and marsh denizens of all kinds...This requires vision...
>
> (NAWMP Update 1998 page vi)

Born in Yorkton, Saskatchewan Rich's childhood was spent in Flin Flon, Manitoba. After receiving a BSc degree in Agriculture from the University of Manitoba, he completed graduate studies in Wildlife Management at Utah State University. For over 30 years, he worked in wildlife and resource management, holding positions of Coordinator in the Canada Land Inventory Program, Director of Fisheries and Wildlife, Director of Parks, and Assistant Deputy Minister of Natural Resources in Manitoba. He was named the first Chief Executive Officer of the Manitoba Habitat Heritage Corporation, seizing the opportunity to expand Manitoba's engagement with the new North American Waterfowl Management Plan. Goulden was active in numerous provincial, national, and international resource management initiatives and served on the founding Board of Directors of Wildlife Habitat Canada.

The PHJV Today

Fast-forward to today, the PHJV is among the most successful of the NAWMP JVs. As of 2023, the partnership had secured 8.6 million acres of habitat, enhanced another 3.1 million acres, and positively influenced 8 million more. By engaging some 17,000 landowners and 566 funding partners since 1987, the JV had invested more than $1.6 billion in conservation.

With a stable but slowly expanding partnership, a succession of five-year implementation plans (among the most frequently updated of all the JVs), decades-long commitments to science, evaluations, and ever-improving decision-support models, the PHJV is arguably the most "delivery tested" of all the JVs. Decades of research including three multi-year Ducks Unlimited Canada studies: PHJV Assessment Study, 1993–2000 (Howerter et al. 2014); Pintail Study, 2005–2007 (Devries et al. 2018); and the Spatial/Temporal Variability Study, 2001–2009 (Bortolotti et al. 2022), greatly improved understanding of hen- and nest-level breeding success related to habitat parameters. These projects have culminated in a dataset that is like no other in wildlife conservation. The data are richest for the five main dabbling duck species breeding in prairie Canada: mallard, gadwall, blue-winged teal, northern shoveler, and northern pintail. These data have allowed DU's Institute for Wetland and Waterfowl Research to build and adapt waterfowl production models that predict the breeding density and breeding success of these species in response to habitat change features in prairie Canada landscapes.

Current PHJV planning tools are the result of 30 years of heavy investments in waterfowl science and landscape change assessments. Their targeting maps and waterfowl productivity models are backed by as much data and tougher testing than plans anywhere on the continent, and that gives the PHJV confidence in estimating the probable value of each new conservation investment (https://phjv.ca/science-and-planning/). For example, the waterfowl return on investment tools that guides program delivery are based on more than 26,979 duck nests found and monitored, 3,214 radio-marked female mallards tracked to examine their preferences and their fates, and 145 waterfowl nesting study sites sampling over 652,000 ha of parklands and grasslands. The Waterfowl Productivity Model and associated tools for costs and risks of loss (Devries et al. 2023) are now used by every organization planning conservation investments in the PHJV.

The PHJV also broadened its area of endeavor. After protracted considerations from 2000 to 2011, the PHJV embraced an expanded role in facilitating conservation in Canada's Western Boreal Forest, adding oversight to a program that had already secured some 12 million acres and influenced 122 million more. Expanding the PHJV science team to include experts and oversight for boreal science, the JV has since 2013 developed separate detailed implementation plans for both the Western Boreal Forest (PHJV 2021b) and the Parkland and Prairies (PHJV 2021a). The JV has also an active policy focus seeking ways to enhance regional habitat quality through improvements in government and corporate policies, important work in both the Boreal region and the prairies.

The PHJV has also engaged experts and processes to add shorebirds, waterbirds, and grassland birds to their habitat actions as incremental funding allows. They are building complementary targeting models based on a steadily improving database for those birds too.

Following the 2012 NAWMP Revision, the JV also added people objectives with a focus on widening support for conservation. The top priority target group is the agricultural community as they control or influence the great bulk of the land in the Prairie Pothole Region. This includes private landowners, land managers, and agricultural industries. Next are governments and the public at large. Details of these initiatives are still developing.

The EHJV

Canada's EHJV was formally created at a meeting of Wildlife Ministers in November 1989. The first comprehensive five-year EHJV plan was completed in June 1990. "The east had a slower start", Jim McCuaig recalled. "With more Provinces to engage, and less previous preparation, the process was more complex". Eventually Raymond Sarrazin from Québec was chosen as the coordinator but he left the job after a year and was replaced by Keith McAloney, from DU Canada's Atlantic Region. Keith served effectively for six years with his unique combination of people skills and program knowledge as the EHJV got up and running. He worked closely with George Finney, Eastern Region CWS Director and Chair of the EHJV Management Board. George had served on the Canada/US NAWMP negotiating team, so he was deeply knowledgeable about NAWMP.

Initially, the JV comprised southeastern Ontario along the shores of Lakes Erie and Ontario, and southern Quebec along both sides of the St. Lawrence River, Anticosti Island, and coastal riparian areas of New Brunswick, Nova Scotia, Prince Edward Island, a few bits of Newfoundland, and two small areas of interior Ontario and Quebec. Later, as an all-bird JV, it expanded to include nearly all of Eastern Canada, some 780 million acres.

Partners proposed a division of effort between breeding (65%), migration (30%), and wintering (5%) habitat, which made biological sense but complicated planning. Breeding black ducks were the primary target species at the outset, but coastal wetlands were valuable for many other species including Canada geese, greater snow geese, canvasbacks, redheads, mallards, wood ducks, ring-necked ducks, and several species of sea ducks connected to both the Atlantic and Mississippi Flyways.

Human impacts on waterfowl and wetlands across the JV included agricultural drainage, urban expansion, declining watershed health, hydroelectric development, acid rain, and more.

Like the PHJV, the EHJV Prospectus included a list of key principles, a summary of goals, objectives, and strategies to attain them positing 648K acres of secured and enhanced waterfowl habitat and some 3.7 million acres of improvements through indirect actions like education and public policy adjustments. In addition to the First-Step projects (Chapter 3), by 1989, EHJV partners had six more projects mostly organized, one in each province.

A more fulsome plan came together in 1990 with specific goals for each province and program type. Responding to perceived threats and opportunities in areas known to be heavily used by waterfowl, these projections were largely best-professional-opinion prescriptions with fewer data and decision-support tools available as developing in the prairies. The spatial footprint of the JV grew with this five-year plan and included some commitment to evaluating the effectiveness of proposed actions. EHJV partners began to assess some initial projects in 1991–1994 and developed the first comprehensive evaluation plan by 1995.

Early on, federal partners urged development of a broader biodiversity focus for EHJV delivery and evaluation, which set the stage for broadening JV actions and an updated prospectus along with an evaluation plan. The EHJV was among the first JVs to recognize and develop strategies to ensure that the waterfowl conservation work was positive, or at least neutral, regarding impacts on threatened or endangered biota.

The EHJV Today

The current EHJV Implementation Plan was finished in 2024 and covers the period through 2030 (Eastern Habitat Joint Venture 2024). That plan notes that, at over 3 million km^2 (1.2 million square miles), the EHJV is the largest of the NAWMP's 22 habitat JVs. This area also supports 68% of Canada's human population, most of which are concentrated in southern Ontario, Québec, and urban centers in the Maritime provinces[1] and holds some 39% of Canada's wetlands across multiple ecological regions (9 Bird Conservation Regions (BCRs); see Chapter 9). These breeding, migration, and nonbreeding habitats support hundreds of bird species and biodiversity broadly.

Some 246 target species have been identified in BCR-specific strategies for the EHJV including 132 landbirds, 54 shorebirds, 28 waterbirds, and 32 waterfowl species. Priority species were selected using population status and trends estimates and the importance of the EHJV area to each species. On that basis, the EHJV identified six high-priority waterfowl species for their latest Plan: American black duck, mallard, wood duck, common goldeneye, Barrow's goldeneye, and the American race Common Eider (hereafter Common Eider). A parallel process was used to identify priority shorebirds, landbirds, and waterbirds. Associated habitat goals were then distributed among the partner provinces.

Avian population monitoring has been improved in the EHJV since the onset of NAWMP. Habitat loss/conversion has been extensive and difficult to monitor across the vast JV. Human development pressures are pervasive. Invasive species have challenged wetland habitats too, the most serious being the non-native European common reed (*Phragmites australis* spp.) followed by water chestnut (*Trapa natans*), Eurasian watermilfoil (*Myriophyllum spicatum*), and purple loosestrife (*Lythrum salicaria*). Each of these species tends to expand rapidly and form dense stands that can lessen the amount of open water in a wetland, reduce nutrients for native plants, and degrade habitat for wildlife. Since around 2013, Phragmites has emerged as the most troublesome invasive plant species in the EHJV. Its rapid expansion in coastal wetlands and along roadways has stimulated extensive research and experimental control. Air and water pollution have also had widespread negative effects on habitats and biodiversity in Eastern Canada.

The suite of JV actions to counteract these challenges includes habitat retention, restoration, and management along with evaluations to test the efficacy of the work. Efforts to affect public policies and legislative fixes to favor positive conservation outcomes are another focus.

Box 4.4 R. Keith McAloney (1948–2016)

Keith McAloney was an important leader in the realization of NAWMP in Eastern Canada. Born in rural Nova Scotia, Keith developed a passion for wetlands and waterfowl. He earned undergraduate and MSc degrees at Mount Allison and Acadia Universities, respectively, and spent his career of over 40 years working for waterfowl conservation.

First hired by DU Canada in Amherst to develop wetland conservation projects, he later became coordinator of the EHJV with the Canadian Wildlife Service in Sackville, N.B., working closely with Regional Director George Finney. Later, Keith returned to DU as manager of conservation programs for Atlantic Canada, before finishing his career with CWS. McAloney was a knowledgeable biologist and a mentor to younger staff. He played multiple roles in implementing NAWMP including a pivotal role in establishing the Sea Duck JV and the Black Duck JV, in addition to the EHJV.

Note

1 Statistics Canada 2021 (https://www12.statcan.gc.ca/census-recensement/2021/dp-pd/prof/index.cfm?Lang=E).

References

Bortolotti, Lauren E., Robert B. Emery, Llwellyn M. Armstrong, and David W. Howerter. 2022. "Landscape composition, climate variability, and their interaction drive waterfowl nest survival in the Canadian Prairies." *Ecosphere* 13: e3908.

Delta Waterfowl and Wetlands Research Station (Editors). 1987. *Evaluation Program. Prairie Habitat Joint Venture. North American Waterfowl Management Plan.* Unpublished Proposal. 27pp.

Devries, James H., Llwellyn M. Armstrong, David W. Howerter, and Robert B. Emery. 2023. "Waterfowl distribution and productivity in the Prairie Pothole Region of Canada: Tools for conservation planning." *Wildlife Monographs* 211: 1–49.

Devries, James H., Robert G. Clark, and Llwellyn M. Armstrong. 2018. "Dynamics of habitat selection in birds: Adaptive response to nest predation depends on multiple factors." *Oecologia* 187: 305–18.

Doherty, Kevin E., David W. Howerter, James H. Devries, and Johann Walker. 2018. The prairie pothole region. Pages 1–10 in C. M. Finlayson, G. R. Milton, R. C. Prentice, and N. C. Davidson (Editors). *The wetland book II: Distribution, description and conservation.* Springer, Dordrecht.

Eastern Habitat Joint Venture. 2024. *Eastern Habitat Joint Venture Implementation Plan 2021–2030: Partners Conserving Habitat for Birds and People.* Eastern Habitat Joint Venture. Environment and Climate Change Canada, Sackville, NB.

Fisher, Robert, W. Ury, and B. Patton. 2011. *Getting to yes.* Penguin Books, New York. 236pp.

Howerter, David W., Michael G. Anderson, James H. Devries, Brian L. Joynt, Llwellyn M. Armstrong, Robert B. Emery, and Todd W. Arnold. 2014. "Variation in mallard vital rates in Canadian Aspen Parklands: The Prairie Habitat Joint Venture assessment." *Wildlife Monographs* 188: 1–37.

Prairie Habitat Joint Venture. 1987. *Prairie Habitat: A Prospectus.* Environment Canada, Edmonton, AB. 32pp.

Prairie Habitat Joint Venture. 2021a. *Prairie Habitat Joint Venture Implementation Plan 2021–2025: The Prairie Parklands*. Report of the Prairie Habitat Joint Venture. Environment Canada, Edmonton, AB.

Prairie Habitat Joint Venture. 2021b. *Prairie Habitat Joint Venture Implementation Plan 2021–2025: The Western Boreal Forest*. Report of the Prairie Habitat Joint Venture. Environment Canada, Edmonton, AB.

U.S. Fish and Wildlife Service, Pacific Flyway Council, Central Flyway Council, Mississippi Flyway Council and Atlantic Flyway Council. 1982. *A National Waterfowl Management Plan for the United States*. U.S. Department of the Interior. Washington, DC. 32pp.

Verbewski, Barry, Patrick Rakowski, and Roger Pederson. 1984. *Waterfowl Action Plan for Manitoba*. Manitoba Waterfowl Technical Advisory Committee. Unpublished Report. 79pp.

5 The First US Habitat Joint Ventures

As Canada was launching its two initial habitat JVs, the United States was establishing five others, completing coverage of the first-priority areas designated in the 1986 Plan (Figure 4.1).

The Lower Mississippi Valley Joint Venture

The first US Joint Venture to launch was the Lower Mississippi Valley Joint Venture (LMVJV) in 1987. Initially envisioned as a combination of the Lower Mississippi Valley and the western Gulf Coast, it quickly made more sense to split these for planning purposes. The LMVJV encompassed the Mississippi River floodplain of some 24 million acres in western KY, TN, and MS, eastern OK, northeastern TX, northern LA, southern MO, and AR. Agricultural development, timber harvest, wetland drainage, and dredging for navigation were the most common impacts on waterfowl habitat abundance and quality. The habitat estate included both public and private lands.

Seth Mott was Chief of Game for Mississippi, in the mid-1980s. He recalls that his first substantive exposure to NAWMP was attending a meeting in Nashville hosted by Gary Myers, where the southern states were asked to identify the "most important areas for new waterfowl conservation projects" (pers. comm. March 2025).

> Everyone there assumed that meant land acquisition. Gary had his staff prepare a large-format map of the southern Mississippi Flyway which was hung on the wall of the meeting room, and we all took turns drawing polygons on the map. Most of these areas came from previous planning efforts for National Wildlife Refuge acquisitions and had helped define the priority areas shown in the 1986 NAWMP but there were some new areas too.

Myers, thinking ahead as usual, had applied for a federal grant to help set up the JV. Half of the money was left with the FWS to help pay for a coordinator and the rest was used to support operational planning.

"Few states, Mott recalled, with the exceptions of Missouri and Tennessee, had money for land acquisition. Prior to the passage of NAWCA, most land acquisition was undertaken by the FWS". At that time, the Assistant Regional Director for Refuges and Wildlife was Harold Benson. Harold was from Wisconsin and spent his early career in the prairies. His outlook was "protect all the land you can now; worry about staffing and management later". Mott recalled that,

> with Benson's support, Refuge folks began aggressively pursuing several of the targets identified at the Nashville meeting. This resulted over the next decade in several new NWRs

DOI: 10.4324/9781003732310-5

> including Cache River, Bald Knob, Tallahatchie, Lake Ophelia, Bayou Cocodrie, and Grande Cote; as well as significant additions to Chickasaw, Lower Hatchie, and Tensas River.

Soon after the Nashville meeting, Gary Myers and Jim Pulliam (then FWS Regional Director for the southeast) began working on creating an LMVJV Management Board. The decision to have separate JVs for the Gulf Coast and LMV had been made by then. Seth added, "I can only speculate this decision was based on the differences on habitat (coastal marsh vs. bottomland hardwoods) and giving each FWS Region (Southeast vs. Southwest) responsibility and money for standing up a joint venture". The administrative boundary between the Mississippi Flyway and Central Flyway may have been considered as well. The LMVJV was being perceived as the "bottomland hardwood" JV, so other states that had bottomland hardwoods outside of the Mississippi Alluvial Valley wanted to be included too. This resulted in the boundary for the JV expanding to include small parts of TX, OK, KY, IL, and IN.

In 1987, Charles Baxter was Project Leader for the Ecological Services Field Office in Vicksburg where he spent much of his time dealing with the Corps of Engineers trying to save bottomland hardwood forests. The week after Christmas, Jim Pulliam and Harold Benson invited him to the FWS Regional headquarters in Atlanta.

> I'll never forget their words, Baxter recalled, and it came from Pulliam who was then the Regional Director. He said 'we don't know what a Joint Venture is, and we don't know what a Joint Venture Coordinator is, but we want you to figure it out.' And like a dummy I said OKAY.

For the next 20 years, Charles served admirably as the LMVJV Coordinator.

Mott recalled,

> as Chief of Game for Mississippi, I had been working with Charles and his staff to implement a provision of the 1985 Farm Bill which allowed the Farmers Home Administration to convey tracts of foreclosed farmland to state or federal agencies for conservation purposes. Throughout the Valley there were tens of thousands of acres of such lands, most all of which were recently cleared of bottomland hardwoods that were too wet for agriculture. Harold Benson was supportive of adding these lands to the Refuge System. He even created the first FWS Wetland Management District in the Southeast to administer and reforest the new tracts. My growing enthusiasm for this kind of work led to me leaving the state agency and accepting the offer to become the assistant LMVJV Coordinator."

As Gary Myers commented much later, "Seth and Charlie did a really good job of figuring how to run a Joint Venture".

The Management Board had a formative meeting at the North American conference in the spring of 1988. Baxter remembered,

> The deep South is not a land of NGOs or public land trusts…so it was the normal players – the state wildlife agencies, Ducks Unlimited, TNC, and the FWS. As a new Coordinator, I had no grounding in the culture of waterfowl conservation. I didn't know Flyway people… but we had some folks that were real leaders, like Ken Babcock (MO), Hugh Bateman (LA), Steve Wilson (AR), and Jim Pulliam and Harold Benson. Those guys were committed, but like me didn't know what a Joint Venture was …but they were committed to being one.

In an interview 18 years later, Baxter quipped,

> I'm a federal employee, but in many respects, I've always worked for Gary Myers, because of his leadership in bird conservation. There was a period when he was providing funding for a full time FWS position! Therein lay much of the strengths of the North American Plan. It kind of prompted and supported that non-traditional thinking. But the initial impetus was on getting an Implementation Plan done.

That was the common focus of all the early Joint Ventures.

Reflecting on the first LMVJV Plan, Baxter admitted that the,

> little 29-page document (October 1990) did not have a bunch of clear, explicit objectives.... It just pointed us in directions like private lands, acquisition, increase carrying capacity of state and federal management areas. And so, the first year was pretty much a traditional, opportunity-based approach to conservation. But at the same time....because of those new things that I saw in the Plan and the imperative to link populations to habitat, we started trying to lay the foundation for – how are we going to move past this opportunistic approach... and actually think about defining... habitat objectives that have some kind of biological linkage to the Plan's population objectives? That was two years after I became Coordinator. And then we started worrying about how can we become more science based in terms of our waterfowl conservation programs.

Initial activities of the JV focused on acquisition and private-land management. Charles reflected,

> While we didn't have a stable source of funding there was ... political support for acquisition in the Mississippi alluvial valley. And so, there were projects that we were able to get going in nearly every state. I felt very strongly that the North American Plan was pointing us toward private lands programs, which we didn't have. The issue there was – how do we develop private lands programs, targeted at waterfowl, and recognizing that it was going to have to be targeted at the agricultural landowners. We would not have been successful without leadership from Ducks Unlimited, specifically David Wesley and Curtis Hopkins. At that point in time, in the South, it was no small amount of risk involved in DU devoting resources on the wintering grounds, and on private lands. These guys that took risks within their organizations, and created some good private lands programs,...particularly in the Mississippi alluvial valley, in Arkansas, Louisiana, and Mississippi. They also brought know how about how to restore wetlands; getting state permits; interceding on behalf of a project, doing some of the leg work that was necessary to get those kinds of projects off the ground. We needed more than money early on.

Importantly, the LMVJV also became a leader in thinking about biological evaluations of JV actions. Baxter felt it was the responsibility of the partnership to create the biological linkage between habitat objectives and population goals.

> I viewed that as part of what I was paid to do and what this partnership that I was coordinating was intended to do. We had to start thinking about difficult things, like... how do you monitor increases in carrying capacity across a 24-million-acre eco-region? One of the early impediments was a tendency, inside the FWS in particular, to see the Plan as a program. In my mind this was not a government program. I mean, the mere concept of Joint Ventures says

> we're going to take what we're doing right now, as a private/state/federal community, and we're going to coordinate it better, and we're going focus it better, and we're going to leverage it better. And so that told me… you don't sit around waiting on new money to show up. You start figuring out what can you do with the… resources that the conservation community has right now, focus it, and leverage it. And that drove much of my thinking as a Coordinator, on the need to establish a strong biological foundation for what we were doing, to get more out of our existing resources.

Thus, with Baxter's leadership, the LMVJV became one of the first Joint Ventures to focus attention and substantial resources on biological evaluations of JV's projects. Baxter maintained his office in Vicksburg, co-located with the offices of the Patuxent Wildlife Research Center, and the National Wetlands Research Center giving the JV staff easy contact with Ken Reinecke and others who were doing groundbreaking research on wetlands, wintering waterfowl, and other bottomland-dependent species. This co-location was fortuitous. Reinecke had been working on the numbers, distribution, and habitat use of wintering mallards. So, the "best science available" for determining where, what kinds, and how much habitat was needed to support NAWMP population objectives was literally a few doors away. Charles and Seth leaned on Ken's knowledge and time to the extent that Reinecke's supervisor once accused him of being "captured by management". Reinecke's published research, along with that of others like Leigh Fredrickson, Mickey Heitmeyer, John Eadie, and Mark Petrie, led to the concept and calculations of "duck use days" as the basis for wintering habitat conservation objectives.

The LMVJV continued to evaluate and refine its objectives, soon adding Chuck Loesch as the first science coordinator and later Bill Uihlein when Chuck moved to the Habitat and Population Evaluation Team (HAPET) office in Bismarck. Over time, additional species were added to the JV population and habitat objectives and the approach spread among other JVs and beyond, leading to the FWS adopting "Strategic Habitat Conservation" as standard practice across the agency. Both the HAPET offices in Bismarck and with Rex Johnson in Fergus Falls were on a similar approach and timeline as the LMVJV.

Later, the LMVJV became an early adopter among the Joint Ventures for an all-bird conservation mission and Baxter again showed his leadership in urging other JVs to do the same (Baxter 2005). The LMVJV had a well-developed landbird plan by 1999 and a shorebird plan by 2000. Baxter credits David Pashley and some innovative FWS Refuge Managers for helping move the JV in this more inclusive direction.

The LMVJV Today

The Joint Venture's ambitious operational plan has a goal of, before 2029, having biologically driven, and partner-vetted population objectives for priority species within all bird guilds throughout its borders. Because of the importance of the Mississippi River as a migratory funnel, the JV estimates that some 60% of North America's migratory birds funnel through the Joint Venture area annually. It also serves as the single most important wintering area for mallards and wood ducks. Their slogan, "Dedicated to bird habitat conservation. Committed to the use of the best science available. Believing in the power of partnership", could well be applied to all of NAWMP.

The JV's early focus area on the 22 million acre Mississippi River floodplain was expanded several years ago to also embrace the 52 million acre West Gulf Coastal Plain and the Ouachitas, an open pine forest/grassland area to the west. Their current goals for that initiative area are first about collaboration and funding; second, to update population goals for all bird guilds there; third, to update habitat objectives for those same species and build effective decision-support tools for conserving them; fourth, to foster habitat project developments; fifth, to monitor and evaluate avian population and habitat statuses; sixth, to test their planning assumptions; and seventh, to carry out the JV's ambitious communication, education and outreach plans. The LMVJV continues to be a leader in all-bird conservation under the NAWMP umbrella.

The Prairie Pothole Joint Venture

Following the successful Flagship projects in North and South Dakota, the Prairie Pothole Joint Venture (PPJV) continued to organize. Although the entire US prairie pothole region was included in its area of responsibility, it was administratively complex including both Region 3 (Minneapolis) and Region 6 (Denver) of the FWS and overlapped the purview of both the Mississippi and Central Flyways and their associated states.

By 1988, ND, SD, MN, IA, and the FWS had agreed to form the JV and, because of the engagement of both FWS Regional Offices, two coordinators were identified – Paul Hartman from the realty division of Region 6 (who was succeeded by Mitch King in 1990) and Jerry Schotzko from Region 3. Steve Lewis (FWS, NCTC interview) recalls that was about the time he came to Denver and began working with both Paul and Jerry to assemble the first implementation plan for the Joint Venture. The JV garnered some seed money to help get started – $100K from The Nature Conservancy and $80K from the Iowa Wildlife Heritage Foundation.

As with the counterpart PHJV in Canada, habitat loss from agricultural intensification was the greatest threat to wetland loss and waterfowl production in the PPJV. In later years, concerns about watershed health, oil production, and climate change have also arisen but agriculture remains the greatest competitor for land use in the region. An underappreciated factor in the ability of both the US and Canadian prairie joint ventures to focus early on known problems was the historical work of the Northern Prairie Wildlife Research Center. Begun in 1967 under the leadership of Harvey Nelson, it was later complemented by the CWS research centre in Saskatoon. These institutions helped lay the science foundations for NAWMP in the 1960s and 1970s with initiatives like the Canadian study areas established near Lousana, AB, Redvers, SK, Minnedosa, MB, and Woodworth, ND, and later on the four-square-mile monitoring plots followed by the FWS wetland management districts. Collectively, this work greatly advanced our understanding of factors affecting waterfowl populations and provided the bases for planning model development.

For example, Al Sargeant, Ray Greenwood, Marsha Sovada, and others at NPWRC learned with studies in ND, and later across Prairie Canada, that red foxes were particularly influential predators on nesting ducks – frequently taking both eggs and incubating hens. They also learned that coyotes tended to displace red foxes and were usually more focused on mammalian prey. Yes, coyotes took duck nests and hens too, but typically at much lower rates than did foxes. Sovada et al. (1995) showed that nest survival of upland-nesting ducks was higher (32%) in areas where coyotes predominated than in areas with mostly red foxes (17%). Both canids were common in six other areas, where nest success averaged 25%.

As noted in Chapter 3, there had long been tensions between the FWS, politicians, and landowners over purchase of lands exclusively for wildlife, as well as residual head-knocking over the controversial Garrison Diversion Project. Early PPJV staff had to navigate this tricky terrain where working lands abut refuges or waterfowl production areas, and various USDA set-aside programs competed for annual cropping. Card Madsen, who joined Nelson's NAWMP office in Region 3, had a long history of working with private land conservation programs, mainly in SD.

He reflected (NCTC oral history, March 2, 2000) on how, in his long experience, private land conservation needed to be approached:

> Our relations in the agricultural community are very personal. They're one on one. People buy their machinery from someone they know. They sell their grain to someone they know… They don't respond well to postcards or mailing things or television ads and just send off and do it. You've got to get to know them personally, and that's been a hallmark of [successful FWS] Private Lands Programs [and] also Wetland Acquisition Programs.

Madsen's insights have proven instructive to workers in every JV where habitat on private lands is important for mission accomplishment.

Long known to be essential to ducks, the PPJV also hosts 45–90% of the continental breeding populations of several grassland bird species that are experiencing steep population declines. The loss of over 70 million acres of grasslands in the PPJV has made grassland birds among the most imperiled guilds in North America (North American Bird Conservation Initiative 2025).

A unique and important feature of the PPJV was the establishment by the FWS of the HAPET teams in 1988 with offices in Bismarck, ND, and Fergus Falls, MN. HAPET was established to monitor waterfowl populations and habitat, assess programs, and develop landscape-scale models to help direct waterfowl conservation efforts in the region. Working in partnership with wetland district managers, the Northern Prairie Wildlife Research Center, and other scientists, it provided vital support for the design, delivery, and evaluation of PPJV programs. Ron Reynolds (Bismarck) and Phil Arnold (Fergus Falls) were chosen to set up and lead the two offices at the outset.

A major early contribution of the group was to lead a partner-funded evaluation of the impact on nesting waterfowl of USDA's Conservation Reserve Program in ND (Reynolds et al. 2001) which resulted in important new insights about the effects of large blocks of perennial grass on nesting success.

Soon after beginning the waterfowl work, HAPET staff expanded their role to include other taxonomic groups (e.g., grassland passerine birds, shorebirds, marsh birds, and threatened and endangered species) as continental planning initiatives developed for those birds too. HAPET staff were early leaders in using Geographic Information System technology to integrate biology with landscape features and develop spatially explicit recruitment models to help guide and evaluate large-scale conservation problems. Two recent fact-filled publications available from the HAPET website sum up more than 30 years of avian habitat research and chart the future for waterfowl habitat work https://www.fws.gov/media/hapet-waterfowlfinal-091323pdf and grassland bird conservation in the PPJV region https://www.fws.gov/media/hapet-grassland-birdsfinal-091323pdf. The current HAPET office leader in Bismarck is Heath Hagy, and the office remains thoroughly engaged in planning for the PPJV and the Service's wetland acquisition program.

The PPJV Today

With the engagement of the FWS HAPET Offices, Northern Prairie Wildlife Research Center, State, and NGO science investments since the early years of NAMWP, the PPJV has substantially increased knowledge of regional waterfowl ecology and developed sophisticated planning tools.

Important fortuitous lessons were learned during the waterfowl population boom that coincided with a wet series of years in the presence of abundant USDA Conservation Program

Reserve perennial grass cover…and the subsequent reversal of population trends with the expiration of millions of acres of cover in later years. This both validated the expectation of what a productive landscape for waterfowl would look like and drove home the vital importance of farm policy as a strategic component of conservation in the Prairie Pothole Region.

The PPJV Implementation Plan of 1989 was revised in 1995. Two changes were to adopt a greater focus on working lands, that is working with farmers and ranchers who control some 90% of the PPJV geography via voluntary, nonregulatory programs to conserve grasslands and wetlands. A second substantial change was to commit in principle to work toward conserving other grassland birds as well as waterfowl. The next revision in 2005 confirmed a focus on wetlands and native grasslands.

The latest implementation plan in 2017 was a massive planning document (Prairie Pothole Joint Venture 2017) with detailed objectives and conservation plans for waterfowl, shorebirds, waterbirds, and landbirds, identifying top-priority species in each guild, and assigning sub-objectives for State "tactical" plans and individual NGO partners. The JV continued to embrace its role in grassland conservation, recognizing that only a small percentage of the original tall, mixed, and short-grass prairies remain and the dire status of several species of native grassland birds. Shorebirds and waterbirds are featured in their plans as well, but the greatest focus remains on breeding ducks in this vital part of the "duck factory". By 2022, in total through various mechanisms, the JV had accomplished 165% of its five-year objectives by conserving 212K acres of wetlands and 350K acres of grasslands (Fields 2022). Nevertheless, losses of both wetlands and grasslands continue in this vital but agriculture-dominated JV. The JV's 2024 Annual Report (https://ppjv.org/assets/docs/resources/2024AnnualReport.pdf) shared that >7 million acres (−19%) of grasslands were recently lost to croplands, mainly due to CRP conversion back to annual tillage. Some 69% of all grasslands had been converted since European settlement, with another 2 million acres anticipated in the next decade. Grassland conservation in the PPJV clearly remains a major conservation challenge.

Central Valley Habitat Joint Venture[1]

The Central Valley Habitat Joint Venture (CVHJV) was formed in 1988 and assembled their first comprehensive implementation plan by 1990. The Valley is a vital wintering area for Pacific Flyway waterfowl, supporting about 60% of the total Flyway population (CVHJV 1990) on a relatively small land base. The Valley was once covered with wetlands, seasonal to permanent in character, that varied with years and seasons, but it has been subject to extreme alteration for agriculture, flood control, navigation, and urban expansion – more than 4 million acres of historical wetlands had been reduced to <300K acres by the 1980s. Although based on waterfowl objectives, from the onset, the JV was designed to also benefit other wetland or riparian-dependant species and broader environmental values. The CVJV website states that, despite dramatic habitat loss, >100 million birds of approximately 400 species still depend on the Central Valley for their survival at some point in their annual cycles.

In 1987, Gary Kramer was the manager at Kern NWR in the San Joaquin Valley. Gary was asked by the Portland Regional Office to consider taking on the job of the CVHJV coordinator. Gary, however, had designs on becoming the Project Leader of the Sacramento NWR Complex, so he agreed to assume the coordinator job for one year. Gary jumped in and helped organize a Management Board of interested organizations and got technical partners working on building quantitative objectives (acres, population numbers) by sub-areas within the Central Valley.

Dave Paullin was hired as Gary's replacement in the spring of '89. Dave describes his arrival in Sacramento as "a deer in the headlights" moment, but much of the initial organizational stuff had been done by Kramer and his first job was to shepherd the plan to completion and get the agreement signed, which he did.

Paullin had been employed in the National Wildlife Refuge system, where he recalls "living and breathing bureaucracy, red-tape, rules, regulations, protocols, and hierarchical lines of authority". Then along came the joint venture and all its "footloose partnership stuff". He recalls (pers. comm. November 2024) that his "supervisors in the Portland Regional Office had little understanding of what we were doing in Sacramento". He heard comments like (1) *you can't do that, that's lobbying*; (2) you *talked to so and so? Wow, who authorized that?;* and (3) *that has never been done before.* The bean counters in Contracting and General Services just rolled their eyes. "Y*ou can't do that"* was one of his favorites because often his retort was *we already did*, followed by stunned silence. In those formative times, taking calculated risks seemed essential and Dave wasn't alone. He noted that,

> those of us in the early days of JV's and NAWMP were an enigma to most bureaucrats in the Service. We were operating way outside their comfort zone. Personally, I loved it. It was challenging but oh, so invigorating both personally and professionally.

Clearly, entrepreneurial leadership was part of the "magic sauce" that helped make the earliest joint ventures successful.

The JV's ambitious first plan in 1990 targeted 80K acres of existing wetlands for protection, 120K acres for restoration and protection, 292K acres for enhancement, and >400K acre feet of water secured for existing wildlife areas. The JV also committed to an ambitious biological evaluation plan. Finalized in 1996 (Miller et al. 1996), that work led to substantial revision of the JV's implementation plans as the caloric needs of wintering waterfowl became better understood.

At the beginning, however, the challenge was building the JV partnership. Although the CVHJV enjoyed the simplicity of dealing with only one State government, California is a state of social and bureaucratic complexity. Water and its use are paramount factors in the Valley, and it is complicated by tensions among agricultural, urban, and wildlife user. Endangered anadromous salmon and other biota require consideration and limit management options. In the Central Valley, wetland-dependent bird habitat is almost entirely "managed", either as semi-permanent or seasonal wetlands or flooded agricultural lands that provide habitat functions part of the year. Ensuring reliable and affordable water supplies for wetland habitat management is one of the JV's greatest challenges.

The saying in CA that *water flows to money* is largely true throughout the Intermountain West. Dave Paullin recalls that in CA it was always about water-based politics and battles for every acre-foot of water. Invariably, wetlands and wildlife were at the end of the train. In some years, a lack of precipitation in California made this difficult situation even worse.

Box 5.1 In California, Water Flows to Money

David Paullin, CVJV Coordinator retired.

Water is a fundamentally limiting resource in the Central Valley of California. I spent time pondering this in my JV years in Sacramento. At the bottom of the priority stack was purchased water for private duck clubs owned or leased by "working guys". These were always shoestring operations with limited budgets. Every dollar was counted and every acre-foot of water judiciously used. The next rung on the ladder was state wildlife areas, federal refuges, and duck clubs owned by wealthier people. These users were willing and able to pay a bit more for water than blue-collar duck clubs but still this was "wildlife water", at or near the bottom of the priority ladder. In CA, agriculture always trumped wildlife. Farmers can and did pay more per acre-foot than folks seeking "wildlife water".

This is where it gets interesting. CA can grow about 200 different crops and agricultural statistics are plentiful – acres planted, tons produced, revenues generated, etc. There has been a steady upward movement to higher value crops. In the earliest days of irrigation, things like native hay and cereal grains took up large acreages but had relatively low value. At this level, there were still tangential values to waterfowl like waste grain and goose pastures. Over time, these crops gave way to row crops (e.g., tomatoes, sugar beets) and these growers could outbid wheat farmers and so water started flowing to row crops first. With row crops, the value to waterfowl was minimal except for corn.

Then, row crops gave way to higher cash crops like orchards – almonds, walnuts, peaches, and pears. The value to waterfowl was near zero and these new growers could outbid row-crop farmers. So, another shift took place – row crops yielded to orchards. Then along came niche crops like grapes, pistachios, kiwi fruit, and avocados that outbid the first orchards. Finally, even avocados, once the most lucrative dollars/acre crop yielded the top position to the cut flower industry, which now trumps everyone else in agriculture because of their profit margins, so water continues to flow toward money.

But it doesn't stop there. Urban use for people can outbid agriculture, even avocados and flowers, so cities outbid agriculture. In the 38 years between 1986 and 2024, the human population of the Central Valley grew by about 13 million people, >60%, demanding a whole lot of space and water! Waterfowl, wetlands, and agricultural land seem to always give way to urban expansion despite efforts to hold the line. Lastly, industries can outbid them all. If you are a tech-titan in Silicon Valley and you are making semiconductor chips or some whiz-bang widgets for the aerospace industry, money is no object. You'll pay whatever it takes to make your widgets even if that means buying up urban water rights. So that's how money (and water) flows in CA in a ladder-like fashion. Wildlife water is at the bottom and hangs by a tenuous thread.

The renamed CVJV describes its mission as "cultivating and supporting partnerships that protect and restore habitat for the benefit of birds, other wildlife, and people". They do this by providing partners with biological, human dimensions, and policy guidance to plan and carry out effective bird habitat conservation. The JV uses the best available science to set population and habitat objectives that are aligned with national and international bird conservation plans and focuses on voluntary conservation. The JV also provides practical guidance for improving bird habitat to policymakers, state and federal agencies, private landowners, local governments, conservation organizations, and corporations. Additionally, the JV helps leverage investments in bird habitats on public and private lands to support healthy bird populations while benefiting local communities.

The CVJV benefited greatly from a strong cadre of waterfowl scientists in central California. As the JV grew, so too did the commitment of others. Delta Waterfowl partnering with the University of California Davis recruited Mickey Heitmeyer as a postdoctoral researcher to study waterfowl biology in the Valley with Dennis Raveling. A world-renowned goose biologist, Raveling mentored many waterfowl-focused graduate students at UC Davis. Raveling was later succeeded by John Eadie who uniquely among academics provided great service to

NAWMP for >20 years. The Northern Prairie Wildlife Research Center also had a field office there led by Dave Gilmer. Waterfowl researchers like Mike Miller, Joe Fleskes, John Takekawa, Dennis Orthmeyer, and Susan De la Cruz all did research useful to either the CVJV or later the San Francisco Bay JV. Ducks Unlimited established an office in Sacramento to focus on Pacific Flyway issues with scientists/managers like Alan Wentz, Fritz Reid, Mickey Heitmeyer, Mark Petrie, and their successors like Matt Kaminski. The California Waterfowl Association added waterfowl expertise in Canadian Bob McLandress (a former Raveling student), Greg Yarris, and Dave Smith. California Fish and Game, especially biologists Dan Connelly, Red Hunt (who served on the North American Wetlands Conservation Act (NAWCA) Council and the Plan Committee), Dan Yparraguirre, and more were hugely supportive of the JV. Humboldt State University (now Cal Poly Humboldt) was also a source of wildlife graduates who went on to serve as Refuge or State Wildlife Management Area managers throughout California. No other joint venture enjoyed more waterfowl experts in more organizations within its boundaries, and this proved highly useful for understanding the needs of wintering and breeding waterfowl in the Valley.

Paullin also credited partner lobbying efforts, both at the State level and in Washington, as vital for JV support. He remembers,

> several California policy people worked effectively at the State level including Dan Chapin (California Waterfowl Association), Glenn Olson (Audubon) and Richard Spotts (Defenders of Wildlife), unusual within his organization in 'commiserating' with duck hunters…but he could see the bigger picture and that to get things done on the ground you had to hitch yourself to the bigger train. It was a totally new world to me, but I learned a great deal from all their efforts.

Dave further reflected,

> while we were geographically small, we had huge problems to overcome largely due to the pressures of >30 million people. Among my Coordinator peers, I liked to say we have lots of problems in CA but fortunately we had lots of tools to deal with them. I had avenues to pots of money and programs in CA that nobody else had.

The Central Valley Joint Venture Today

The Central Valley Joint Venture remains one of NAWMP's success stories, despite serious threats of water availability, climate change, and competing land use. Science investments by the JV and associated agencies and universities allowed rigorous re-evaluations of population and habitat goals in 2020 (Central Valley Joint Venture 2020) and their commitment to adaptive management continues to help guide their conservation work. The JV presently has developed, in priority order, objectives for nonbreeding waterfowl, breeding waterfowl (mainly mallards, gadwall, and cinnamon teal), nonbreeding shorebirds, breeding shorebirds (mainly avocets, black-necked stilts, and killdeer), waterbirds, riparian landbirds, and grassland/oak savannah landbirds.

Specific plans are in place for five river basin complexes. From north to south, these are the Sacramento basin, Yolo-Delta, Suisun Marsh, San Joaquin, and Tulare, and these include specific objectives for each of the bird groups and habitat interventions available to the JV. Water availability remains the key to wetland habitat in the Valley and JV scenario planning runs the gamut from full historical supply to 75% less. Climate change threatens to exacerbate this problem with the likelihood of more variable weather in the future, ranging from deluge to drought.

The CVJV partnership remains strong and diverse including as contributors to this latest revision 11 nongovernment partners, four state and six federal agencies, plus seven university or business contributors. Work on public policy (federal, state, local) remains a productive focus for the partnership, as does work with farmers, especially flood management of harvested rice fields (see Appendix D). Both efforts remain important supplements to management of secured wetland habitats. Like many JVs, the Central Valley is still expanding work in human dimensions with waterfowl hunters, farmers, nonhunting recreationists, and urban residents, in that order, the current focus of their outreach and engagement efforts.

Gulf Coast Joint Venture

Based on the 1986 NAWMP objectives of sustaining 62 million breeding ducks and winter populations of 6 million geese, the Gulf Coast was another area identified for immediate attention. In the initial Plan, however, the Lower Mississippi Valley and the western Gulf Coast were lumped together, encompassing the main wintering areas of both the Mississippi and Central Flyways, two FWS Regions, and numerous states. Following an early practical decision by the partners to create two complementary adjacent JVs, an internal FWS proposal for the Gulf Coast Joint Venture (GCJV) was approved on December 22, 1988, and funded with $200,000 in federal aid funds, endorsed by the IAFWA.

Before NAWMP, Jerry Johnson, like Charles Baxter, was serving as an Ecological Services Field Supervisor but, in Texas, attached to the Albuquerque Region. It was a generalist kind of role, not particularly concerned with waterfowl. With only a day or so to think about it, Jerry was offered the GCJV Coordinator position, but after conversations with Ross Melinchuk (SK), Harvey Nelson, and others he agreed to jump in and remained the Gulf Coast Coordinator until retiring in 1993. Johnson recalled (NCTC interview) first contacting every organization that might possibly become involved.

> We had an organizational meeting for the Joint Venture...hosted by Louisiana Department of Wildlife and Fisheries at Rockefeller Refuge, ...and the place was absolutely packed. I remember there were a lot of questions and few answers but everybody was enthusiastic about getting this thing started.

That was also Jerry's introduction to all the people in the potential joint venture, expecting that "most of them had never seen or heard of me before".

Soon after, the Management Board was selected with representative from all the states, two FWS Regions, and a few private organizations like Audubon and Ducks Unlimited. Hugh Bateman (LA) was the first President of the Management Board and became a friend and great asset to Johnson. Jerry remembers him as "a very good leader...who had a way of working with people and getting through issues". "Harold Benson was a great help to me too".

"After a couple of months, Jerry reflected, planning stagnated for a while, especially with the FWS Regions, perhaps because folks were expecting new money that didn't appear right away". Fortunately, the state and NGO people on the Management Board kept things moving. One of the first things the JV did was change its boundaries. "We weren't given a precise boundary, Jerry recalls, but we wanted to extend it into the coastal prairies, so we could cover the entire rice prairies". These adjustments helped them identify a half-dozen priority landscapes with common issues that could be administered by local partnerships including business interests. For instance, the Mobile Bay (AL) initiative was developed as an important partnership with Scott Paper Company. Other federal agencies, including the Corps of Engineers, Environmental Protection Agency, and the Soil Conservation Service, eventually joined the JV as well. This flexibility allowed for the Joint Ventures to develop diverse partnerships and modes of operation were, to Jerry, one of the most important keys to NAWMP success across the regions.

Dave Case's notes from a 1988 JV "round up" meeting captured Johnson's description of their JV planning process as "an opportunity to think", which was widely appreciated by other JVs as well. Yet, the group expressed a desire for some generic models for the future, national step-down objectives, ideas for how to meld local issues with national issues, and the need for an evaluation program.

Some 4.25 million acres of wetlands of significance to waterfowl once existed within the GCJV with about 80% of that now in private ownership. Rice agriculture had long been important to wintering birds in the region, but by 1987 with shifts in global markets only about 516K acres remained of what had been 1.23 million acres of rice in 1950. Besides wintering waterfowl, the region is valuable for trans-Gulf migrant birds in both fall and spring, and home to resident breeding mottled ducks. Hundreds of other species of birds, fish, shellfish, reptiles, and more also depend on the coastal marshes of the region. Coastal marsh loss has been severe due to land subsidence, sea-level rise, hurricane damage, and human conversion for other uses.

The JV's current plans (https://www.gcjv.org/resources) suggest 386K additional acres of mallard and northern pintail habitat are needed in GCJV areas through a combination of acquisition, easements, restoration, or creation of wetlands and better management of associated agricultural lands. The JV organized in six initiative areas for detailed planning for population and habitat objectives and a tailored suite of actions. The planning areas comprised:

Mobile Bay (AL)
Coastal Mississippi Wetlands
Mississippi River Coast Wetlands (LA)
Chenier Plain (TX and LA)
Texas mid-coast
Laguna Madre (TX)

Based on historic distribution patterns for 1983–1988, overall winter population objectives were set for nine common species of ducks and for four species/populations of geese. For each initiative area, acreage objectives under each strategy were also set, costs estimated, and small teams appointed to oversee the work. Under the coordinator and the Management Board, two technical teams were struck: (1) monitoring evaluation and research and (2) communications.

The GCJV Today

The GCJV remains today a partnership of federal and state agencies and private organizations. It remains one of the most important wintering and migration areas in North America for both Central and Mississippi Flyways birds. The latest stepping down of NAWMP goals suggests the need for the GCJV to support some 13 million ducks and 1.1 million geese to ensure an adequate spring flight from the region.

The GCJV continues to deliver habitat programs in each of the six planning areas listed above. They have developed plans in each of these areas for each of four bird guilds: waterfowl, shorebirds, waterbirds, and landbirds (https://www.gcjv.org/biological_planning). They have selected as priority waterfowl species northern pintails, gadwall, lesser scaup, redheads, blue-winged teal, and mottled ducks.

The JV has special responsibility for mottled ducks that breed and winter almost entirely within the GCJV. Since the early 1990s, based on wintering surveys, mottled ducks declined from around 150K birds in the early 1990s to around 50K in recent years. In the last decade, research has added greatly to the understanding of habitat suitability for breeding mottled ducks (Krainyk and Ballard 2014; Krainyk et al. 2019), resulting in a much-improved conservation plan for the species. The GCJV also recommended moving to a breeding population objective for the species which was accepted in the 2024 NAWMP Update. The new objective derived from breeding surveys is an aspirational 212K individuals, up from an average of 126K for the 2011–2021 surveys.

Atlantic Coast Joint Venture

Development of the Atlantic Coast Joint Venture (ACJV) was arguably the most complex undertaking of the first seven JVs. Initially, comprising 13 states (Fl, GA, VT, and Puerto Rico were added later to align the JV with the Atlantic Flyway), multiple federal agencies, and private interests, bringing the first plan together was a massive job. Dick Dyer, the first ACJV Coordinator hired in 1988, had the job of wrangling all these partners to create a first JV implementation plan. Governed by a policy committee of 19 including ten state directors, three FWS officials, and six NGOS, it was both a great partnership and a complex proposition. Region 5 (Hadley, MA) of the FWS was designated to help stand up this JV.

The partnership was founded with the broad goal of protecting, restoring, and enhancing coastal marshes to benefit birds, other wildlife, and people throughout the ACJV area. Threats of habitat loss and degradation due to urbanization, industrial development, agriculture, and sea-level rise were and remain the major threats to coastal habitats.

As with most of the initial joint ventures, when local experts began to look critically at the habitat objectives listed in the 1986 Plan, they chose to expand them. The first ACJV plan (Atlantic Coast Joint Venture 1989) called for protecting or enhancing 879K acres of wetlands and upland buffers from ME to SC before 2000. It also recommended enhancing or improving another 166K acres of existing federal and state refuges and management areas. The Plan also recognized a role that strengthened state and federal policies and regulations could play in adding to wetland protection, but such work was slow to develop.

Focus areas for initial attention were identified within each State and plans included acreage targets for specific strategies in each area. Special focus was included for black duck wintering habitat in recognition of the species high priority under NAWMP, although planners acknowledged they did not know how much of what habitat was needed to support the desired wintering population of 385K birds. A connection with the science-based black duck joint venture (Chapter 7) was envisioned as a path to resolving this uncertainty about habitat needs. Unfortunately for this JV, much of the initial BDJV research focused on breeding birds and the development of black duck adaptive harvest management models. Connections with the Sea Duck JV were needed too but took longer to develop as more initial attention was paid to Pacific coast sea duck populations.

Uniquely among the JVs, after a few years, the boundaries of this JV evolved to match that of the Atlantic Flyway. This helped facilitate coordination between the FWY Council and the JV, and the JV frequently engaged the Flyway Technical Committee as peer reviewers for its plans. This alignment has proven useful as noted in the 2007 Assessment (Chapter 15). The JV revised its waterfowl implementation plan in 2005 after 17 years – an unusually long gap.

The ACJV Today

Coastal marsh and saltmarsh remain the key focal systems for ACJV conservation. From 2019 to 2022, the JV invested in detailed biological planning for three focal species, each representing different elevational gradients of coastal marsh systems. These focal species, American black duck, black rail, and saltmarsh sparrow and their preferred habitats have become the flagships of the JV enterprise, and the JV provides annual updates on accomplishments and developments for each of these initiatives. No other habitat joint ventures have taken this approach, and it will be interesting to track their progress over time. The JV further developed coastal marsh restoration recommendations for ten states, by geographic location and technique, that provide guidance for state-specific partnership actions.

In summary, by early 1989, planning was well underway by the original seven joint ventures identified in the 1986 Plan and interest in developing similar regional partnerships was growing across the United States and Canada. A valuable migratory birds joint ventures website (https://mbjv.org/) offers a present overview of all the JV partnerships, including their geography, brief history, current leaders, website links, and much more for readers who wish to explore further the work of any of the habitat or species joint ventures.

The Dance between the National Offices and the Joint Ventures

Without direct organizational authority, one of the early challenges for NAWMP was developing effective working relationship between the regional Joint Venture partnerships and the National Plan offices that also stewarded the relationship between countries. Jim McCuaig who succeeded Jim Patterson and briefly Lee Munn, as Canadian Director of the North American Plan was faced with getting the Plan running, getting systems designed for annual financial cycles, helping figure out how joint ventures would function, smoothing the relationships between Ottawa and the Joint Ventures, and nurturing the relationship between Canada and the United States. Also vital was developing the annual Treasury Board submissions for Canadian federal funding. Those submissions began with the First-Step projects and provided base funding for all the Canadian habitat and species JVs. In total, this was a formidable set of tasks.

Jim recalls that an immediate challenge was building effective relationships between the Canadian Wildlife Service headquarters and the Canadian Wildlife Service Regions. Trust needed to be strengthened there. Mostly, Jim recalls that meant "doing what he said he would do, doing what the Plan required us to do, and doing the things that only Ottawa could do". In other words, both parties "stayed in their lanes" and developed trust. One advantage in Canada was that CWS Regional Directors got directly involved by chairing the JV management Boards and most of them had biological backgrounds. Establishing trust was vitally important as PHJV Chairperson Gerry McKeating also recalled (pers. comm. June 2025) and given the size and diversity of Canada that took work. McCuaig credits Jim Patterson for years of liaison work with the provinces prior to establishing the JVs. That ensured that "none of this NAWMP stuff" was a surprise to the provinces or other partners. So, when a JV planning group met, the participants had been working together for some time.

Another piece of the puzzle was accomplishment tracking. "Quite a challenge", as McCuaig noted, but one ably handled by Danielle Thibeault of the Ottawa office. Danielle led a cross-Canada team that included the CWS regions, provinces, delivery agencies, and other partners to develop the Canadian NAWMP National Tracking System that has effectively tallied NAWMP accomplishments in Canada since 1986. Compatibility with partner tracking systems, matching apples to apples, and avoiding double counting by project partners were all substantive challenges to resolve.

In the United States, similar challenges arose between the FWS Regional Office structure and North American office in Minneapolis, initially led by Harvey Nelson, and later in Washington with Bob Streeter. In 1988, Bob was the Deputy Executive Director of the NAWMP but, in 1990, with the NAWCA to also manage, Bob was asked to come to Washington and become the Coordinator for the NAWCA Council. The two US offices were tightly linked but still separate. Then when Nelson retired, Bob Streeter also became the Executive Director of the Plan.

One of Streeter's memories of a first meeting in the field in 1988 was, "how much enthusiasm there was around the table, even though nobody fully knew what was going on yet". None of us knew what a joint venture was supposed to be, but people were out there trying to create these "things" with minimal guidance from the 1986 Plan. There was no cookbook, so it was "an evolving process, a negotiation", as Bob remembers it, both at the JV level but also at the North American office and with the FWS Regional Directors. Paullin recalled, "collectively we (the JV coordinators) were the enthusiasm boys. Let's get together and 'build a car!'. Yeah, great! But what's a car?"

Bob Streeter noted that having specific goals and objectives was important in positioning NAWMP for success. "True, the population goals and the habitat goals were not linked [at the beginning], but they were the best that could be done when the Plan was written". Then, the second key thing was, "implementation was not the responsibility of any one organization, or any one group, or any one budget. It was a partnership that the sum was stronger than the pieces". And as time went on, "the strength of those partnerships developed at every level – bureaucratic level, political level, implementation level". Streeter recalled that it unfolded organically as the months passed.

There was sometimes conflict, between Regional Directors and the Plan Office, or between the Plan Office and the JV Coordinators, because "the Coordinators really had to dance to the tune of the partners". Bob noted

> Although they were paid by the FWS, and there was some operational money coming through the Service, they had to get all those partners to the table, and get something out of each one, and make sure they all got something back. And the organizations just evolved. It was through hard work, debate and desire, rather than just simple logic getting to the way it had to be.

One of the things that helped make it work, Streeter felt, was that there was clearly a crisis in waterfowl numbers.

> Whatever we had been doing wasn't enough and everyone wanted to see the situation improve. And [FWS Director] Frank Dunkle, though not widely popular, did believe in the North American Plan. He knew that it was important to do something…and he was convinced that the North American Plan was a way that might get it done.

It's not surprising that the "dance" was challenging. Political scientists have tried to describe the unusual governance relationships in NAWMP (Wilson 2016). The complexity, fluidity, and nonhierarchical structure of the enterprise defied succinct description. Central hierarchical control had been supplanted by governance in a "complicated network comprising supranational, national, and sub-national actors – public and private. Such policy networks reflect a changed relationship between state and society", concluded Wilson. No wonder we duck biologists encountered a few bumps along the way!

Life of an Early JV Coordinator – An Interview with Dave Paullin

The cadre of first JV coordinators had an interesting time. As Dave Paullin recounted to me (pers. comm. November 2024), most of the US coordinators had come from the refuge system or realty system of the FWS where they toiled within a single agency often limited in their actions by frameworks of policies and procedures and rather strict lines of communication. In Canada, by 1990, both coordinators (Sandy Macaulay and Keith McAloney), while working through CWS Regional Offices, had been seconded from long service in DU Canada and accustomed to a different internal culture than the federal government.

As Paullin described it, the operating world of a JV coordinator could not have been more different than that of a FWS mid-level administrator. The JV job was largely about networking, about people, about recruiting, and finessing arrangements among diverse players. Coordinators had to be nimble, entrepreneurial, and inclined to act rather than ask permission. It was a job in which some people thrived, but others found it difficult. Unlike in the federal service, Paullin felt as though he did not work for any single person. Rather, he was a cheerleader, advocate, and salesperson for every partner and every program. He was a JV employee.

The coordinators talked a lot among themselves while seeking to develop their JVs. Site visits with each other were encouraged and an important part of the learning process. Paullin recalls Steve Moran, newly chosen to coordinate the Rainwater Basin JV in 1995, coming to Sacramento, driving around, and talking about JV's and how to do things. "He listened and asked lots of good questions", Dave remembered. In the years following, "Steve and I became buds, and he told me how valuable some little pearl of wisdom was that I had shared with him on that tour". In turn, Dave learned things from Steve, who was a disciple for "friend raising" rather than fundraising, critical in the RWBJV with a high proportion of private land ownership. "It fit, Dave reflects, and it was vitally important to what made JV's work".

In the early days, Paullin said, "my fellow JVC's and I were scurrying around trying to learn the ropes and finalize our Implementation Plans". Clearly, it was a situation where it was hard to see the forest for the trees. Fortunately, Harvey Nelson and later Bob Streeter organized in-person coordinator meetings twice a year and they would change venues each time. "It was a great opportunity to see new places and how others were getting things done". "We always had 'show and tell' time, a chance to share our latest victories, challenges, and setbacks". Paullin recalls Charlie Baxter (LMVJV) coming up to him several times just shaking his head in disbelief at all the hoops we had to jump through in CA to make things happen. Mississippi and California were opposite environments in many respects. "We each had different tools, different partners, different politics, and different ecosystems but we learned from one another". Paullin recalled, "We charged ahead, often far ahead of our headlights. It was fun!" While learning from their JVC network, it still seemed that each JV had its own unique story about how they gave birth to their enterprises. Unfortunately, there was little contact across the international border in those early years, largely the result of difficulty in securing "foreign travel" permission. These cross-border connections became stronger in the late 1990s and beyond with a joint annual meeting and group calls in between.

As time went on, several folks in various JVs moved around and transferred knowledge from one region to another. Dave Smith, for example, started out in California and ended up as the coordinator of the Intermountain West JV. Josh Vest worked for a time with the Intermountain group and now leads the technical work of the PPJV. Jerry Schotzko worked both with the early PPJV and the early Upper Mississippi River/Great Lakes JV. Ken Sambor worked with PPJV partners in ND before becoming the coordinator for the Northern Great Plains. Keith McAloney was coordinator for both the Sea Duck JV and the Eastern Habitat JV. Deanna Dixon served both the Arctic Goose JV and the Prairie Habitat JV (for > 30 years!), and Tasha Sargeant led both the Canadian Intermountain JV and the Canadian side of the Pacific Birds JV. Many other technical leaders in the JVs from nongovernment organizations also shifted locations to the benefit of the enterprise.

Dave Paullin, about eight years into his role as Coordinator of the CVJV, reflected on the start-up challenges and what made his and other JVs work (pers. Comm. June 2024). In a presentation for JV colleagues, he offered these points associated with success:

- A clear vision simply stated.
- Well-crafted, science-based quantitative objectives and a game plan.
- Partners who passionately embrace a common vision.
- The whole must be greater than the sum of its parts, otherwise why partner?
- Park the egos at the door – individually and corporately.
- Have faith that money gravitates to a good idea. Don't get hung up early about funding. Funding comes to good ideas and passionate partners.

Dave saw his key role as nurturing the partnership and by all accounts he did that successfully for a decade.

Funding Support for the Joint Ventures

The work of standing up joint ventures, of course, required funding. In the FWS, initial funds for the North American office came from the Director's account but did not cover the JV coordinators. Streeter thinks their first support came from Regional Office budgets. Their costs were first appropriated by Congress as line items in FY1988 or 1989. In Canada, federal funds covered the national office and coordinator costs.

The Plan Committee needed to get moving too but was also "trying to figure out what needed to be done" in Streeter's recollections. Some useful work was done by Canadian staff and US staff by presenting things to the Committee and seeking forward guidance. The PC did make some specific early recommendations, for example, in sequencing of the next joint ventures.

With the energy and excitement building around the first habitat joint ventures, proponents in both Canada and the United States were anxious to assemble and present additional JV partnerships. I take up the emergence of these in the next chapter.

Note

1 Renamed Central Valley Joint Venture (CVJV) before the 2006 revised implementation plan.

References

Atlantic Coast Joint Venture. 1989. *Atlantic Coast Joint Venture Plan*. USFWS, Hadley, MA. 106pp.

Baxter, Charles K. 2005. "Defining the demands and meeting the challenges of integrated bird conservation." Pages 7–12 in C. John Ralph and Terrell D. Rich (Editors) *Bird Conservation Implementation and Integration in the Americas: Proceedings of the Third International Partners in Flight Conference,* 2002 March 20–24. Asilomar, CA. Volumes 1 and 2. USDA Forest Service General Technical Report PSW-GTR-191. Albany, CA.

Central Valley Habitat Joint Venture. 1990. *Implementation Plan: A Component of the North American Waterfowl Management Plan*. Sacramento, CA. 102pp. (https://www.centralvalleyjointventure.org/wpcontent/uploads/2024/02/cvjv_implementation_plan.pdf).

Central Valley Joint Venture. 2020. *Central Valley Joint Venture 2020 Implementation Plan*. U.S. Fish and Wildlife Service. Sacramento, CA. Available from: www.centralvalleyjointventure.org

Fields, Sean. 2022. *35 Years of Conservation: Prairie Pothole Joint Venture*. 922 Bootlegger Trail, Great Falls, MT. (https://ppjv.org/assets/pdf/PPJV_35_Years_of_Conservation).

Krainyk, Anastasia and Bart M. Ballard 2014. *Decision Support Tool: Prioritization of Mottled Duck (Anas fulvigula) Habitat for Conservation and Management in the Western Gulf Coast*. Gulf Coast Joint Venture. 128pp.

Krainyk, Anastasia, Bart M. Ballard, Michael G. Brasher, Barry C. Wilson, Michael W. Parr, and Cynthia K. Edwards. 2019. "Decision support tool: Mottled duck habitat management and conservation in the Western Gulf Coast." *Journal of Environmental Management* 230: 43–52.

Lancaster, Joel, D. Tim Anderson, Michael G. Brasher, Warren C. Conway, Steven J. DeMaso, Jena A. Moon, Kevin M. Ringelman, and Barry C. Wilson. 2023. *Gulf Coast Joint Venture Mottled Duck Conservation Plan Update*. Gulf Coast 1034 Joint Venture, Lafayette, LA. 75pp.

Lower Mississippi Valley Joint Venture Management Board. 2023. *Lower Mississippi Valley Joint Venture Operational Plan 2024–2029*. Vicksburg, MI. 27pp.

Miller, Michael R., Robert McLandress, and Frederick A. Reid. 1996. *Central Valley Habitat Joint Venture Evaluation Plan*. NAWMP, Sacramento, CA. 67pp.

North American Bird Conservation Initiative. 2025. *The State of the Birds, United States of America, 2025*. StateoftheBirds.org. (https://www.stateofthebirds.org/2025/wp-content/uploads/2025/03/state-of-the-birds-sotb-2025-spreads.pdf).

Prairie Pothole Joint Venture. 2017. *Prairie Pothole Joint Venture Implementation Plan*. Sean P. Fields (Editor). U.S. Fish and Wildlife Service, Denver, CO.

Reynolds, Ron E., Terry L. Shaffer, Randy W. Renner, Wayne E. Newton, and Bruce D. J. Batt. 2001. "Impact of the conservation reserve program on Duck recruitment in the U.S. Prairie pothole region." *Journal of Wildlife Management* 65: 765–80.

Sovada, Marsha A., Allen B. Sargeant, and James W. Grier. 1995. "Differential effects of coyotes and red foxes on duck nesting success." *Journal of Wildlife Management* 59: 1–9.

Wilson, Jeremy. 2006. "New" and "old" modes of environmental governance: the evolution of the North American waterfowl bird policy regime." *Western Political Science Association Conference*, Albuquerque, NM. 30pp.

6 Maturation and Proliferation of Habitat Joint Ventures

With the enthusiasm and progress shown by the initial habitat Joint Ventures and anticipation of new funding through NAWCA, interest in the creation of additional JVs was high as the 1980s ended. Space doesn't permit a fulsome description of each of the now 22 habitat JVs but I will touch on each of these briefly as they are making important contributions to waterfowl conservation and to other conservation initiatives that came together on the coattails of NAWMP. The Migratory Bird Joint Venture website provides a timeline for the establishment of each joint venture: https://mbjv.org/who-we-are/jv-timeline/ and useful links to all the joint venture websites. A map of JV boundaries as of 2024 is found in Appendix F (www.routledge.com/9781041224389).

Playa Lakes Joint Venture

The Playa Lakes are a unique region of eastern NM, CO, the Panhandle of TX, and western OK, KS, and NE. Much of the land there overlies the Ogallala Aquifer and is a region of little topographic relief, with scattered shallow lakes important to many species of migrating and wintering waterfowl. Playas – locally called mud holes, buffalo wallows, or lagoons – are the most numerous wetlands in the region, with more than 80,000 of them scattered across the six states. These round, shallow depressional wetlands are lined with clay soil, which collects and holds water from rainstorms and runoff, creating temporary lakes. Species wintering there include mallards, northern pintails, sandhill cranes, plus many migrating species like blue-winged teal, white-rumped, semipalmated, and Baird's sandpipers to name a few. Fingers of streams and associated riparian habitats also crosscut the region on mostly west-to-east paths. Water availability for irrigation and other human use has often conflicted with wildlife habitat needs in the region.

An early addition to the JV family under the leadership of Harvey Miller (FWS) from 1989 to 1992, the Playa Lakes region was of concern because of widespread drainage and agricultural modifications of wetlands. It was also the first JV to develop major funding from a corporate partner, Phillips Petroleum, that had commercial interests in the area. Later, under the long-term leadership of Mike Carter (2001–2024), the JV has been innovative in blending the needs of water for wildlife and people. Research on waterfowl use and especially regional hydrology has proven important to Playa Lakes Joint Venture (PLJV) progress. Habitat conservation that also benefits people has been a key to progress. The Joint Venture also invested in social studies that have helped guide their work. Teaming with D.J. Case & Associates, the PLJV undertook focus groups and surveys of residents which pointed to widespread water quantity concerns related to the shrinking Ogallala Aquifer. Armed with hydrological studies demonstrating the importance of playas for recharging the aquifer, Carter and his team built creative new partnerships among water and wildlife interests that moved their work forward.

DOI: 10.4324/9781003732310-6

The PLJV's stated purpose today is to find solutions to habitat conservation challenges in the western Great Plains so that birds and other wildlife, agricultural producers, and local communities all prosper in this landscape. The range of partnerships illustrated on their website is uniquely broad (https://pljv.org/). The JV's current focus is on building partnerships, conservation design for working lands and public lands, and helping to site energy developments in a bird-friendly manner. Range management is another focus striving to benefit numerous grassland birds. Their objective criteria for including any species as a priority are that more than 8% of the global population can be found within the short-grass and mixed-grass conservation regions (BCRs 18 and 19; https://nabci-us.org/resources/bird-conservation-regions-map/) and that the species has been declining or of unknown population status there, or the species is a game bird important to state agencies.

Pacific Birds Habitat Joint Venture (originally the Pacific Coast Joint Venture)

The PBHJV has traveled a unique path, beginning as an international joint venture with US and Canadian co-chairs and a binational technical committee. Carey Smith served as the US coordinator from 1991 to 2012 (http://www.pacificbirds.org/); three Canadian co-chairs served briefly until Trish Hayes settled in for a run of years beginning in 1995. At the outset, CWS Regional Director Art Martell was an influential advocate for adding this west coast program to Canada's family of JVs. Fortunately, he found broad support elsewhere in Canada for providing a portion of Canadian NAWMP funding to the Pacific Coast. Martell was also an early advocate of expanding the focus of the partnership to more than waterfowl, including shorebirds, seabirds, and many other animals that rely on intertidal habitats. Jim McCuaig recalls that the first idea was for a coastal JV stretching from Alaska to Baja, but jurisdictionally that was judged to be unworkable in 1990.

The JV focuses on coastal habitats, primarily northwards from San Francisco Bay to northern British Columbia. As the JV evolved, Alaska and Hawaii were added to the partnership. The JV updated its ten-year US strategic plan in 2024 (https://pacificbirds.org/wp-content/uploads/2024/07/PBHJV-Coastal-Wetlands-Plan-7.7.24_reduced-for-web.pdf). Dozens of collaborators were engaged, led by PBHJV staff Laura Farwell, Monica Iglecia, and Sara Evans-Peters. Their restated goal is "to conserve coastal wetlands thereby meeting the habitat needs of migratory birds and other native fish and wildlife as human communities adapt to rising sea levels, a changing economy, and other social and ecological changes". For the next ten years, Pacific Birds is prioritizing conservation of coastal habitats and Western forests within their boundary area. They are also developing a new plan for conservation in Hawaii.

Pacific Birds-Canada published a complementary BC Implementation Plan (2020–2030) in 2021. Led by principal authors Bruce Harrison, Andrew Huang, Kathleen Moore, and Jocelyn Wood, it details priorities, strategies, and objectives for coastal bird and habitat conservation north of the US border. Top-tier habitats for protection under both plans include (a) tidal wetlands, intertidal mudflats, and sandspits; (b) eelgrass meadows vital for Pacific Brant; (c) freshwater wetlands and lakes for ducks, shorebirds, and waterbirds; and (d) bird-friendly working lands and waters, principally in aquacultural and agricultural landscapes. Major focal areas are the Fraser River Delta in British Columbia, the Puget Sound region in Washington, and several others.

An important feature of the Pacific Coast plan is building on partnerships with indigenous peoples and governments. Indigenous control of land and related roles in resource management are expanding in both countries, and they are vital partners. An explicit strategy of the BC plan is to help build coastal habitat and community resilience to climate change.

Another vital connection for PBHJV is with the Arctic Goose and Sea Duck Joint Ventures (Chapter 7) whose members and research projects offer helpful guidance for the design and evaluation of coastal habitat projects.

Upper Mississippi River and Great Lakes Joint Venture

The states along the Upper Mississippi River and the Great Lakes provide both breeding and migration habitats for an array of ducks, geese, and swans. The JV covers all of MI and WI; eastern MN, KS, and NE; eastern and southern IA; northern MO and IL; and western OH. Major habitats include riparian areas, agricultural lands, forested lands, urban areas, coastal wetlands, and large lakes, all of which face numerous conservation challenges. Planning in the region began in 1988 but the partnership was not consummated until September 1993. Jerry Schotzko (FWS Region 3) facilitated early planning for the JV until Jim Leach became the JV coordinator in 1993 (https://umgljv.org/).

For many years, Greg Souliere, Barb Avers, and other technical contributors to the JV have been among the leaders in advancing JV evaluation plans and biologically based conservation planning for both migrating and breeding waterfowl. The JV adheres to a framework of Strategic Habitat Conservation (https://www.fws.gov/policy-library/052fw1) that involves shifting toward more accountable and adaptive actions driven by research and monitoring results. The JV's overall planning goal is "to guide regional conservation that results in habitat to support populations of priority waterfowl species and related social values, consistent with continental bird conservation goals". Regional waterfowl population and habitat trends, and assessments of factors likely to limit population growth, provide their biological planning foundation.

Adding to the complexity of conservation planning here is that over 60 million people live in the JV region, which also accounted for about 30% (300,000) of all US waterfowl hunters during 1999–2013, and more annual waterfowl hunter-days afield than any other JV region (Raftovich et al. 2014). The presence of five NABCI Bird Conservation Regions evidences the ecological complexity of the JV as well.

Since the last comprehensive implementation plan in 2007, individual plans have been updated for waterfowl (2017), waterbirds (2018), and landbirds (2020). This JV faces diverse conservation challenges with important breeding areas, extensive migration habitats, and some wintering habitats. Based on the 2004 continental species prioritization (NAWMP 2004b) and the weightings for five BCRs, the 2017 waterfowl plan chose a handful of focal species for breeding ducks (mallard, wood duck, blue-winged teal, and ring-necked duck) and nonbreeding species (wood duck, gadwall, northern pintail, green-winged teal, canvasback, and lesser scaup). State-specific BCR plans were subsequently established with objectives for habitat retention, restoration, and enhancements.

Some recent creative projects have developed around flood abatement and watershed nutrient management that also provide waterfowl habitat. Another is a program to spatially prioritize new conservation areas based on a joint assessment of biological values and human recreation potential for those sites (Soulliere et al. 2017). In Michigan, Barb Avers and her team have studied what users of coastal wildlife management areas value, published in 2023 (https://www.canr.msu.edu/news/research-project-evaluates-what-users-of-coastal-wildlife-management-areas-value-msg23-triezenberg23-1). The Upper Mississippi River and Great Lakes Joint Venture (UMGLV) has for many years been one of the most innovative and science-based joint ventures in an ecologically and socially complex region.

Rainwater Basin Joint Venture

Nebraska's Rainwater Basin is a small but biologically significant mid-continent habitat for millions of migrating ducks, geese, cranes, and shorebirds in both spring and fall. It is a flat to gently rolling loess plain adjacent to the Platte River Valley, with numerous shallow playa wetlands, reliant on rainfall and snow melt for water. Much of the land and the few remnant wetlands are in private ownership. This JV's intimate relationship with agricultural producers is notable. In fact, its stated mission is "to support our agricultural communities while providing habitat conservation for millions of migratory waterfowl, shorebirds, and other wildlife through cooperation and sound science" (https://www.rwbjv.org/).

Established in 1992 (Gersib et al. 1992), the plan famously and legitimately depicted the Rainwater Basin as the narrow waist in the hourglass of the Central Flyway. Wetland loss since European settlement had been severe. Nearly 95K wetland acres in 1900 were reduced to fewer than 21K acres by 1983. Most of those basins were converted to agricultural use (Nebraska Game and Parks Commission 1984). That loss of wetlands contributed to concentrating birds which increased the risk of disease transmission, particularly for avian cholera.

The Rainwater Basin Joint Venture partnership began with little funding but was able to hire coordinator Steve Moran by 1995. With Steve's leadership, partners focused heavily on relationship building in a region where ducks and farmers historically were often at odds. Conservation actions during the initial years focused on protecting, restoring, and enhancing wetlands on both public and private land to support migrating waterfowl. The combination of Moran's partnership-building skills with Andy Bishop's technical expertise and networking served the Joint Venture well.

With the coming of other bird initiatives (Chapter 9), the JV began to expand its conservation efforts to include other habitats within Nebraska's portions of Bird Conservation Regions 11 and 19, roughly quintupling its spatial extent and increasing the private land portion of the JV to ~97% (https://www.rwbjv.org/wp-content/uploads/2022-Annual-Report-Final.pdf). The expanded area, primarily the Nebraska Sandhills, supports several hundred thousand breeding waterfowl and JV partners are working to understand what specific areas within this expansive landscape are most important to breeding waterfowl and most compatible with cattle production, the major agricultural practice in the Sandhills.

Now led by coordinator Andy Bishop, with a diverse advisory board of landowners, federal agencies, NGOs, and corporations, and armed with plans for waterfowl, shorebird, and landbird habitats, the JV delivers best-practices guidance for cropping and grazing systems, wetland management practices, and still some easements and perpetual protection projects. Linkages with USDA Farm Programs have been a valuable part of their strategies for many years.

For waterfowl, early investments in science led to the development of a bioenergetics model for the nonbreeding period that estimated foraging needs of waterfowl using the RWB. The JV estimated that the RWB would need to provide 4.4 billion kcals of wetland-derived food resources, requiring some 62,500 acres of functioning wetland habitat from a combination of public lands, private lands enrolled in long-term conservation programs, private lands in short-term conservation agreements, and nonprogram wetland acres. One novel useful strategy has been voluntary acquisitions of "roundouts", the privately owned portions of wetlands currently in split (public and private) ownership.

Intermountain West Joint Venture

The Intermountain West Joint Venture shares some important features with the RWB, namely a focus on partnerships with agricultural producers in an area increasingly stressed by limited availability of water. The JV includes eastern WA and OR, northeastern and east-central CA,

ID, NV, northern AZ, UT, western and southeastern WY, western MT, western NM, and western CO. Jim Cole (US Forest Service), with early NAWMP experience in California, largely crystallized the partnership during 1994 (https://iwjv.org/).

Brad Bortner, who in 1995 was the FWS Region 1 Chief of Migratory Birds, had funding for the CVJV, but was already moving some of that to support the new Pacific Coast Joint Venture (PCJV). Brad and John Cornely got permission to hold a scoping meeting in Salt Lake City but were told that the FWS would not fund another JV out of the already stretched Region 1 allocation of JV funding. Bortner recalled that everyone was excited about the idea, but the FWS made it clear that they would need other partners to fund any JV positions. Soon, the Bureau of Land Management, Forest Service, and Bureau of Reclamation agreed to fund the JV start-up costs and Jim Cole was hired. Cole also joined the lobbying efforts and, after a few years, JV appropriations were increased allowing FWS funding for both the PCJV and the Intermountain West Joint Venture (IWJV). Bortner (pers. comm. May 2025) noted that Cole and Cornely had important roles in that success.

Box 6.1 The Association of Joint Venture Management Boards

David A. Smith, former NAWWO Executive Director, recalled that this group was born out of a conundrum of two colliding developments. First, the success of the NAWMP joint venture partnerships was viewed by many as a desirable model for conservation, but, at the same time, FWS funding for joint venture staff and science coordinators was stagnant. JV management boards with diverse memberships soon recognized that private landowners, State agencies, and NGO partners were much better suited to lobbying their Congressional delegations for funding than to rely solely on Federal agency appropriations. This realization happened as JVs were expanding, thus bringing more states and more partners to bear on engaging Congressional delegations than at the outset of NAWMP.

Smith remembered (pers. comm. May 2025) several years when diverse contingents of JV management board members from around the country would converge on Washington in the spring. They brought success stories from conservation activities back home to share with members of Congress and their staffs demonstrating that great leverage of non-federal funds could be obtained for their partnership projects. Sometimes habitats were restored on private lands, sometimes on a state wildlife management area, or a National Wildlife Refuge. Importantly, these projects were always cooperative partnerships among Federal, State, and private entities and, as such, were politically appealing to members of Congress of all political stripes. JV management boards were able to share stories demonstrating the leveraging of relatively small federal investments in projects with oversized accomplishments and never as the result of land condemnation or lawsuits. These stories were effective in persuading earmarked JV funding.

For intermountain wildlife, as for most human endeavors, availability of good water is vital and climate change seems to be exacerbating that. Human land use also changed the landscape for wildlife over the last century. Invasive species in grasslands and uncontrolled fires are ascendant problems as well. The IWJV website today declares that they "believe in forging productive, lasting relationships built on trust and respect with a diversity of partners, fostering innovations in our work, and producing tangible, on-the-ground results for habitat conservation across private and public lands in the West" (https://iwjv.org/). Their vision is succinctly "An Intermountain West where people, birds, and other wildlife thrive". Focal habitats include many state- and federal-owned wetlands but also extensive working lands where the JV seeks to enable those lands to continue providing food, fiber, and forest products for people while sustaining bird habitats in a voluntary, nonregulatory manner.

Dave W. Smith became the JV's innovative, partner-focused coordinator in 1997 and continues in that role today. The JV completed a comprehensive implementation plan in 2013 with primary attention to waterfowl, shorebirds, and other waterbirds. By building several explicit decision support tools, it sought to tighten the linkages between emerging conservation science and conservation delivery. The IWJV updated its strategic plan in 2023 (https://iwjv.org/wp-content/uploads/2023/04/Strategic-Plan-2023-version-1.pdf). Each year, this is supplemented with annual operating plans also available on the IWJV website. These are intended to keep the JV adapting to changing circumstances and opportunities. Like most other NAWMP JVs, it was originally a waterfowl and wetland-focused enterprise, but has expanded to encompass all birds and their grassland, shrubland, and forested habitats.

Much of the IWJV's work is accomplished through a network of conservation managers and implementers called the State Conservation Partnerships. These seem to be enabling meaningful conservation outcomes at regional scales. Much of the work of JV staff centers on translating scientific results to private partners, enabling them to interpret and apply those insights in the field, merged with local and traditional knowledge. IWJV declares that "The people of the West are our greatest asset" and they work that way.

San Francisco Bay Joint Venture

Established in 1996, the San Francisco Bay Joint Venture (SFBJV) is a diverse partnership working to protect, restore, increase, and enhance wetland habitats throughout the San Francisco Bay region to benefit birds, fish, and other wildlife. Planning and partner assembly began with Nancy Schaefer as the initial Coordinator, building on the experience of several organizations already engaged in the Central Valley Joint Venture and other groups seeking to conserve Bay Area habitats (https://sfbayjv.org/).

This unique JV is the smallest, geographically, but one of the most complex socially and ecologically with a high human density, multiple threatened or endangered species, high species diversity, and very expensive land. Focused on nine counties that surround San Franciso Bay and six watersheds, it was, as the JV began, home to some 3.1 million people and a human population growing at a rate of nearly 1%/year. The Bay supports large numbers of Pacific Flyway waterfowl, particularly diving ducks, myriad shorebirds, and more. As a WSHRN site of hemispheric importance, San Francisco Bay has long been recognized as important to shorebirds and other waterbirds, as well as waterfowl.

In the mid-1800s, the Bay was ringed by an estimated 190K acres of tidal wetlands, 50K acres of tidal flats, 85K acres of seasonal wetlands, and 5K acres of riparian habitats in what are now known as San Francisco Bay, San Pablo Bay, and Suisun Bay (Steere and Schaefer 1999). By 1997, a state of the estuary report estimated that only 3% of the original wetland area was in pristine condition, and that human development had destroyed or profoundly altered 80% of tidal marsh and 40% of the mudflats. Also, about 70% of the freshwater flows that historically entered the Bay from the Sacramento and San Joaquin River systems were being diverted for agricultural or urban use. Most of the vanished marshes had been filled for urban development or diked for salt production or agriculture, and some 200 invasive species dwelled in the region competing with native plants and animals. This most significant estuary on the US Pacific Coast had indeed been greatly diminished.

The JVs' first implementation plan in 1999 included ambitious goals for protecting, restoring, and enhancing Bay habitats, with additional attention to seasonal wetlands, embedded creeks, and lakes. Accomplishments of the first 15 years of the JV are celebrated in a booklet by Huning et al. (2014) and a 2024 fact sheet can be found at https://drive.google.com/file/d/1BOaZN1Td-QO4QTzGozkgMQ_eTS2Yz_tT/view.

Partnerships and people have been the key to progress in this JV. Its eclectic management board of 25 includes national nongovernment organizations, local private groups, corporations, and private citizens as well as 11 federal, state, and local government agencies. Their 2022 Implementation Plan boasts >100 organizations as JV partners and funders to date. The JV is indeed about building and supporting diverse coalitions of people to act for conservation. The 2022 Plan titled "Restoring the Estuary to Benefit Wildlife and People" (https://sfbayjv.org/conservation/planning/), authored by a long list of scientists, administrators, and advocates, charts a science-supported vision of what is needed and hopefully achievable to restore much of the ecological integrity of the Bay. The JV offers through its partners a wide array of expertise. Current projects range from physical marsh restoration to advocacy for promising public policies. The SFB website offers, "With a shared passion for birds and their wetland habitats, we believe by working together, our partnership has the power to positively impact birds, other wildlife and people…". Partnerships can be powerful when people share a common vision.

Canadian Intermountain Joint Venture

Conterminous with the US IWJV and rounding out the important waterfowl habitats of the settled region of southern Canada, the Canadian Intermountain Joint Venture (CIJV) encompasses one of Canada's most biodiverse regions, boasting some 373 species of birds, >1,500 species of plants, 29 species of amphibians and reptiles, and 94 species of mammals occupying wetland, grassland, forested, and alpine habitats. An average of 1.45 million waterfowl breeds here annually (https://cijv.ca/).

It also supports one of the fastest growing human populations in Canada with about 1 million people living and working within its boundaries. People living there rely mostly on resource-based industries such as crop agriculture, ranching, forestry, mining, and tourism for their livelihoods. The area is renowned for recreational opportunities like fishing, biking, golfing, hunting, birding, and sightseeing. It also supports growing retirement communities in its river valleys, and these human activities affect wildlife habitats. Other issues, such as wildfires, Mountain Pine Beetle infestations, and climate change with strains on water quality and supply, underscore the need for conservation actions. The region includes several indigenous lands as well.

After initial stirrings in 2000, the JV was formally established in 2003, as Canada's first "all-bird" joint venture. The private-public partners of the CIJV work with landowners and governments to conserve bird habitat. Higher priority habitats include wetlands, grasslands, and riparian areas that are critical for breeding and transiting migratory birds. The JV sees itself as an important continental link in habitat connectivity of birds, plants, and other wildlife.

Led by a compact six-member steering committee, the JV nonetheless numbered >1000 landowner partners by 2023. Krista de Groot (CWS) was the JV's first coordinator coming on during planning in 2001. Seven others followed before Tasha Sargeant served from 2010 to 2018, followed by Andrew Huang. Since 2003, CIJV partners have secured, enhanced, or supported stewardship on nearly 400K hectares (1 million acres) within the CIJV's 500K-km^2 expanse.

Initially, the JV had relatively little waterfowl data with which to build conservation plans, but that gap has been reduced and around 2005 the JV formally committed to adaptive management as its operating approach. The JV's Implementation Plan was last updated in 2021 (Harrison et al. 2021) and now features a greater focus on landowner engagement to promote wildlife-friendly management practices on private lands.

Changing Formulas for Canada

With the approach taken by the Canadian NAWCA Council to distribute funding across Canada based on biological need (Chapter 3) and vesting the regional joint venture partnerships with sorting out the expenditure priorities within each region, the coming of new JVs required adjustments in the Council's business. Jim McCuaig recalls,

> Originally there were two Joint Ventures in Canada, the Prairie Habitat and the Eastern Habitat Joint Ventures, and it had long-been decided that 80% of the resources were to go to the Prairies where we perceived most of the [waterfowl] problems to be.
>
> Then when Art Martel and I, with Marv Plenert and others in the United States, began developing the Pacific Coast Joint Venture, it was important for us in Canada, we really did need to be coast-to-coast. It became a hard jaunt to persuade, particularly the Prairie Provinces, to lower their number. I needed to get a third number, 80/20 had to turn into 3/77/20.

It was arbitrary but rational.

Years later, I served as the technical lead on a multiagency NAWCC Allocation Task Force (Colpitts et al. 2004) that attempted a more rigorous biological review of this allocation strategy with the addition of all birds in the calculus and the new CIJV. We reviewed historical allocations and estimated BCR-specific importance to wetland-associated waterfowl, landbirds, shorebirds, and other waterbirds. We added historical habitat loss rates, estimated future threats, likelihood of US matching funds, and a feasibility score. These scores were developed for each of 12 BCRs and aggregated. The results largely confirmed the wisdom of the original expert opinion allocation formula, but minor adjustments were recommended to provide justifiable new resources to the CIJV and the Western Boreal Forest (Table 6.1).

Table 6.1 NAWCC Allocation Changes Among Canadian JVs that Continued Until 2025, Colpitts et al. (2004).

Joint Venture	*Current Allocation (%)*	*New Allocation (%)*
Pacific Coast Joint Venture (PCJV)	6	7
Canadian Intermountain Joint Venture (CIJV)	0	3
Western Boreal Forest (WBF)	0	3
Prairie Habitat Joint Venture (PHJV)	77	70
Eastern Habitat Joint Venture (EHJV)	17	17

Box 6.2 Early Communications and the NAWMP Logo

By Robert Streeter, USFWS-retired

The FWS Director in 1988 was Frank Dunkle and he was committed to making the Plan a success. He asked each employee to have a role in supporting it and he redirected resources from various offices to help make it happen. Dunkle called me in February 1988 with “an offer I couldn’t refuse”. I was reassigned from my position in the research arm of the Service to assist Harvey Nelson in organizing the US North American office, establishing a communications program, supporting the Plan Committee, coordinating and encouraging the US Joint Ventures, and working with our Canadian counterparts.

Things seemed to be moving fast, helping activate the US Joint Ventures, organizing and supporting the US Plan Committee and the International Plan Committee, pushing for funding through the FWS, supporting the US Implementation Board, developing a communications program to keep all informed and energized, and traveling to cheerlead the Plan at various meetings, and more. The development of the logo seemed to happen on the edges, but I think it proved to be an important edge!

The idea of creating a logo for the NAWMP was floated at a Plan Committee meeting in 1988. The US Office took on the project. Harvey Nelson knew a graphic designer who had created a recent John Deere logo with the tag line “Nothing Runs Like a Deere”, which he liked. We contracted the artist, who subsequently produced various logos that featured wetlands, stylized eggs with a wetland juxtaposed, and duck and marsh designs. Various draft logos were presented to Plan Committee members on multiple occasions but were rejected. It became the “find a rock” game! When the stylized pintail reflected on water design was eventually presented, the Plan Committee was either worn down or it was a design that finally grabbed them. It became a unanimous choice and the enduring symbol for NAWMP across North America. The logo was first used publicly in the December 29, 1989, Joint Canadian-US progress report, “A new beginning” (Figure 6.1a).

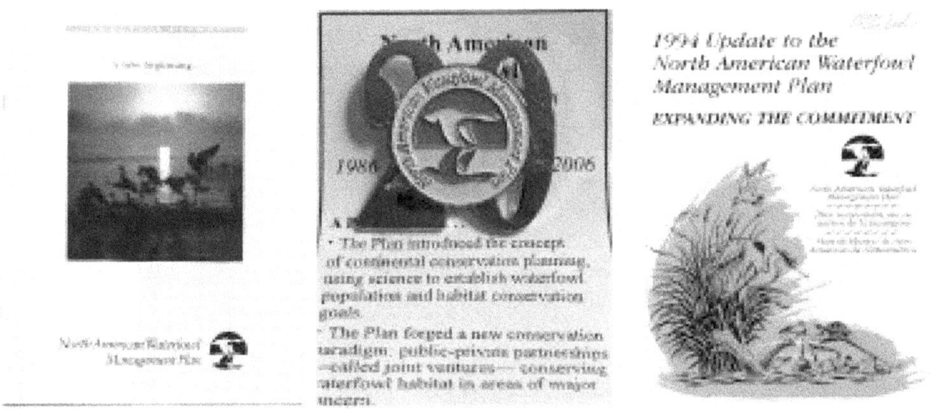

(a) (b) (c)

Figure 6.1 First uses of the NAWMP logo 1989a, 20th Anniversary lapel pin, and first use of trilingual names 1994. (Photos by the author.)

After joining Harvey in Minneapolis, I called on the resources of my former Office of Information Transfer. Bill Seitz and I came up with the initial Plan newsletter, *Waterfowl 2000. Quacks, Honks and Whistles – Sounds from the North American Waterfowl Plan.* It was a crude, four-page, black and white, photocopied newsletter with short stories from the five US Joint Venture Coordinators. It also informed the JVs about deliberations and actions of the US Plan Committee, the US Implementation Board, and the International Plan Committee. I was able to contract with a young writer, Kathryn Hoeman, to keep this newsletter going. Another result of the Dunkle push was the assignment of a seasoned Service Public Affairs specialist, Inez Conner, to promote the Plan and publicize its activities. Inez, Kathryn, and Canadian office staff morphed "Waterfowl 2000" into a more polished format to keep partners informed.

Inez and the JVs started using the logo in various ways, but there were never any official standards for how it was to be used or with what accompanying words. The only thing copywritten was the image itself. When Jim McCuaig joined the Canadian Plan Office, he and his communications folks coordinated with Inez to develop the NAWMP white and blue enamel pin (Figure 6.1b) and began adding French to the tag for some Canadian applications. I think it was with the 1994 update that the words appeared in English, French, and Spanish together for the first time (Figure 6.1c).

References

Colpitts, Lorne, Andrew de Vries, Michael Anderson, Len Ugarenko, Deanna Dixon, Karen Loftus, B. Morgan, Alison Grose, Reginald Melanson, Kenneth Cox, Danielle Bridgett, Ian Barnett, and Douglas Chekay. 2004. *Final Report on NAWCA Funding Allocations.* North American Wetland Council Canada. Ottawa. 47pp.

Gersib, Richard A., Kenneth F. Dinan, Jon D. Kauffeld, Michael D. Onnen, P. Joe Gabig, John E. Cornely, Gerald E. Jasmer, Joseph M. Hyland, and Kenneth J. Strom. 1992. *Rainwater Basin Joint Venture Implementation Plan.* Nebraska Game and Parks Commission, Lincoln, NE. 56pp.

Harrison, Bruce, Andrew Huang, Kathleeen Moore, and J. Wood. 2021. *CIJV Implementation Plan 2020–2030.* (https://cijv.ca/cijvca-content/uploads/2022/11/CIJV-Implementation-Plan-2020-2030.pdf).

Huning, Beth, Caroline Warner, John Cain, Laurel Marcus, Renee Spenst, Brad Olson, Michelle O'Herron, Nancy Schaefer, John Bourgeois, Eric Mruz, Lorraine Parsons, Carolyn Shoulders, Hattie Brown, Wendy Eliot, Susan E. W. Dela Cruz, and Marily Latte. 2014. *The San Francisco Bay Joint Venture: Celebrating 15 Years of Partnerships Protecting Wetlands and Wildlife.* San Francisco Bay Joint Venture. Fairfax, CA. 39pp.

Nebraska Game and Parks Commission. 1984. *Survey of Habitat Work Plan K-83 W-15-R-40.* Nebraska Game and Parks Commission, Lincoln. 13pp.

North American Waterfowl Management Plan, Plan Committee. 2004b. *North American Waterfowl Management Plan 2004. Implementation Framework: Strengthening the Biological Foundation.* Canadian Wildlife Service, U.S. Fish and Wildlife Service, Secretaria de Medio Ambiente y Recursos Naturales. 106pp.

Raftovich, Robert, V., Stephen C. Chandler, and Khristi A. Wilkins. 2014. *Migratory Bird Hunting Activity and Harvest During the 2012–2013 and 2013–2014 Hunting Seasons.* U.S. Fish and Wildlife Service, Laurel, MD. 68pp.

Soulliere, Greg. J., M. A. Al-Saffar, John M. Coluccy, Robert J. Gates, Heath M. Hagy, John W. Simpson, Jacob N. Straub, R. L. Pierce, Michael W. Eichholz, and Dave R. Luukkonen. 2017. *Upper Mississippi River and Great Lakes Region Joint Venture Waterfowl Habitat Conservation Strategy – 2017 Revision.* U.S. Fish and Wildlife Service. Bloomington, MN. 170pp.

Steere, J.T. and N. Schaeffer. 1999. *Restoring the Estuary: An Implementation Strategy for the San Francisco Bay Joint Venture.* San Francisco Bay Joint Venture, Oakland, CA. 124pp.

7 The Species Joint Ventures – Progress and Futures

As the 1986 Plan came together, some waterfowl species seemed to require different attention than the work envisioned for the first habitat Joint Ventures. The "species" Joint Ventures were conceived to help fill some of these gaps in our understanding and ultimately provide guidance to appropriate habitat JVs or other management authorities.

American black duck (hereafter black duck) populations had declined but the extent and primary causes of the decline were unclear. Mallard displacement or genetic introgression were concerns, but so too were changes occurring to breeding and wintering habitats. It seemed that a better understanding of black duck population dynamics and the ecological factors affecting productivity and survival were prerequisites for confident management actions.

Sea ducks comprise a diverse array of species breeding across a wide range of largely undeveloped habitats but mostly wintering in human-impacted coastal waters. These species were less well enumerated than most other ducks. So as with black ducks, it seemed that the greatest need was to better understand sea ducks' population statuses and trends and factors affecting those trends. For several species, as with some northern-breeding goose populations, indigenous communities in Alaska and Canada had traditional interests in hunting these species and there were growing expectations for co-management of the birds.

Arctic Geese also breed mainly beyond the footprint of human development, but winter near human concentrations in both coastal and interior habitats. Again, it was unclear what habitat conservation efforts might most affect their populations. Most goose populations were tightly managed by the federal governments and four Flyways with the distribution and occurrence of subspecies or flocks well identified and harvest managed jointly by the affected jurisdictions.

Years later, in response to sharply declining trends in certain duck populations, two additional species focus groups were formed – the Pintail Action Group (2003) and the Scaup Action Team (2007) (Chapter 8). These were self-formed by concerned scientists and managers who developed formal prospectuses seeking endorsement from the NAWMP community but without the formal structure and staff support that attended the original species JVs.

Arctic Goose Joint Venture

Canadian and US federal agencies and the Flyway Council representatives began planning for this joint venture in 1987, producing a formal prospectus in 1991 (https://www.agjv.ca/wp-content/uploads/2017/11/aprospectus.pdf). In considering their challenges, they focused on inadequate information on the breeding distributions of many Arctic-nesting goose species, recent changes in population sizes, and changes in winter distributions, all requiring coordinated research by multiple jurisdictions. The broad purpose envisioned for the JV was to establish the science foundation for management.

DOI: 10.4324/9781003732310-7

In the 1970s, long before NAWMP, goose biologists and managers had established the North American Arctic Goose Conference and Workshop, a science/management forum held at three- to four-year intervals (most recently October 2025). That experience helped set the stage for the Arctic Goose Joint Venture (AGJV), by connecting scientists and keeping Arctic Geese visible on agency agendas.

The JV created an oversight structure with 16 members, mostly federal or state government people, that acted as a management board with a slightly smaller group comprising a supporting technical committee. Ray Alisauskas (CWS) coordinated the initial planning work, followed by Paul Gregoire for a couple of years, before Deanna Dixon assumed the role in 1994 and continued to serve until 2024 – the longest service record of any Joint Venture coordinator under NAWMP (Box 7.1).

The AGJV and Black Duck Joint Venture (BDJV) were the first international joint ventures, so the logistics of establishing these partnerships had the added complication of involving two countries. For the AGJV, both the Management Board and Technical Committee are co-chaired internationally. There are two US federal agencies (USFWS and USGS) and all four Flyways represented on both the Management Board and the Technical Committee. In Canada, federal representation comes from different administrative regions. Provincial representation has fluctuated. There are also NGO members from both countries. For a few years, the JV also included Mexico, but their participation varied.

At the onset, the AGJV discussed high-priority research topics and how the JV might improve coordination, information sharing, and resource sharing to address those needs. It emphasized science to support federal policies, inform Flyway management of populations and advise habitat planning for other JVs. The JV also served as a forum to discuss numeric goals with CWS, FWS, and the Flyways for 16 populations of geese that breed mostly north of the boreal forest in Canada and Alaska. An early challenge was population delineation. Although goose populations were probably best identified by breeding-ground affiliations, sometimes it was easier to estimate numbers in wintering concentrations. Genetic research that helped elucidate population structures and things like the relationship of Canada geese to cackling geese and snow geese to Ross's geese were particularly helpful. Scientists would study each situation and recommend an approach, but the federal agencies made the final determinations in consultation with the Flyways. Resolving the legal separation of science and policy roles also arose during the difficult debates around overabundant snow geese (below).

The partnership concept has obviously worked for the AGJV. Fast-forward to 2024, when the JVs fact sheet notes 109 funding partners to date, 155 major projects completed, and, in 2023, $435K in FWS funding leveraging another $1.25 million in research support from States and Canadian sources. Today, the AGJV facilitates research and monitoring for seven species and 24 populations of geese that breed in the Arctic and migrate throughout the continent. The JV works cooperatively to coordinate approaches to meeting high-priority information needs. The partnership approach has proven especially valuable for conducting Arctic research where field logistics are costly.

The greatest challenge addressed by the AGJV was the unusual problem of overabundant white geese whose populations had benefited from agricultural expansion in their primary migrating and wintering haunts. The complex history of that vexing issue is presented in timeline format on the AGJV website (https://www.agjv.ca/light-goose-timeline/) spanning the time from the 1916 Migratory Bird Treaty through 2018.

The challenge of overabundant white geese with attendant impacts on their Arctic breeding habitat spurred the AGJV to form an Arctic Goose Habitat Working Group in 1995. Their report,

Arctic Ecosystems in Peril (Batt et al. 1997), was a landmark report that urged the responsible federal and Flyway officials to act to reduce white goose populations. Subsequent unprecedented and controversial actions to do this by greatly expanded hunting opportunities for snow geese, including spring hunting and generous bag limits, unfortunately, had limited impact on controlling goose populations (Alisauskas et al. 2011).

The AGJV usually updated its strategic plan every four years after 1990 but has more recently moved to a review every two to three years to align with their research funding cycle. The latest plan in 2022 (https://www.agjv.ca/wp-content/uploads/2022/05/AGJV-Strategic-Plan-2022-FINAL.pdf) chronicles 30 years of progress in better understanding 24 specific populations of geese. Focus areas recommended in the 2022 Plan were habitat degradation caused by populations of snow and Ross's geese, evaluation and improvement of population delineations, impacts of climate change and resource development on Arctic Geese, ecology of brant and emperor geese, evaluation and improvements of harvest estimates, and improvements in population monitoring.

A wonderful resource available on the AGJV website is an illustrated history of the main Arctic Goose research camps in Canada (https://www.agjv.ca/history-of-goose-field-camps-in-northern-canada/). Highly recommended!

Black Duck Joint Venture

As NAWMP was forming, American black ducks posed a special challenge. The species had long been revered by hunters as the wiliest of ducks and was the most abundant duck across much of Eastern North America from Labrador to the Gulf Coast. Surveys only partially covered its breeding range but those data, as well as winter indices, suggested that the species had declined substantially since the 1950s. Yet, the causes of the decline were uncertain (Bailey et al. 1988). Ideas included changes in land use (e.g., forestry practices) on the breeding grounds, displacement or genetic introgression by mallards (especially game farm mallards in the Atlantic Flyway), hunting mortality, leucocytozoan infections, or inadequacy of wintering habitat.

Planners concluded that black ducks needed special attention and that developing better information on population trends and factors affecting the species' population dynamics was urgent. Planning for the BDJV began in 1989, led by the federal governments, states of the Atlantic Flyway, and some university researchers. Kathy Dickson (CWS) signed on as the JV coordinator and continued in that role through 1995, followed by Brigitte Collins from 1996 to 2013. Additional coordinators have included Trish Edwards, Tania Morias, Kristin Bianchini, Kristina Hick, and Pat Devers (Science Coordinator). The JV published its first prospectus in 1991 (CWS and USFWS 1991) followed by a strategic plan (Black Duck Joint Venture 1993). These proffered three initiatives broadly described as research, surveys, and banding. The list of research needs was extensive, covering recruitment rates, causes and extent of mortality, habitat selection and management, and mallard/black duck interactions. Banding needed to be expanded across more geographic locations, and enhanced communication between harvest managers in Canada and the United States seemed desirable. An expanded survey program was envisioned including helicopter surveys across random blocks on forested breeding habitat, fixed-wing surveys over more southern areas, and ground surveys in Prince Edward Island. Prior to the BDJV, eastern waterfowl breeding areas did not have a population survey in place to match what existed for the prairie region. One reason was difficult terrain. Accordingly, an early priority was to evaluate the pros and cons of helicopter-based vs. fixed-wing-based surveys for eastern landscapes.

The BDJV has accomplished many of its initial goals and is presently focused on understanding the linkages between population dynamics and habitat circumstances to provide guidance for growing the continental population. The JV has:

1. Established long-term monitoring programs for breeding populations, expanded preseason banding, and improved knowledge of black duck ecology. Witness the operational Eastern Waterfowl Breeding Population and Habitat Survey (Zimmerman et al. 2012).
2. Supported the development of the Black Duck International Harvest Strategy that was first implemented in 2013 (USFWS 2025).
3. Begun to link black duck population dynamics to habitat (Robinson et al. 2016) with the development of decision support tools in the Atlantic Flyway (e.g., Hartley et al. 2020).

Development of a cross-border adaptive harvest management strategy had not been achieved before for any species. A group of federal, state, and Flyway managers settled on three goals (1) maintain a black duck population that meets legal mandates and provides consumptive and nonconsumptive use commensurate with habitat carrying capacity; (2) maintain societal values associated with the hunting tradition; and (3) maintain equitable access to the black duck resource in Canada and the United States. The harvest strategy is based on two hypotheses: (a) that black duck population growth is limited by competition with mallards during the breeding season and (b) that black duck population growth is limited by harvest that is additive to natural mortality. Annual updating of model weights leads to recommended regulatory choices for the United States and Canada individually.

The latest iteration of the BDJV implementation plan (BDJV Management Board, 2015) was focused on developing and implementing decision tools to guide habitat management at multiple spatial scales. This was done with the objective of increasing continental carrying capacity and thus growing the population of black ducks. The Plan also embraced a framework of strategic habitat conservation (Chapter 14) to focus on explicit management decisions while tailoring its monitoring and research programs to reduce key uncertainties that influence those decisions.

A possible role of human harvest in affecting black duck numbers remains uncertain, as does the evidence supporting black duck density-dependent reproduction vs. survival, so the new management framework will contrast support for these models by comparing model predictions and observed system states and updating model weights among three competing density-dependent hypotheses:

1. The black duck population is limited by density-dependent productivity based on the amount of available breeding habitat;
2. The black duck population is limited by density-dependent post-hunting seasonal survival based on the amount of available nonbreeding habitat; and
3. The black duck population is limited by density-dependent productivity based on the amount of available nonbreeding habitat.

The development and implementation of habitat projects often take three to five years to complete. Therefore, the time step for the BDJV decision framework was set at five years. Aggressive monitoring and research programs were launched to support this adaptive framework.

In 2025, Kristina Hick (CWS) and Patrick Devers (USFWS) serve as the BDJV Coordinator and Science Coordinator, respectively, and the JV is in the process of updating their strategic plan.

Box 7.1 Deanna Dixon – Reflections of the Longest Serving JV Coordinator

In 1990, I was hired by the Canadian Wildlife Service to assist with a new program just getting underway. I was hooked immediately, so what began as a short-term contract ended up as a full career working on NAWMP.

I served as the Coordinator for both the Prairie Habitat Joint Venture and AGJV for 29 and 30 years, respectively. It was the vision of NAWMP and the partnerships that motivated me, not any single organization, including the one I worked for. That helped me considerably as I think the most important aspect of my job was to assist in nurturing this partnership, what I came to think of as our NAWMP family.

I was hired by Gerald McKeating, who soon became the Director of CWS, Prairie and Northern Region, the Chair of the PHJV, and the Canadian Co-Chair of the AGJV. He taught me to reach out and engage people in discussions, respect differing views, assist in developing a collective voice, and most importantly build trust needed for a partnership to succeed.

I have much admiration for the early visionaries of NAWMP. Developing a clear and concise continental message of urgency that could be stepped down to science-based goals and objectives and delivered through untested, unknown, and relatively autonomous regional partnerships was nothing short of incredible. This was long before email (never mind Zoom!), so communication consisted of in-person meetings, phone calls, mail, and faxing. I can't fathom the number of meetings required to create the Plan. But in truth, it probably couldn't have happened any other way. Forging partnerships and building trust really needs to be done in person.

NAWMP has been a success for almost 40 years, so it's been interesting to reflect on a few of the factors that really made the JVs "sing". I'd say it started with those early leaders, people who reached far beyond their day to day to envision and strive for something better. Through much of the 1990s, we were "singing", with an exceptional number of leaders involved on both the AGJV and PHJV Boards, including the Chairs of both boards, the Chairs of their technical and science teams, plus leaders on other committees as well. There was also a strong team at CWS national headquarters which helped build the national alliance. There was a tremendous amount of energy and enthusiasm, with members devoting countless hours to attend meetings, traveling almost constantly, and providing energy to roll out the plans.

There are naturally ebbs and flows in partnerships with turnover. Several times during the later decades we have seen that same level of energy as the 1990s. Those times aligned with the strong engagement of members and, in particular, the number of "leaders" involved. However, there have been notable shifts over time, from strong board leadership, to more of a board oversight role with the science/technical and other committees taking the lead. It is neat to see new members, initially quiet, get excited and really jump in as they realize what a great program they landed in.

The AGJV is a small but mighty JV. It's importance to geese continentally can be seen in the interest and dedication of its members. They are in the room, not because they are filling a role, but because they really want to be there. It is one of the least funded JVs, and yet the passion for the work is consistently high. Members participate in-person, and unique synergy exists between the Board and Technical Committee as they mostly attend each other's meetings!

One limitation to the successes of the JVs has been the relatively low and sporadic investment in communications and marketing. Both of my JVs fully recognized the need and benefits, but it has been difficult to maintain consistent effort. When they have made the investment, the results have been remarkable. The most interesting example is how the AGJV undertook a massive communications strategy and media campaign to educate the public about snow geese, a bird relatively unknown to many, and gained public acceptance and support to significantly reduce their populations.

The PHJV's importance to waterfowl and habitat conservation is well understood in our NAWMP community, and the JV's achievements have been outstanding. But outreach beyond the NAWMP partnership hasn't been consistent. The JV had a couple of hard lessons, one in the mid-2000s and another more recently, that a lack of communication can result in unnecessary misunderstandings or mistrust. The more recent issue was with Max McGraw Wildlife Foundation, which led to an extensive, two-year review of NAWCA spending in Prairie Canada. The main finding of that review was that the significant accomplishments of the PHJV had been underreported. Its success story could and should have been better told. At the same time, those wetlands that had not been protected were dwindling. With some assistance from Max McGraw, the PHJV has recently stepped up the resources for communications, which will be important to gaining more public understanding and broader support for stemming the continued loss of wetlands.

Alongside strong leadership and communications, I think big strides have been made in times of controversy, and while having fun, and these aren't mutually exclusive! There wasn't always agreement among members, and I recall some heated discussions in both JVs. Respecting divergent views and finding a way to hear and understand them were and are very important in maintaining a partnership. It was those discussions, and watching key leaders get creative to reach a positive outcome, that pushed the JVs forward in a positive and productive way. The daytime discussions could be long, arduous, and sometimes intense, but it was often the fun at the end of the day – shooting the breeze while watching a flock of birds, dinners or special events, or just a beer in the bar – that really did wonders for increasing understanding, building friendships, and developing trust in ways that can't be achieved in formal meetings. Building relationships in person has been key to the health of these long-term partnerships. One example – embracing the Western Boreal Forest into the PHJV was a long and arduous process with varied opinions, locally, nationally, and internationally, about if and how it should be done. The discussions needed to include partners beyond the PHJV and consider administrative, policy, science, and financial factors. It was hard work and stressful for the partners to reach a decision but led to an outcome that was accepted and respected to this day.

In the AGJV, the scope and magnitude of the snow goose overabundance issue was enormously complicated, with so many factors and viewpoints to be considered. The AGJV's leadership through the establishment of several scientific working groups managed to develop a complete picture of the issues through a series of reports and led to unprecedented changes to management of those populations. The AGJV is still considered a leader by government agencies in facilitating science and making recommendations on goose management.

I think the development of trust among partners was the most important outcome from these and many other discussions over the years, which has brought so much strength to the large and diverse NAWMP partnership.

Mike Anderson asked me, why did I take on the coordination of two JVs? That was unheard-of. Well, how could I not? To be asked was such an honor! And to get to work with so many dedicated people has been the highlight of my career. Some people may think of partnerships as formal endeavors, limited to official representatives on boards or committees. Truly, the number of people involved in NAWMP is too vast to count. I sometimes wonder if those early planners could have possibly envisioned that on any given day we could reach out and request help from an NGO program delivery manager, a university research scientist, a communications specialist from a US joint venture, a wildlife director from a province, a US state biologist, a federal GIS specialist, on and on, all in the name of NAWMP. The partnership reach of NAWMP is amazing and what a fulfilling career it has been for me to be a part of it.

Sea Duck Joint Venture

Sea ducks (Tribe *Mergini*) comprise a group of diving ducks occupying fresh or saltwater habitats that share numerous anatomical and behavioral characteristics (Baldassarre and Bolen 2006). Most species are uncommon quarry for hunters although many indigenous communities, especially in northern Canada and Alaska, rely on sea ducks for meat, eggs, and in some cases down and feathers. Sea ducks have long been popular with birders for their striking plumage and their elaborate courtship behavior. The annual ranges of some North American species include Greenland on the east or Russia to the west. Some species use almost entirely coastal waters throughout their lives; some breed in freshwater wetlands but winter in coastal zones. Loss of habitat, pollution, coastal development for commerce, nearshore wind turbines, and harvest of concentrated colonial species were known threats. Most importantly for conservation purposes, many of these species were poorly known to science and estimations of their population status and trends are problematic. Two Alaskan species, the spectacled eider (1992) and the Steller's eider (1997), had been listed as threatened by the FWS.

NAWMP planners realized that addressing the paucity of information on sea duck distribution, population status and trends, habitat requirements, and what might be done to secure the future of eiders, scoters, goldeneyes, buffleheads, long-tailed duck, harlequin duck, and mergansers was the greatest immediate need for this group of birds. Much of the impetus to launch the JV came from CWS in Atlantic Canada (George Finney, Keith MacAloney, and others) along with Alaskans like Tom Rothe concerned about declining eiders and harlequin ducks affected by oil pollution. Delineation of populations was essential. With so little banding data available, molecular genetics proved helpful especially with species with disjunct east-west distributions like harlequin ducks and Barrow's goldeneyes. Once research could answer some of the basic population questions, then connecting with agencies to help develop operational surveys and working with coastal-zone Joint Ventures to identify, protect, and possibly enhance sea duck habitats would be natural next steps.

With competing demands on agencies, including the formation of other JVs, it wasn't until 1999 that the Sea Duck Joint Venture (SDJV) was formally established but its strategic planning documents have been updated regularly since then (e.g., SDJV 2008, 2022). The partners' vision was that "sustainable populations of North American sea ducks are maintained throughout their ranges". The first management board of 13 included Co-chairs George Finney (CWS) and Jon Andrew (FWS). The agreement also created two Coastal Scientific Advisory teams, each with one Arctic region representative and a combined Continental Technical Team (CTT) with links to the management board. Initially, dual JV coordinators were designated; the Atlantic coordinator was located in Canada and the Pacific coordinator was located in the United States. Keith McAloney (CWS) was retained in 1998 to lead the technical work and had a strong hand in getting the JV launched. He remained in place through 2001. Tim Bowman (FWS) assumed the role in 2002–2017. Presently, Kate Martin (FWS) serves as the overall JV Coordinator.

The JV has evolved since its inception; there are no longer separate coastal scientific teams. The CTT now serves a comprehensive advisory role and includes staff representing organizations on the Management Board and at-large members with appropriate scientific expertise. Members on the Board and CTT span the continent. New members include the Upper Mississippi River Great Lakes JV, recognizing that the Great Lakes region is essential for wintering sea ducks, and that conservation of coastal areas that are largely managed by state and federal agencies requires more participation from those partners. Kyle Spragens (WA) and Megan Ross (CWS) serve as technical committee co-chairs while Todd Sanders (FWS) and Becky Whittam (CWS) co-chair the management board.

The JV's latest strategic plan update was completed in 2022 for the decade 2022–2031 (https://seaduckjv.org/wp-content/uploads/2022/06/SDJV-Strategic-Plan-2022-2031-FINAL.pdf.) and it outlines strategies for achieving the partners' vision over the next decade. Priorities include (1) focusing on species of management concern, (2) assessing population status and distribution, (3) improving harvest management, (4) conserving important habitats, (5) better understanding consequences of environmental change, and (6) mitigating effects of industrial development. Their approach for achieving these things will continue to be supporting research that informs management decisions, improving curation of existing data sets, and increasing partnerships, awareness, and engagement in sea duck conservation.

One major accomplishment has been compiling a comprehensive Atlas of important sea duck habitat sites throughout North America (Bowman et al. 2022). For 85 sites, the Atlas provides information on recent sea duck abundance, temporal usage, and potential concerns. Rigorously vetted, with relevant citations, it offers strong recommendations to the NAWMP community, especially the habitat JVs, about needs for sea duck habitat conservation.

An ongoing focus for the JV is delineating population units for management purposes and seeking ways to monitor those populations. As of the 2024 Update (Appendix C in NAWMP Committee 2024), there were still eight species or populations (of 25) with no adequate population estimates available, and 18 for which the SDJV cannot recommend a population goal. The need for continued research is obvious as is the challenge for managing these "data-poor" species. As summarized in the latest SDJV strategic plan, recent satellite telemetry studies of surf scoters, black scoters, long-tailed ducks, king eiders, common eiders, and harlequin ducks have enabled helpful strides in delineating the annual cycle ranges of several populations. The 2022 Plan provides an interim ranking process for species/populations of management concern and the human impacts that may be most germane to each of them.

Since its inception, the SDJV has supported over 175 research and monitoring projects (https://seaduckjv.org/sea-duck-joint-venture-projects/). In their 2022 report to the Plan Committee, the JV noted that, in the latest five years, the 20 major projects supported involved 50 partner organizations with matching funds varying annually between 22 and 130% of SDJV funding. Also, to support NAWMP's "people" goals (NAWPEP Chapter 22), the JV established a Student Fellowship Program in 2021, which had as of 2024 supported 14 graduate students working on sea duck projects (https://seaduckjv.org/funding-opportunities/student-fellowship-program/; https://seaduckjv.org/funding-opportunities/recipients-of-sdjv-student-fellowship-awards/).

Box 7.2 Planning for the Future on Long-term Study Areas

Much of what we know about northern geese and sea ducks has emerged from a handful of long-term field studies in northern Canada and Alaska, mostly supported by the federal governments, Flyway partners, or national grant programs like NSERC, NSF, or Canada's Polar Continental Shelf Program. A major challenge for any long-term study is managing the succession of principal investigators and field staff when initial leaders and local indigenous team members move on. It's not unlike the challenge the entire Plan Community faces as decades pass. Every long-term field study risks the loss of momentum and value when early leaders retire, and every team faces the challenge of mitigating annual turnover in personnel.

Recognizing this, a recent program developed by the SDJV (https://seaduckjv.org/funding-opportunities/student-fellowship-program/) in collaboration with Ducks Unlimited, Inc. is an attempt to recruit new biologists to focus on sea ducks. The SDJV Fellowship program offers support to graduate researchers with an interest in such studies. Three or four awards have been made annually since 2022. The long-term goal of the program is to increase the number of skilled early-career professionals interested in sea duck research and conservation.

A promising complementary program has been developed to engage local indigenous partners by Grant Gilchrist of Environment and Climate Change Canada, working with community leaders at his long-term Belcher Islands Archipelago eider study area. Inuit youth live and work with western science-trained biologists conducting fieldwork at their remote summer camps. By experiencing life and work in a research camp, these indigenous youth are exposed to new education and employment ideas and offered career consulting advice if desired. In return, the southern staff and students gain field assistance and traditional knowledge of the land. It's a small-scale program with support from ECCC and the Polar Continental Shelf Program but offers a promising model ripe for replication elsewhere.

A Summary Thought

Envisioned at the onset of NAWMP as a means of providing needed information to guide conservation of poorly understood species, these species' Joint Ventures have been largely successful. Logistical challenges of working in wilderness areas, limited funding, and human resources have constrained the pace of progress from what was hoped at the onset, but clearly all of them have been useful. Black duck population and habitat management are on firmer scientific footing. Many goose and sea duck populations are better understood. Challenges remain for many places and populations, but the work of these groups remains vital to the NAWMP enterprise.

References

Alisauskas, Ray T., Robert F. Rockwell, Kevin W. Dufour, Evan G. Cooch, Guthrie Zimmerman, Kiel L. Drake, James O. Leafloor, Timothy J. Moser, and Eric T. Reed. 2011. "Harvest, survival, and abundance of midcontinent lesser snow geese relative to population reduction efforts." *Wildlife Monographs* 179: 1–42. †

Bailey, Robert O. and the Eastern Canada Waterfowl Advisory Council. 1988. *The Black Duck in Eastern Canada: A National Overview.* Canadian Wildlife Service, Ottawa, ON. 24pp.

Baldassarre, Guy A. and Eric G. Bolen. 2006. *Waterfowl Ecology and Management.* 2nd Edition. Krieger Publishing Company. Malabar, FL. 567pp.

Batt, Bruce D. J., Editor. 1997. *Arctic Ecosystems in Peril: Report of the Arctic Goose Habitat Working Group.* Arctic Goose Joint Venture Special Publication. U.S. Fish and Wildlife Service, Washington, DC. and Canadian Wildlife Service, Ottawa, Ontario. 120pp.

Black Duck Joint Venture. 1993. *Strategic Plan.* Canadian Wildlife Service. Nepean, ON. 7pp.

Black Duck Joint Venture Management Board. 2015. *The Black Duck Joint Venture Strategic Plan, 2015–2019.* Canadian Wildlife Service, Gatineau Quebec. 37pp.

Bowman, Timothy. D., James L. Churchill, Christine Lepage, Shannon S. Badzinski, Scott G. Gilliland, Nic R. McLellan, and Emily Silverman. 2022. *Atlas of Sea Duck Key Habitat Sites in North America.* Sea Duck Joint Venture March 2022. 286pp. (https://seaduckjv.org/science-resources/sea-duck-key-habitat-sites-atlas/).

Canadian Wildlife Service and U.S. Fish and Wildlife Service. 1991. *The Black Duck: A Species of International Concern.* Ottawa, ON. 9pp.

Hartley, Mitch J. and Aimee J. Weldon, eds. 2020. *American Black Duck Conservation Plan.* Atlantic Coast Joint Venture. 60pp. www.acjv.org/documents/ACJV-American-Black-Duck-Conservation-Plan.pdf

NAWMP Committee. 2024. *North American Waterfowl Management Plan 2024 Update: Expanding the Partnership.* U.S. Fish and Wildlife Service. Washington, DC. 50pp.

Robinson, Orin J., Conor P. McGowan, Patrick K. Devers, Rodney W. Brook, Min Huang, Malcolm Jones, Daniel G. McAuley, and Guthrie Zimmerman. 2016. "A full annual cycle modeling framework for American black Ducks." *Natural Resource Modeling* 29: 159–74. (https://doi.org/10.1111/nrm.12088).

Sea Duck Joint Venture Management Board. 2008. *Sea Duck Joint Venture Strategic Plan 2008–2012.* USFWS, Anchorage, AK; CWS, Sackville, NB. 95pp.

Sea Duck Joint Venture. 2022. *Sea Duck Joint Venture Strategic Plan 2022–2031.* U.S. Fish and Wildlife Service, Anchorage, AK; Canadian Wildlife Service, Whitehorse, YT. 38pp.

U.S. Fish and Wildlife Service. 2023. *Adaptive Harvest Management: 2024 Hunting Season.* U.S. Department of Interior, Washington, DC. 76pp. Available online at (https://fws.gov/project/adaptive-harvest-management).

U.S. Fish and Wildlife Service. 2025. *Adaptive Harvest Management: 2026 Hunting Season.* U.S. Department of Interior, Washington, DC. 85pp. (https://fws.gov/project/adaptive-harvest-management).

Zimmerman, Guthrie S. John R. Sauer, William A. Link, and Mark Otto. 2012. "Composite analysis of black duck breeding population surveys in eastern North America." *Journal of Wildlife Management* 76: 1165–176.

8 The Species Action Groups

In the early 2000s, concern was building about the population statuses of northern pintails and lesser and greater scaup (e.g., Miller and Duncan 1999; Afton and Anderson 2001). Discussions began about whether NAWMP ought to add new species joint ventures to address these issues as had been done for black ducks, sea ducks, and Arctic Geese. The Plan Committee recognized, however, that the existing species JVs remained essential and underfunded. A simpler approach seemed desirable.

The NSST (Chapter 13) suggested, and the Plan Committee agreed, that it might be better to try a different model, one that would require less process, infrastructure, and administrative funding than another formal JV. Two groups of interested scientists came together to propose first a Pintal Action Group (PAG as it became known) in 2003 and a Scaup Action Team (SAT) in 2007. The prospectuses elaborated the concerns of the scientific and management communities and set objectives for the teams. The common idea was to encourage and coordinate research on these species to better understand what was affecting their population trends and then to convey what was learned to harvest managers and to the habitat JVs already in place to take remedial actions.

The Pintail Action Group

Northern Pintails (hereafter, pintails) had displayed sharp population declines during the 1980s, reaching lows of 1.8 million birds in 1991 and again in 2002. Historically, when wetlands counted during May duck surveys had increased, so had pintail numbers. But during 1996 and 1997 with abundant wetlands, pintail numbers grew only modestly and remained 36% below the NAWMP goal of 5.6 million birds. In contrast, almost all other prairie-nesting dabbling duck populations increased strongly in those years. This failure of pintails to respond as expected to better pond conditions heightened concerns of waterfowl managers and suggested that the problems facing pintails may be different or more severe than previously thought (Miller and Duncan 1999). Yet, little recent management activity or research has been focused on pintails, particularly on the breeding grounds.

The Joint Flyway Councils in the summer of 2000 passed a resolution requesting that the FWS organize a workshop to probe the decline. The workshop was held in Sacramento in March 2001. Researchers and managers debated the pintail "problem" and synthesized information about the species (Miller et al. 2001). Potential causes of the decline were identified, management actions recommended, and research needs outlined. Proponents argued that a coordinated, multi-JV approach was required and that some kind of pintail technical group might be able to mobilize and channel efforts among NAWMP partners without creating redundant administrative structures.

DOI: 10.4324/9781003732310-8

Pintails have a broad geographic range and although poor breeding success on the prairies seemed likely to be an important factor limiting population recovery, hypotheses involving other regions and times of the year, or harvest pressure, could not be ruled out. As recently as the 1970s, 60% or more of the continental population of pintails bred in southern Canada. By the early 1980s, the numbers of pintails settling there had dropped, while other breeding areas remained relatively stable. Additional research and management trials were needed. Maintenance of existing high-quality habitats on wintering, staging, and northern breeding areas (e.g., Alaska) was also seen as important to ensure that any recruitment gains on the Canadian prairie would not be compromised by habitat degradation elsewhere. Concerns for the maintenance of winter habitats in California's Central Valley, rice-field habitats from Arkansas to Texas, and the west coast of Mexico remained, along with migration habitats in the Rainwater Basin of Nebraska and the Klamath Basin of California and Oregon.

The Pintail Action Group was envisioned as a partnership of Canadian and US Federal, State, and Provincial agencies, nongovernment organizations, and university scientists that would help identify and coordinate actions within and among multiple JVs to address management and research needs. Topics of initial focus were:

- Developing an adaptive harvest management framework for pintails.
- Improving the geographic scope and precision of population estimates.
- Improving pintail survival estimates and identifying sources of heterogeneity.
- Pintail nesting ecology and sources of variation in recruitment rates.
- Factors affecting pintail settling patterns in spring.

Having eschewed a formal JV structure, the PAG was established as a subgroup of the NAWMP Science Support Team. The Mission of the Pintail Action Group was *to advocate and support planning, coordination, and evaluation of northern pintail management and research actions among the NAWMP Joint Ventures, Flyways, government agencies, and other organizations*.

The PAG would consist of biologists from JVs important to pintails, government agencies, and nongovernmental groups. It would meet in person once annually in conjunction with the NSST or other meetings, establish a website, and share information electronically. Government agencies, nongovernmental organizations, and other partners would contribute funding for the habitat conservation work, evaluation, monitoring, and research needed to support the recovery of the species. Direct expenses for the PAG were expected to be limited to meeting expenses and communication materials.

The prospectus for the new group was developed through 2002, mainly by a writing team of Karla Guyn, Michael Miller, Dave Duncan, Robert Clark, and me. With endorsement of the NSST, the Plan Committee approved the prospectus in the summer of 2003. Joe Fleskes (USGS, California) was chosen as the first Chair through 2005.[1] The group published its first annual report in 2004, did so annually through 2012, and then again in 2015 and 2019 (www.pintailactiongroup.com).

The meetings and newsletters served effectively for sharing information about pintail research, as networking opportunities across agencies and JVs, for brainstorming new initiatives, and to a lesser degree connecting with habitat managers. I was a member of the PAG for several years, though not as a pintail expert but as a link to the NSST and the Plan Committee. I think the networking of researchers was helpful in advancing pintail science and that many individual members were also usefully engaged within their regional JVs in sharing what was learned for conservation planning purposes. However, the PAG last met in person in 2019 just before the COVID pandemic. Chair Mitch Weegman and the group prepared a comprehensive update for the Plan Committee in 2021, and, though research has continued, the group itself has been largely inactive since then. Some interesting progress was made recently through a long-term analysis of banding data by Bartzen and Dufour (2017) and spatial and temporal variation in pintail productivity by Mattsson et al. (2017).

These and other data enabled an integrated population modeling effort led by postdoctoral fellow Qing Zhao et al. (2019) who concluded that agricultural intensification in the PPR resulted in lower pintail productivity and higher sensitivity of pintail productivity to variation in pond numbers (a climate index). Thus, it seems pintail populations have become more vulnerable to climate variation under intensified land use, which suggests that future conservation strategies must adapt to these altered relationships.

Next steps with the IPM modeling should perhaps add harvest as a component. One overarching lesson from all this work on pintails is emphasizing the importance of large-scale, long-term monitoring of birds and landscapes. Operational programs (counts, banding) inform habitat conservation in space and time, and while operational monitoring programs could be improved, they remain vital to analytical efforts and adapting conservation plans.

Scientists affiliated with the PAG are discovering how populations respond demographically to the direct and interacting effects of climate and land use change, a valuable advance. They have also revealed common patterns and drivers of productivity within the prairie dabbling duck community, the importance of ponds (and winter wheat) to partly offset agricultural intensification effects thus reinforcing the importance of limiting drainage. Continuation of these broad monitoring programs (counts, banding, etc.) has potential to help estimate the effects of planned higher bag limits for pintails in the United States beginning in 2025.

As of autumn 2025, the PAG has not met for several years nor assembled a substantial progress report since 2021. With a track record of improving our understanding of pintail population dynamics through integrated research, with policy and conservation implications, it would be unfortunate if their work isn't revitalized.

The Scaup Action Team

Quickly following the creation of the PAG, managers and biologists with growing concerns about scaup began crafting plans for a parallel working group. The combined breeding-season populations of greater scaup and lesser scaup (hereafter, scaup) declined from estimates of 5 to >7 million birds in the 1950s and 1970s to a record low of 3.25 million birds in 2006, 48% below the goal of 6.3 million scaup set in the 1986 NAWMP.

There was much uncertainty about factors limiting scaup populations; so appropriate management actions were unclear. Scaup have the most widespread distribution of all North American diving ducks, extending from the northern tundra in summer to southern Mexico in winter. This continental range challenged attempts to understand both the nature of constraints on the populations and the temporal and geographic scope of their influences. The western boreal forest, where 70% of breeding scaup are counted, had experienced the fastest rates of decline (Afton and Anderson 2001) but it remained unclear what and where recent limitations were operating.

The desire to recover scaup populations led to new research and two scientific workshops (1998 and 2006) that addressed hypotheses about factors constraining the population and delineation of research and management needs (Austin et al. 2000). Workshop participants in 2006 concluded that dialog between researchers and managers was vital for focusing research and helping to frame management questions, but that gathering only once every seven years was inadequate. Scaup issues merited a more frequent forum to foster communication and coordination, integrate new results, discuss emerging issues, and keep scaup at the forefront of management agencies. Thus, a SAT under the auspices of NAWMP was proposed modeled closely after the PAG.

Participants at the 2006 Workshop developed a list of factors thought to be limiting scaup populations. These were subsequently condensed and captured as desired action items:

- Examining historical spatial and temporal patterns of population change and associated climatic and landscape characteristics.
- Quantifying demographic rates in the Western Boreal Forest and analyses of habitat changes there that may be affecting vital rates.
- Estimating breeding propensity and identifying sources of variation in same.
- Identifying factors influencing adult survival.
- Examining winter or spring migration habitats with possible implications for survival and cross-seasonal effects.
- Evaluating estimation biases in population surveys.
- Identifying initiatives that might be relevant for habitat JVs.

Proponents believed that close coordination among projects might be beneficial as might novel partnerships with climatologists, hydrologists, and landscape ecologists. A broad, strategic approach to scaup research, permitting landscape-level inferences, seemed desirable.

Thus, the SAT's mission was *to help reverse the decline of scaup populations by advocating and supporting the planning, coordination, conduct, and assessment of scaup research and management actions among the NAWMP Joint Ventures, Flyways, government agencies, and other interested parties.*

Following the lead of the Pintail Action Group, scaup advocates recommended a similar structure and operational model with the SAT functioning as a subgroup of the NSST. One difference was a strong recommendation to secure the services of a dedicated coordinator to steadily advance the work of the SAT, the PAG, and perhaps other high-priority waterfowl science and conservation initiatives not presently covered by JV or national NAWMP staff.

The SAT Prospectus was finished following the workshop in 2006[2] and approved by the NSST and then the Plan Committee in 2007. Unfortunately, when the PC endorsed the SAT, they specifically excluded any commitment to a staff coordinator, which had lasting ramifications for progress of the SAT.

The SAT started strong with in-person meetings in 2008 and 2009. Jane Austin (USGS) was elected the first Chair; and Steve Cordts (Minnesota DNR) succeeded Jane in 2010. The Team produced a first information-packed newsletter in 2009. Jane (USGS) and Mark Koneff (USFWS) acquired seed funds to help develop a Scaup Conservation Plan and hosted two workshops to help develop it. Their approach was to apply principles of structured decision-making and develop a framework for recovering scaup populations. The first session (Minneapolis in April 2009) focused on breeding season matters and the second (Memphis in September) on migration and wintering areas and human dimensions. Austin, Koneff, and Scott Boomer followed up, advancing some candidate decision support models in 2011, but further work on the framework stalled. Although nothing concrete emerged, the scaup team did start thinking about how to link duck-hunter dynamics with scaup numbers, including integrating management costs and benefits. This was novel thinking at the time. Fast-forward to 2025 and the National Flyway Council has approved setting a closed-season threshold on scaup of 2.5 million birds, a cautionary step but without the benefit of any experimental structure in place to ensure learning.

Although some research continued, the Action Team fell quiet after 2011. When the SAT and PAG reported jointly to the Plan Committee in 2016, the SAT had little to report. That same year, Todd Arnold et al. (2016) completed an informative analysis of banding data from 1951 to 2011 and concluded that declining fecundity was the most likely explanation for population declines during 1975–2005 and that hunting mortality probably played only a minor role in affecting population dynamics of lesser scaup.

Reflections on the Action Groups

Bob Clark instigated a joint meeting of the PAG and the SAT at the North American Duck Symposium in 2017. The hope was to revitalize the SAT or at least consider why the pintail group seemed to be succeeding while the scaup group had stalled. I think there were probably three main differences in play. First, the SAT did not include active JV participants ready to apply on the land what was being learned. The JV structure is less strong throughout most of the scaup breeding range and the birds are widespread in the nonbreeding period as well. Habitat improvement actions were not evident. Second, unlike with pintails, there are relatively fewer hunters that actively target scaup or make huge investments (think duck clubs in the Central Valley) in providing habitat for scaup. Most scaup utilize bigger waters and public lands. Third, with some modest recovery of scaup numbers between 2009 and 2017, the sense of urgency about the status of scaup began to wane, reinforced by the conclusion of Arnold et al. (2016) that human harvest was probably not materially driving scaup numbers. Lastly, with few people studying scaup, after the original advocates of the SAT retired, there seemed not to be a critical mass of scaup-engaged successors to ensure continuity.

The PAG and SAT were experimental in nature – something different than a formal JV. What can be learned from our collective experience with the action team model? Might it also be valuable to take stock of the science needs that remain for these species and others that merit special attention? Bob Clark (CWS), who like me served on both groups, offered these reflections (personal communication, April 2025):

- The teams provided useful forums for engaging technical and management staff in problem-solving, both at the outset and on a regular basis thereafter until the COVID pandemic disrupted in-person meetings.
- The initial work of reviewing existing data and building hypotheses linking demographic rates and a wide range of environmental factors was useful. This included potential habitat management actions.
- The process inspired research on several high-priority information needs, mostly completed by teams of collaborating individuals and institutions, in many cases with modest funding.
- For pintails, studies identified where management actions might be most effective (e.g., increase nesting success in prairie breeding areas, retain rice production on wintering areas [Weegman et al. 2024]) and advocated for experimental harvest management.
- For scaup, results highlighted impacts of environmental drivers on productivity in boreal systems (Ross et al. 2015), where direct management options are limited; rather, climate policy efforts might be most effective over the long term. Other findings suggest that conserving important migration and wintering habitats, including coastal marshes and large lake/river systems, will remain important objectives (e.g., Hall et al. 2024).
- The NAWMP partnership has access to remarkable data for many species and a technical pool of talented individuals who are enthusiastic and able to generate and test ideas to help guide

management actions. Many graduate students and postdoctoral fellows were trained via these initiatives. This capacity-building represents another important outcome for NAWMP.

- Looking ahead, maintaining and improving operational programs (e.g., surveys, banding) remain critical for identifying patterns and causes of future population changes, to help evaluate experimental harvest management and to identify management actions that could benefit species of concern.

Box 8.1 The Bird on the Cover of the 1986 Plan

Sometime in 1988, I posed the question to Harvey Nelson, "What about the bird on the cover?". I was deeply engaged in canvasback research at the time and despite my enthusiasm for NAWMP I had hoped for more prominent treatment of canvasbacks. In 1985 as the Plan was firming up, the May survey estimated 376K breeding birds vs. the average of 542K through the 1970s. For a bird long held in high esteem by hunters, epicureans, and artists alike, this seemed concerning. Why not a canvasback JV?

Harvey indulged me and a few other canvasback enthusiasts and organized a meeting to talk this through. Carl Korschgen who was working on canvasbacks and other diving ducks on the Upper Mississippi River around La Crosse WI was another advocate. Jerry Serie (FWS), Dave Sharp, Matt Perry, and a few others were interested in the notion as well. So, we met in Minneapolis to consider whether advocating for another species JV to the Plan Committee might make sense.

After a thorough debate, we concluded, somewhat reluctantly in my case, that the added structure and cost of a formal species JV probably were not warranted. Unlike with Black Ducks, Sea Ducks, and some Arctic Goose populations, most canvasbacks nested in the mid-continent region where they likely were well surveyed. There was already a good deal of research underway on the species on breeding, migration, and wintering areas, and the researchers involved were already communicating with one another. The people involved in research on the breeding grounds were already connected to the emerging JVs there (PPJV and PHJV). So too, it seemed likely to become the case for the still developing UMRGLJV, ACJV, GCJV and a proposed SFBJV. Another interesting twist with canvasbacks is the special entanglement they have with the parasitic and congeneric redhead. It made little sense to us to single out canvasbacks with their population dynamics so entwined with redheads. Thus, the canvasback remained on the cover, but the JV was one that never flew.

Notes

1 Subsequent Chairs were Jim Devries (DU Canada) 2007–2009, David Haukos (USGS and Kansas State University) 2010–2012, Bob Clark (CWS) 2013–2015, Bart Ballard (Texas A&M University) 2016–2019, and Mitch Weegman (University of Saskatchewan) 2020–2025.

2 Writing team members were Jane Austin, Alan Afton, Michael Anderson, John Barclay, Robert Clark, Mark Koneff, and Stuart Slattery.

References

Afton, Alan D. and Michael G. Anderson. 2001. "Declining scaup populations: A retrospective analysis of long-term population and harvest survey data." *Journal of Wildlife Management* 65: 781–96.

Arnold, Todd W., Alan D. Afton, Michael J. Anteau, David N. Koons, and Christopher A. Nicolai. 2016. "Temporal variation in survival and recovery rates of lesser scaup." *Journal of Wildlife Management* 80: 850–61.

Austin, Jane E., Alan D. Afton, Michael G. Anderson, Robert G. Clark, Christine M. Custer, Jeffrey S. Lawrence, J. Bruce Pollard, and James K. Ringelman. 2000. "Declining scaup populations: Issues, hypotheses, and research needs." *Wildlife Society Bulletin* 28: 254–63.

Austin, Jane E., Michael J. Anteau, John S. Barclay, G. Scott Boomer, Frank C. Rohwer, and Stuart M. Slattery. 2006. *Declining Scaup Populations: Reassessment of the Issues, Hypotheses, and Research Directions.* Consensus report from the Second Scaup Workshop, 17–19 January 2006, Bismarck, ND.

Bartzen, Blake A. and Kevin W. Dufour. 2017. "Northern Pintail (Anas acuta) survival, recovery, and harvest rates derived from 55 years of banding in Prairie Canada, 1960–2014." *Avian Conservation and Ecology* 12, no. 2: 1–11.

Hall, Laurie A., Christopher J. Latty, Jeffrey M. Warren, John Y. Takekawa, and Susan E. W. De la Cruz. 2024. "Contrasting migratory chronology and routes of Lesser Scaup: Implications of different migration strategies in a broadly distributed species." *Journal of Field Ornithology* 95: 1–9.

Lesser Scaup, Greater Scaup – Prospectus for an Action Team. Available from: (https://www.researchgate.net/publication/312168782_Lesser_Scaup_Greater_Scaup_-_Prospectus_for_an_Action_Team#fullTextFileContent). [accessed Feb 12 2025].

Mattsson, Brady J., Michael C. Runge, James H. Devries, G. Scott Boomer, John M. Eadie, David A. Haukos, Joseph P. Fleskes, David N. Koons, Wayne E. Thogmartin, and Robert G. Clark. 2017. "A modeling framework for integrated harvest and habitat management of North American waterfowl: Case-study of northern pintail metapopulation dynamics." *Ecological Modelling* 225: 146–58.

Miller, Michael R. and David C. Duncan. 1999. "The northern pintail in North America: Status and conservation needs of a struggling population." *Wildlife Society Bulletin* 27: 788–800.

Miller, Michael R., David C. Duncan, Karla Guyn, Paul L. Flint, and Jane E. Austin. 2001. *The Northern Pintail in North America: The Problem and Prescription for Recovery.* Northern Pintail Workshop. Sacramento, CA.

Ross, Beth. E., Mevin B. Hooten, Jean-Michel Devink, and David N. Koons. 2015. "Combined effects of climate, predation, and density dependence on Greater and Lesser Scaup population dynamics." *Ecological Applications* 25, no. 6: 1606–17.

Weegman, Mitch D., James H. Devries, Robert G. Clark, David W. Howerter, Daniel Gibson, J. Patrick Donnelly, and Todd W. Arnold. 2024. "Ecological and anthropogenic drivers of waterfowl productivity are synchronous across species, space, and time." *Ecological Applications* 34, no. 4: e2979.

Zhao, Quing. I, Todd W. Arnold, James H. Devries, David W. Howerter, Robert G. Clark, and Mitch D. Weegman. 2019. "Land-use change increases climatic vulnerability of migratory birds: Insights from integrated population modelling." *Journal Animal Ecology* 88: 1625–37.

9 "Birds are Just Like Ducks" – Other Bird Plans Take Wing

Following the waterfowl plan, other visionaries soon created complementary frameworks for conserving landbirds, shorebirds, waterbirds, and more. Within a few years, this development fundamentally altered the approach to avian conservation in the Western Hemisphere and the breadth and business of the Migratory Bird Joint Ventures (JVs).

Each of these national/international partnerships that coalesced during the 18 years following NAWMP owes some of their inspiration to the waterfowl plan, although concerns about the status of neotropical migrant birds preceded NAWMP by many years. This commonality of purpose in bird conservation strategies and practice led Gary Myers, then head of the Tennessee Wildlife Resources Agency, to famously exclaim, "Birds are just like ducks!" That seemed to reflect Myers' understanding that landscape-scale conservation was what had to be achieved for all birds. Bob Ford observed (pers. comm. July 2025) that this philosophy came with a business side as well, recognizing that state agencies needed support from wildlife watchers as well as hunters.

Most, but not all, of the plans below focus on a guild of migratory birds. Some are associated with wetland habitats, but several are not. Nearly all latched onto the expanding network of Migratory Bird JVs as a primary mechanism for project delivery. At the same time, the original seven NAWMP JVs (Figure 4.1) grew to national coverage in the United States, and much expanded geographic areas in continental Canada and Mexico (Appendix F online, www.routledge.com/9781041224389). Several of the latest JVs came together to address conservation needs mostly unrelated to waterfowl (Chapter 10).

Partners in Flight

The first substantive coalition to emerge after NAWMP was Partners in Flight (PIF). Launched in 1990, it offered a vision for landbird conservation parallel to the wetland-focused NAWMP partnership. The roots of PIF, however, date back to a seminal symposium at the Smithsonian Institution in 1966, "The Avifauna of Northern Latin America" (Buechner and Buechner 1970), where scientists shared concerns about declining birds and the possibility of cross-seasonal and cross-border influences on their populations. Terry Rich, who served as the PIF National Coordinator from 2000 to 2014, explained that the 41 attending ornithologists reviewed the ecological changes underway in Latin America – principally urbanization, industrialization, agricultural expansion, rapid population growth, and the related risks posed to birds. But in 1966, there were no systematic data with which to assess the changes that many ornithologists felt were happening.

In the absence of systematic data, there is always a risk, known in conservation biology as the shifting baseline syndrome (Soga and Gaston 2018), wherein people's reference points for

DOI: 10.4324/9781003732310-9

environmental conditions are gradually lowered over time. In the absence of past information or personal experience with historical conditions, each new generation tends to accept the situation in which they were raised as normal or baseline, thus forming an obstacle to objective consideration of long-term trends. Fortunately, soon after that 1966 Symposium, the US Bureau of Sport Fisheries and Wildlife (a previous name for the USFWS) launched an annual Breeding Bird Survey (BBS) that in time became the foundation for PIF's species' vulnerability assessments. The brainchild of Chandler Robbins, that survey continues today. Later, major symposia on landbird concerns were held in 1977, also at the Smithsonian, and in 1989 at Manomet Bird Observatory. By 1989, leaders of the BBS reported alarming changes in neotropical bird population indices since the 1970s (Robbins et al. 1989). For many groups of birds, those declines have continued (Rosenberg et al. 2019).

Also in 1989, Amos Eno at NFWF commissioned a study of neotropical migratory bird needs in Caribbean and Latin American countries, with a focus on the winter habitats of species declining as breeding birds in the United States (Stangel and Eno 1992). Initially proposed as a Neotropical Migratory Bird Conservation Program, the architects envisioned cooperation of an array of public and private organizations throughout the ranges of the target species. NFWF created a restricted fund to help get the initiative underway. Another contemporary voice calling attention to declining neotropical migrants was Russell Greenberg, then director of the Smithsonian Migratory Bird Center (Askins et al. 1990; Greenberg 1991). John Terborgh's influential 1989 book, "Where have all the Birds Gone?", did much to popularize the growing scientific concerns about bird declines.

A first organizational workshop for the Neotropical Program was held in Atlanta in 1990 sponsored by NFWF, USDA Forest Service, USFWS, the National Wildlife Federation, seven states, 17 nongovernment organizations, and a few participants from Canada and Mexico. Envisioning a program parallel in many ways to NAWMP, participants advanced a "framework for cooperation" theme. They emerged with an initial committee structure, and a first list of high-priority research, monitoring, protection, and education actions. They also designated a technical working group.

Rebranded as "Partners in Flight – Aves de las Americas", the partnership's vision was to address conservation of habitats on both breeding and nonbreeding areas. The added notion of "keeping common species common" was intended to benefit species and ecosystems before they become endangered, while also helping species in decline. Federal fiscal year 1991 saw the first appropriated dollars from USFWS and USDA Forest Service. Eno credits Bob Nelson at the US Forest Service as an immensely helpful partner at the onset. PIF also began to organize regional working groups with federal and NGO cochairs.

Other US federal agencies participating in PIF included the Department of Defense, Bureau of Land Management, and US AID. Peter Stangel was hired by NFWF to help launch a grant program and Megan Hill was appointed to oversee international grants from Guatemala City. Years later (2009), IAFWA organized its "Southern Wings" program to facilitate transferring funds from States and others to support international work.

Initial goals of the PIF program were to determine the status and causes of population changes of neotropical migrants; to maintain stable populations and habitats of mainly forest and grassland birds; and to enhance declining populations and habitats. Coordinated monitoring, research, and education were envisioned as main mechanisms for achieving this. The program was hugely ambitious, encompassing 355 species of birds in 62 families, a vastly larger suite of species and variety of habitats compared to the taxonomic scope of the waterfowl plan. PIF was originally created for all birds except waterfowl, but not long after, the shorebird advocates and waterbird group sought their own branded initiatives, in part because they could not agree on details of an early vulnerability assessment framework (now Panjabi et al. 2021) that could work for all birds (Bob Ford pers. comm., July 2025).

As with any complex venture, Stangel and Eno (1992) noted, challenges with achieving consensus soon emerged as well as growing needs for funding, taxonomic focus, travel, and finding a balance between the needs of individual organizations for identity while pursuing mutual goals. Gradually, these were managed successfully despite the "loose" organizational model of the initiative (Ralph and Rich 2002). Terry Rich (pers. comm., June 2025) recalls early discussions about a legal charter, bylaws, and even making PIF a 501(c)(3) charitable company, but this was set aside in favor of a fluid partnership that would encourage entities to contribute when and where they could and minimize formal processes.

The complex evolution of the "other bird" initiatives deserves its own book-length treatment. I can only mention a few critical steps here: the prioritization process, NFWF funding and the regional coordination structure, major funding commitments, a comprehensive PIF plan, the decision to cleave away the shorebird and waterbird plans, and the coming of NABCI as an umbrella concept for all the bird plans.

PIF Species Vulnerability Assessments

Chuck Hunter (FWS) was a key figure in development of the PIF species vulnerability assessments, a vital set of tools for targeting and prioritizing conservation investments. As a young professional, Hunter developed experience with bird surveys in the Southwest United States and had noticed that temporal trends in numbers of resident and migratory riparian species seemed to be diverging. After moving to a new job with the FWS in Atlanta, he began exploring how BBS data might be used to help assess conservation needs for landbirds (mostly) and among different areas, not only in the Southwest but continentally. Hunter recalls (https://partnersinflight.org/resources/one-biologists-journey-hunter/) that Danny Bystrak and Sam Droege provided him with trend data on every species by physiographic strata. "They sent me a hard copy printout that allowed me to have all this information in front of me as I looked for patterns. I wasn't much for computers yet, would rather eyeball it all first", he confessed. He started formulating a system based upon the factors he had first considered in Arizona along with those recommended by others including Brian Millsap's team in Florida (Millsap et al. 1990). Hunter developed scoring criteria for each of seven factors (on a scale of 1–5) and assigned preliminary scores for each species in each physiographic area. This provided an initial scoring of conservation needs for each species in each physiographic area.

David Pashley, then the Southeast Region PIF cochair, recognized the value of the process Hunter had developed and reached out to other PIF Regional Working Groups, including Mike Carter, then Chairman of the Western Working Group, on the need for a standard species' vulnerability assessment. Work on the framework expanded and at the first PIF Conference in Colorado in 1992, Hunter, Pashley, Carter, and Keith Barker presented a paper on the nascent system (Finch and Stangel 1993). Continuing work on the system, now including Ken Rosenburg in the group, led to formal publication eight years later (Carter et al. 2000). That system and the related database have continued to undergo updates but are still housed with the renamed Bird Conservancy of the Rockies and include all birds of interest to PIF across the Western Hemisphere. For an update on the Avian Conservation Assessment Database Handbook, see Panjabi et al. 2019 (https://pif.birdconservancy.org/wp-content/uploads/2021/03/PIF-ACAD-Handbook-Version-2019.pdf).

PIF Coordination

In the mid-1990s, PIF secured the first multistate grant from IAFWA to fund four staff positions to help develop PIF activities in the United States. Terry Rich, in his draft history of PIF, credits Naomi Edelson, for vision and persistence in helping that happen. At the same time, the American Bird Conservancy (ABC) agreed to support David Pashley as the first US National Coordinator to lead the group. Three such grants over seven years supported regional coordinators Carol Beardmore (West), Ken Rosenberg (NE), Dean Demarest (SE), and Jane Fitzgerald (Midwest) who collectively gave a vital boost to the enterprise. In 1993, in Veracruz, MX the first of a series of international meetings focused on Mexico and Canada brought together more ecologists and agencies, helping to spur the development of PIF beyond the United States.

As planning continued, people primarily interested in shorebirds and waterbirds had difficulty fitting their groups of birds into the evolving PIF prioritization matrix (Rich and Ford, pers. comm., July 2025). This led eventually to decisions by those groups to set up separately identified initiatives in 2000/2001 (see below). While this allowed narrower focus on smaller groups of species, the split increased coordination challenges. The FWS subsequently hired national coordinators for PIF and each of the two new initiatives.

In Canada, PIF leadership was assumed initially by the Canadian Wildlife Service and national plans emerged more slowly. A Canadian PIF working group published a Framework for Landbird Conservation in Canada in 1996. It highlighted habitat loss and fragmentation from agriculture, forestry, urbanization, linear developments and climate change, toxic chemical use, non-native species expansion, and human-caused mortality. The group proposed working with the existing NAWMP partnerships as a mechanism for conservation delivery, as well as with COSEWIC (Committee on the status of endangered wildlife in Canada), but the group was still forming as NABCI began to take shape. Thus, in a short while, PIF was embraced as one of the four pillars of the North American Bird Conservation Initiative (see below) along with initiatives for waterfowl, waterbirds, and shorebirds rather than as a stand-alone program. In Canada, its leadership was subsumed under the broader initiative. CWS staff have been notable leaders in PIF Canada. Steve Wendt, who also worked for many years on NAWMP, was honored with one of PIF's Lifetime Achievement Awards upon his retirement in 2017. Presently, there are three PIF Regional Working Groups active in Canada: a Boreal Working Group, a British Columbia Working Group, and the international Eastern Working Group. Bird Studies Canada has also been working with the Prairie Habitat JV and the new Central Grasslands Consortium (https://www.grasslandsroadmap.org/) to create a new initiative to help stem the loss of grasslands in Canada (https://www.birdscanada.org/bird-science/grassland-birds-at-risk).

A North American Landbird Conservation Plan

Support for the US PIF Regional Coordinator positions ended in 2002, leaving only the National Coordinator, Terry Rich, in place. A third International PIF Conference featuring the Migratory Bird JVs, NABCI, NAWMP, and others was staged at Asilomar CA in 2002. I remember it as a high-energy event pervaded by a sense of positive possibilities (Ralph and Rich 2005). Stepping off from advances reported there, PIF partners produced the first comprehensive, quantitative North American Landbird Conservation Plan (Rich et al. 2004). It included a "Watch List" of 100 species of greatest concern, population estimates and objectives, monitoring needs, and more. While assembling the plan, Rich recalled, partners realized that it made little sense to address only the original focal group of PIF, namely Nearctic-Neotropical migratory landbirds.

Many habitats also contained short-distance migrants and residents, and waterbirds and shorebirds as well as landbirds, so the scope of the PIF plan expanded accordingly, excepting only waterfowl that had their own well-developed plan then being updated.

Another landmark publication, "Saving Our Shared Birds – The Partners in Flight Tri-National Vision for Landbird Conservation" (Berlanga et al. 2010), published in English, Spanish, and French, featured all the landbirds that occur in Mexico, the United States, and Canada. Remarkably, >200 species rely on habitats in all three nations every year.

Rich concluded in his historical sketch that PIF, through its wide network of scientists, activists, birders, and other conservationists, had made enormous gains in understanding the biology and ecology of birds, in building partnerships, in producing state-of-the-art vulnerability assessments, and in getting the word out about bird conservation. But it's also clear that many of the problems facing birds today remain much the same.

Bob Ford, another long-serving PIF coordinator, observed something that distinguished PIF from many endangered species programs – its landscape/ecosystem base and the considerable attention paid to "keeping common species common" as well as addressing the needs of declining species. "Our work included research, large-scale conservation actions, policy initiatives and more" (pers. comm. July 2025). Bob fondly recalled

> spending time with knowledgeable local ecologists to better understand the natural community structure and composition, what disturbance factors drive the dynamics of the community, how the pattern of these communities fit into a large landscape context, and most importantly, how bird populations respond to the current content and context of these places.

It is a formula that hopefully leads to more sustainable, long-term progress than piecemeal approaches. For more on PIF history see https://partnersinflight.org/tag/pif%E2%80%91history/

Today, PIF draws upon the resources of federal and state agencies, nongovernment conservation organizations, academia, industry, and landowner partners to implement conservation for neotropical migrants and other birds. Various plans typically have both public and private land objectives where voluntary incentives might keep working lands working for people while also protecting birds.

Box 9.1 David Pashley: PIF and NABCI

David Pashley was born in Victoria, British Columbia, migrated to California as a youth, received his B.S. from Humboldt State University, and his Ph.D. from Louisiana State University. Early in his career, David was with the Science and Stewardship program of The Nature Conservancy in Louisiana where he was instrumental in working with NFWF and the US Forest Service to help launch what became PIF. At the 1992 PIF Conference in Colorado, he and others presented a proposed PIF species prioritization scheme, which in time led to the foundation for landbird conservation planning. In 1995, David joined the newly formed American Bird Conservancy (ABC) and ABC agreed to support him as the first US PIF National Coordinator. He worked with IAFWA to secure funding to hire Regional Coordinators to implement the new plan. This was a critical step in bringing operational capacity to the partnership.

David also helped develop NABCI's Bird Conservation Regions (BCRs), in conjunction with Canada and Mexico. He remained as PIF National Coordinator through 1999. In 2000, he became the US National Coordinator for NABCI and was deeply committed to expanding the role of US JV in "all bird" conservation. He also helped stimulate the formation of several new JVs focused mainly on landbirds (Chapter 10). In their tribute to David after his passing, ABC remembered that David sometimes pushed faster and further toward all-bird planning than others were ready to travel, but in the end, major progress was achieved.

During his time at ABC, David coauthored several important documents including the first (2009) and subsequent State of the Birds reports. He was awarded a PIF Lifetime Achievement Award and remained active until forced to step down due to illness in 2018. ABC described him as "a relentless advocate for bird conservation throughout the Western Hemisphere" and "a master of inclusion with an incredible ability to forge partnerships". The David Pashley Award was created by the US NABCI Committee to honor one of its founding members and recognize excellence and significant contributions by US NABCI participants.

The North American Bird Conservation Initiative's Broad Umbrella

The Commission for Environmental Cooperation (CEC) was established by the United States, Canada, and Mexico under the 1994 North American Agreement on Environmental Cooperation that complemented the North American Free Trade Agreement. In 1996, the CEC Council passed a resolution establishing an advisory group to develop a comprehensive North American strategy for the conservation of birds (www.NABCI.net). The nine-member group sketched out the framework for what could be established. They had frequent meetings leading to products including the Important Bird Areas project, the delineation of BCRs, and a strategy for approval by the CEC Ministers.

Steve Wendt (CWS) was one of the first delegates from Canada. He recalled (pers. comm., 2025) the first international stakeholder meeting was in Puebla, Mexico in 1998. where a mapping team began developing a spatial framework for bird conservation in North America. A first product in 1998 was a list of 150 "Important Birds Areas" in North America, followed in 1999 by a strategy and action plan, which set forth goals, objectives, and actions for the conservation of all birds on the continent.

By March 1999, they further established NABCI BCRs – ecoregions with similar bird communities, habitats, and resource management issues to assist with planning the all-bird work. These BCRs were congruent, or nearly so, with several of the previously described NAWMP JV boundaries (https://nabci-us.org/resources/bird-conservation-regions-map/). Wendt, however, reckons that the second large NABCI meeting in 2001 in Queretaro, Mexico, was the real birth of the full Initiative, with formal committees established and national coordinators in place for the three countries. It had been a five-year process. Administrative arrangements followed within each country, including the appointment of national NABCI coordinators, Art Martell (CWS) for Canada, David Pashley (ABC) for the United States, and Humberto Berlanga (CONABIO) for Mexico.

Much as with NAWMP in 1986, in 2005, the three countries signed a Declaration of Intent for "The Conservation of North American Birds and their Habitat" and conservation planning began for many of the individual BCRs. One of the first undertakings was a comprehensive review of all of Canada's avian monitoring programs seeking to better apportion efforts for all-bird priorities (Avian Monitoring Review Steering Committee 2012).

With NABCI fledged as an umbrella framework, plans for individual bird groups proliferated. PIF, with a focus on landbirds, had launched before NABCI but fit easily under the NABCI banner while retaining its own identity with several regional working groups and staff support (https://partnersinflight.org/).

Importantly, with the coming of NABCI and articulation of the BCRs, the FWS declared in February 1999 that all the NAWMP JVs in the United States would henceforth be all-bird JVs, although their engagement in waterfowl compared with other bird science and conservation activities would vary greatly. With encouragement from NABCI Canada, the Canadian JV partnerships moved in that direction too but at their individual pace. There was considerable angst among NAWMP partners over this merger (Chapters 11 and 12). Mexican partners readily agreed with the combination. In the United States and Canada, leaders like Gary Myers, Duane Shroufe, William Molini, Steve Wendt, Art Martell, Gerry McKeating, and Rod Fowler were effective voices that saw long-term benefit in some form of union, but others saw mostly competition for resources.

Both Canada and the United States created national NABCI Councils to provide policy oversight and guidance to the partnerships within their countries. In Canada, the new NABCI Council became the umbrella body over all the individual bird plans including NAWMP and its associated JVs, PIF, and in time the Canadian shorebird and waterbird initiatives.

Environment Canada led the drafting of all-bird conservation strategies in each of Canada's BCRs by integrating new and existing strategies into an all-bird framework (https://www.birdscanada.org/bird-science/nabci-bird-conservation-regions). These all-bird conservation guides continue to inform conservation planning across Canada and help guide Canadian support for conservation work in other countries important to Canada's migrant birds.

State of the Birds Reports

Led by the US NABCI Committee, ABC, AFWA, Cornell Lab of Ornithology, Klamath Bird Observatory, National Audubon Society, The Nature Conservancy, USFWS, and USGS a first comprehensive report on the status of birds in the United States was released in 2009, followed by a parallel report for Canada in 2012. Relying on various data sources, they reviewed population trends since 1970 for groups of waterfowl, raptors, shorebirds, grassland birds, and associated habitats. Wetland birds were contrasted with grassland birds, arid-landbirds, forest birds (Eastern, Western, and Boreal), and arctic alpine. The report thus drew attention to those ecosystems and species most in need of help (https://archive.stateofthebirds.org/).

Subsequent reports in 2010 (with a focus on climate change), 2011 (with a focus on public lands in the United States), and 2013 (with a focus on private lands in the United States) focused on special topics. The 2014 report was again a comprehensive one. The State of Canada's Birds report was repeated in 2019. The latest reports in the series (see also Chapter 24) were published for Canada (Birds Canada 2024) and the United States (NABCI 2025) with slightly different groupings of birds. The Canadian report noted continued positive trends for waterfowl, waterbirds, and birds of prey over the past 54 years while grassland birds, aerial insectivores, and several shorebird species have declined markedly. The US report noted that several sea duck species were not on the same positive trend as many other waterfowl species and that grassland birds, arid-landbirds, and some shorebirds were doing poorly. Hawaiian forest birds were also of high concern.

Box 9.2 Humberto Berlanga

Humberto Berlanga has been engaged in virtually every bird conservation initiative in Mexico for more than three decades. He graduated from the National Autonomous University of Mexico where he remained to study the ecology of tropical bird communities. Later, he represented Mexico in international forums including CITES, RAMSAR, and the Convention on Biological Diversity and in 1997, became the copresident of the migratory bird board of the Trilateral Commission for Wildlife and Ecosystem Conservation of México, Canada, and the United States. Humberto served as Mexico's Cochair of the NAWMP Committee during the years leading to the 1998 Update. He engaged early in the PIF program focused on neotropical migrants and became the first Mexican National Coordinator of NABCI while based at CONABIO (the National Commission for the Knowledge and Use of Biodiversity). In 2009, he received the "Myers Award" for his contributions to the conservation of North American birds and in 2016 a leadership award in bird conservation from PIF. Humberto is currently promoting regional alliances for bird conservation including two cross-border JVs, the development of information tools like eBird Mexico, the Mexico Avian Knowledge Network, and the growth of bird monitoring networks across that country.

The Neotropical Migratory Bird Conservation Act

This US federal legislation was established in 2000 to help fund conservation of approximately 340 species of migratory birds that winter in Latin America and the Caribbean, and summer in North America. It created a competitive grants program administered by the US Secretary of the Interior, through the USFWS. Initially appropriated at $5 million per year, at least 75% of the funds were to be spent outside of the United States and grants initially required a nonfederal match of $3 to every $1 of federal money (2:1 in 2024), although in-kind contributions could be counted toward match. Habitat protection, education, research, monitoring, and capacity building were eligible expenses. In April 2024, the "Migratory Birds of the Americas Conservation Enhancements Act of 2023" reauthorized and slightly amended the Neotropical Migratory Bird Conservation Act (NMBCA).

Neotropical migrants face an increasingly complex range of threats, including human development pressures, invasive species, avian diseases, and a changing climate. As of 2024, grants under the Act have affected some 3 million acres across 40 US states and >30 Latin American and Caribbean countries.

Echoes of NAWCA are evident here with the matching funding approach. As with the wetlands program, progress in neotropical conservation has required building public/private partnerships and leveraging nongovernmental contributions. Since the NMBCA was enacted in 2000 until the spring of 2024, the United States had invested more than $30.4 million in 296 projects, with nonfederal partners contributing >$4 for every dollar invested by the federal government.

Western Hemisphere Shorebird Reserve Network

The Western Hemisphere Shorebird Reserve Network (WHSRN) was another contemporary initiative with NAWMP and PIF. Planning for WHSRN began in 1985, in response to worrying declines in the populations of several shorebird species (https://whsrn.org/about-whsrn/history/).

The notion of fostering "stepping stone" protected sites for migrating shorebirds seems to have derived from experience with aerial shorebird surveys of coastal South America undertaken by Guy Morrison and Ken Ross of the Canadian Wildlife Service, and international shorebird surveys coordinated by Brian Harrington of the Manomet Bird Observatory (later Manomet Center for Conservation Science). George Finney (CWS Regional Director), who was also one of the architects of NAWMP, proposed a voluntary network approach which was further developed by Morrison and other shorebird researchers including J.P. (Pete) Myers then at the Philadelphia Academy of Natural Sciences.

Myers and Pete McLain (New Jersey Department of Fish, Game, and Wildlife) presented the concept of a network of conserved sites for shorebirds to the IAFWA annual meeting in 1984. Thus, the concept of WHSRN was brought to the attention of state wildlife directors and NGOs such as Audubon, World Wildlife Fund, and others. McLain nominated Delaware Bay as the first WHSRN site. Famously important for Red Knots and Horseshoe Crabs, it was dedicated on May 21, 1986, followed by Canada's Bay of Fundy designation in 1987. By 2024, WHSRN was active in 20 countries where 123 sites totaling 39 million acres have been designated as places to provide migratory connectivity for shorebirds. Some 450 partners have been involved in achieving these successes. The degree of protection for those sites varies widely, depending on the laws and mechanisms available in each jurisdiction, but at minimum, these are publicly profiled as habitats of international importance.

Shorebird Conservation Plans

Building on the WHSRN network, moving away from PIF, but under the umbrella of NABCI, Canada and the United States completed national shorebird conservation plans (Donaldson et al. 2000; Brown et al. 2001). Noting widespread declines in shorebird populations (Morrison et al. 2001), these plans were directed at preventing further losses, sustaining the distribution, diversity, and abundance of all species, and restoring those of threatened or endangered status.

Canada has recorded 75 species of shorebirds, including 47 that breed or occur regularly there. The Canadian shorebird working group is positioned under NABCI Canada, with its own science support team to advise delivery agents on conservation work. Gary Donaldson served as the shorebird plan coordinator in Canada at the time of its writing. As of 2022, Cynthia Pekarik (CWS) serves in that role. Subsequent regional plans were developed in Canada including for the North in 2006 and the Prairies in 2017 (https://www.phjv.ca/wp-content/uploads/2017/10/Prairie-Canada-Shorebird-Conservation-Plan.pdf).

Assembled by a broad coalition of partners, the ambitious US shorebird plan offered three goals: (1) at a regional scale, to ensure that adequate quantity and quality of habitats are maintained to support the shorebirds that use it; (2) at a national scale, to stabilize populations of all shorebird species known or suspected to be in decline; and (3) at the hemispheric scale, to restore and maintain the populations of all shorebird species in the Western Hemisphere through cooperative international efforts. Three working groups lead the effort at the national level, overseen by a US Partnership Council. A research and monitoring group focuses on population estimation, identification of research needed to guide conservation, and funding. A habitat group works with regional habitat delivery partners like the JVs, and an education group reaches out to the public about shorebirds and their needs. Eleven regional teams work under the national group.

The FWS established new full-time positions for individuals to coordinate PIF, the shorebird initiative, and the waterbird initiative in 2000–2001. Terry Rich recalled this as another critical step in the evolution of bird conservation in the United States. Hemispheric objectives, of course, implied coordinated planning and actions among Canada, the United States, Mexico, the Caribbean, Central America, South America, and, for a few species, Pacific Oceana and East Asia.

North American Waterbird Conservation Plan

Close on the heels of the shorebird initiatives and engaging some of the same biologists, Canada and the United States prepared plans for the conservation of waterbirds not covered by NAWMP or the shorebird plans. Their ambitions spanned 29 nations of North America, Central America, the Caribbean, and the northern coast of South America. It included 23 families and 210 species of seabirds, freshwater wetland birds, colonial waterbirds, waders, rails, cranes, and more. Coleaders James Kushlan and Melanie Steinkamp, both with USGS, led a group of 23 international authors from 14 organizations assembling the first plan (Kushlan et al. 2002). The Plan's intention was by, "weaving together cultures, opinions, resources and science to achieve sustainable waterbird populations and appropriately manage waterbird habitat throughout the entirety of their ranges".

Water birds as categorized here are dependent on aquatic habitats ranging from swamps to streams, marshes, coastal zones, and pelagic waters which makes them vulnerable to many threats facing water globally, especially human impacts on their habitats. Moreover, about 80% of the species considered in the Plan are colonial nesters, which can increase the risk of catastrophic events affecting many individuals. Other species are secretive, dispersed, and especially difficult to count (e.g., rails) resulting in uncertainty about their status.

Priorities identified going forward were gaining better information on population status and trends, better understanding of factors affecting population change, and execution of site-specific conservation projects. Plans called for a Waterbird Conservation Council to help advance the work recommended in the Plan, but to use existing agencies, habitat JVs, and other partnerships to accomplish the work. Like with NAWMP, where adequate information existed, the planners sought to develop population goals, habitat goals, and education goals, all of which were developed in considerable detail.

A Canadian-specific plan, titled "Wings Over Water", was developed in parallel with the North American plan and finished a few months thereafter. Involving 13 families and 93 species of birds, Canada's plan, understandably, had a greater emphasis on seabirds. Canada has the longest coastline of any nation on earth with a total length of 243K km (151K miles) and thus tens of millions of breeding seabirds. Canadian planners, led by CWS scientists Bob Milko (National Wings over Water coordinator) and Loney Dickson (Chief of the Nongame Division, Prairie, and Northern Region), developed a list of top-priority research and conservation needs (Milko et al. 2003). About 30% of Canadian waterbirds appeared to be in decline with another 10% or so too poorly known to assess their status. Under the umbrella of NABCI, implementation of the waterbird work in Canada is overseen by a voluntary National Working Group of partners supported by a technical team. As in the United States, work is being undertaken by partner scientists, existing habitat JVs, or new project-specific coalitions but the waterbird initiative has eschewed development of new delivery structures.

A first regional spin-off plan appeared just a year later (Beyersbergen et al. 2004), a joint effort of the Prairie Pothole JV in the United States and the Prairie Habitat JV in Canada. Led by authors from CWS and FWS, the Northern Prairie and Parkland Waterbird Conservation Plan sought to identify key species and habitats in need of urgent attention. Loss of wetlands and associated grasslands appears to be a problem faced by most prairie waterbirds. Thirty-nine species breeding in the region plus migrating whooping cranes were included in the plan. Focus issues included better estimates of population trends, better understanding of factors presently affecting survival and reproduction, development of spatially explicit habitat suitability models, and connections of the region to waterbird habitats elsewhere. Development of wildlife-friendly agricultural programs is a priority in both countries as is integration of conservation planning for waterbirds with the waterfowl efforts already underway. The prairie plans include prioritization of landscapes and species based on population status and trends. Although BBS data are similar in both countries, wetland inventories and landcover data differ north and south of the 49th parallel which limits integration of national datasets.

Other Partnerships

Influenced by the NAWMP model, some nonmigratory conservation partnerships also developed as the migratory bird plans proliferated. These included an international North American Grouse Partnership in 1999 (https://www.grousepartners.org/) with a special focus on species dependent on prairie and sage ecosystems, and a (United States) National Fish Habitat Partnership, supported by the Association of Fish and Wildlife Agencies (https://www.fishwildlife.org/afwa-inspires/national-fish-habitat-partnership) and others. The fish plan was born in 2001 when an *ad hoc* group explored the notion of developing partnership efforts for fish on the scale done for waterfowl. AFWA supported staffing for the National Fish Habitat Partnership and serves on the Board. Since 2006, the Partnership has supported >850 projects benefiting fish habitat in all 50 states under 20 regional partnerships.

The NAWMP model has indeed been emulated time and time again and comprises one of the Plan's great legacies to conservation. Other initiatives have adapted the central idea of self-formed regional partnerships, connected at national or international levels, with nuances to better fit their target birds, supporters, and political realities. There is much to celebrate about the evolution of bird conservation over the past 40 years!

References

Askins, Robert, Russell S. Greenberg, and James F. Lynch. 1990. "Declines in birds of eastern deciduous forests." *Current Ornithology* 7: 1—57.

Avian Monitoring Review Steering Committee. 2012. *Environment Canada Avian Monitoring Review – Final Report*. Environment Canada, Ottawa ON, xii + 170 pages + 3 appendices.

Berlanga, Humberto, Judith A. Kennedy, Terrell D. Rich, Maria del Coro Arizmendi, Connor J. Beardmore, Peter J. Blancher, Gregory S. Butcher, Andrew R. Couturier, Ashley A. Dayer, Dean W. Demarest, Wendy E. Easton, Markus Gustafson, Eduardo Iñigo-Elias, Elizabeth A. Krebs, Arvind O. Panjabi, Vincente Rodriguez Contreras, Kenneth V. Rosenberg, Janet M. Ruth, Eduardo Santana Castellon, Rosa Maria Vidal, and Thomas C. Will. 2010. *Saving Our Shared Birds: Partners in Flight Tri-National Vision for Landbird conservation*. Cornell Lab of Ornithology. Ithaca, NY. 52pp.

Beyersbergen, Gerard W., Neal D. Niemuth, and Michael R. Norton. 2004. *Northern Prairie & Parkland Waterbird Conservation Plan. A Plan Associated with the Waterbird Conservation for the Americas Initiative*. Published by the Prairie Pothole Joint Venture, Denver, Colorado. 183pp.

Birds Canada and Environment and Climate Change Canada. 2024. *The State of Canada's Birds*. (https://naturecounts.ca/nc/socb-epoc/main.jsp).

Brown, Stephen, Catherine Hickey, Brian Harrington, and Robert Gill (Editors). 2001. *United States Shorebird Conservation Plan*. 2nd edition. Manomet Center for Conservation Science, Manomet, MA. 61pp. (http//shorebirdplan.org).

Buechner, Helmut K., and Jimmie H. Buechner (Editors). 1970. "The Avifauna of Northern Latin America: A Symposium Held at the Smithsonian Institution13–15 April 1966." *Smithsonian Contributions to Zoology* No. 26. Smithsonian Institution. Washington, DC.

Carter, Michael F., William C. Hunter, David N. Pashley, and Kenneth V. Rosenberg. 2000. "Setting conservation priorities for landbirds in the United States: The partners in flight approach." *The Auk* 117: 541–48.

Donaldson, Gary, Colleen Hyslop, Guy Morrison, Loney Dickson, and Ian Davidson (Editors). 2000. *Canadian Shorebird Conservation Plan.* Canadian Wildlife Service. Ottawa, ON. 27pp. (https://publications.gc.ca/collections/collection_2011/ec/CW69-15-5-2000-eng.pdf).

Finch, Deborah M. and Peter W. Stangel (Editors). 1993. "Status and Management of Neotropical Migratory Birds." *USDA Forest Service, General Technology Report RM-229.* Fort Collins, CO. 422pp.

Greenberg, Russell S. 1991. *Birds Over Troubled Forests.* Migratory Bird Center, Washington, DC.

Kushlan, James A., Melanie J. Steinkamp, Katherine C. Parsons, Jack Capp, Martin Acosta Cruz, Malcolm Coulter, Ian Davidson, Loney Dickson, Naomi Edelson, Richard Elliot, R. Michael Erwin, Scott Hatch, Stephen Kress, Robert Milko, Steve Miller, Kyra Mills, Richard Paul, Roberto Phillips, Jorge E. Saliva, Bill Sydeman, John Trapp, Jennifer Wheeler, and Kent Wohl. 2002. *Waterbird Conservation for the Americas: The North American Waterbird Conservation Plan. Version 1.* Waterbird Conservation for the Americas. Washington, DC. 78pp. (https://www.fws.gov/sites/default/files/documents/north-america-waterbird-conservation-plan.pdf).

Milko, Robert J., Loney Dickson, Robert Elliot, and Gary Donaldson. 2003. *Wings over Water: Canada's Waterbird Conservation Plan.* Environment Canada. Ottawa, ON. 32pp. (https://nabci.net/wp-content/uploads/WOW-2003.pdf).

Millsap, Brian A., Jeffrey A. Gore, Douglas E. Runde, and Susan I. Cerulean. 1990. "Setting priorities for the conservation of fish and wildlife species in Florida." *Wildlife Monographs* 111: 1–57.

Morrison. R. I. Guy, Yves Aubry, Rob W. Butler, Gerry W. Beyersbergen, Garry M. Donaldson, Cheri L. Gratto-Trevor, Peter W. Hicklin, Vicky H. Johnston, and R. Ken Ross. 2001. "Declines in North American shorebird populations." *Wader Study Group Bulletin* 94: 34–38.

North American Bird Conservation Initiative, U.S. Committee. 2009. *The State of the Birds, United States of America, 2009.* U.S. Department of Interior. Washington, DC. 36pp.

North American Bird Conservation Initiative Canada. 2012. *The State of Canada's Birds, 2012.* Environment Canada, Ottawa, Canada. 36pp.

North American Bird Conservation Initiative. 2025. *The State of the Birds, United States of America, 2025.* StateoftheBirds.org.

Panjabi, Arvind O., Wendy E. Easton, Peter J. Blancher, Allison E. Shaw, Brad Andres, Carol J. Beardmore, Alaine F. Camfield, Dean W. Demarest, Randy Dettmers, Marcel A. Gahbauer, Rebecca H. Keller, Kenneth V. Rosenberg, and Tom Will. 2021. *Avian Conservation Assessment Database Handbook, Version 2021.* Partners in Flight Technical Series No. 8.2. (https://pif.birdconservancy.org/acad.handbook.pdf).

Partners in Flight Canada. 1996. *A Framework for Landbird Conservation in Canada.* Environment Canada. Hull, Quebec, 23pp.

Ralph, C. John and Terrell D. Rich. (Editors). 2005. "Bird conservation implementation and integration in the Americas: Proceedings of the Third International Partners in Flight Conference, 2002 March 20–24. Asilomar, CA. Volumes 1 and 2." *USDA Forest Service General Technical Report PSW-GTR-191.* Albany, CA.

Rich, Terrell D., Carol J. Beardmore, Humberto Berlanga, Peter J. Blancher, Michael S. W. Bradstreet, Greg S. Butcher, Dean W. Demarest, Erica H. Dunn, W. Chuck Hunter, Eduardo E. Iñigo-Elias, Judith A. Kennedy, Arthur M. Martell, Arvind O. Panjabi, David N. Pashley, Kenneth V. Rosenberg, Christopher M. Rustay, J. Steven Wendt, and Tom C. Will. 2004. *Partners in Flight North American Landbird Conservation Plan.* Cornell Lab of Ornithology. Ithaca, NY.

Robbins, Chandler S., John R. Sauer, Russell S. Greenberg, and Sam Droege. 1989. "Population declines in North American birds that migrate to the neotropics." *Proceedings National Academy Science* 86: 7658–62.

Rosenberg, Kenneth V, Adriaan M Dokter, Peter J Blancher, Adam C Smith, Paul A Smith, Jessica C Stanton, Arvind Panjabi, Laura Helft, Michael Parr, and Peter P Marra. 2019. "Decline of the North American Avifauna." *Science* 366, no. 6461: 120–24.

Soga, Masashi and Kevin J. Gaston. 2018. "Shifting baseline syndrome: Causes, consequences, and implications." *Frontiers in Ecology and Environment* 16: 222–30.

Stangel, Peter W. and Amos S. Eno. 1992. "Conservation on a grand scale." *Transactions of the North American Wildlife and Natural Resources Conference* 57: 648–56.

Terborgh, John. 1989. *Where have all the birds gone? Essays on the biology and conservation of birds that migrate to the American Tropics.* Princeton University Press. Princeton, NJ. 207pp.

Vogt, William. 1970. The avifauna in a changing ecosystem, "The Avifauna of Northern Latin America: A symposium held at the smithsonian institution13–15 April 1966." Pages 8–16 in H. K. Buechner and J. H. Buechner (Editors). *Smithsonian Contributions to Zoology No. 26*. Smithsonian Institution. Washington, DC.

10 Habitat Joint Ventures for Other Birds

With NAWMP experience growing and the starburst of new avian conservation plans in hand, a next natural step was expanding the habitat Joint Ventures to include ecosystems important mainly for birds other than waterfowl. The federal governments with responsibility for migratory birds were assumed to be players. Some states with mature nongame programs quickly stepped up too, but the new partnerships needed to engage a different set of nongovernment organizations, corporate partners, and private donors to support these new enterprises.

By the early 2000s, all the NAWMP Joint Ventures had embraced to varying degrees an all-bird conservation mission. Early adopters, even before the 1999 FWS declaration, had begun to develop partnerships and initiatives for "more than ducks". In Canada, the EHJV was an early adopter as was the new Canadian Intermountain JV. In the United States, the Lower Mississippi and Prairie Pothole JVs began early to engage in multi-taxon planning. I recall Charles Baxter, LMVJV Coordinator, not only speaking passionately in a keynote address to the PIF conference in 2002 advocating this expanded mission for the NAWMP Joint Ventures, but also articulating the challenges of integrating planning and delivery for multiple bird guilds.

Movement toward all-bird integration proceeded haltingly among the JVs. The Playa Lakes JV provided an interesting example. The PLJV had existed through its first 12 years with no dedicated FWS funding, instead drawing from GCJV to fund a half-time Coordinator. The management board, not happy with the situation, began working directly with Congress to increase funding for all JVs. A few other JVs (IWJV, RWBJV) with similar funding challenges joined them. Repeated visits to Capitol Hill by Ross Melinchuk and Barth Crouch succeeded. For PLJV, this resulted in a meager but steady ramp-up of federal funding from 2001 ($220K) to 2006 ($660K). The PLJV Board also assessed itself $5K per seat for administrative costs. With these funds in hand, the Board began recrafting the JV as an all-bird JV and hired Mike Carter as its full-time Coordinator. Carter was the founder of Rocky Mountain Bird Observatory and was very active with Partners in Flight. He became the first Coordinator hired and specifically tasked with developing an all-bird JV. Carter, with Brian Sullivan (waterfowl) and Christopher Rustay (landbirds), developed databases and GIS tools to assist with habitat targeting.

Sonoran Joint Venture

The Sonoran Joint Venture (SJV) was a hybrid creation in two respects. It was a cross-border JV, the first between the United States and Mexico. It was also conceived from the beginning as a multi-bird Joint Venture with both waterfowl and other bird components.

In 1999, at the request of The Nature Conservancy of Arizona, the Arizona Game and Fish Department, and the FWS convened a meeting of interested agencies from the United States

DOI: 10.4324/9781003732310-10

and Mexico to discuss developing a partnership for bird conservation across the Sonoran Desert region (SJV strategic plan 2003. https://sonoranjv.org/downloads/SJV_StrategicPlan_v2023-05.pdf).

Several factors made this a timely development including the 1998 NAWMP Update, the advent of PIF and later NABCI, and Audubon was about to publish its influential atlas, *Areas de Importancia para la Conservación de las Aves en México*. The new SJV was established soon afterward, housed in the FWS Southwest Region in Tucson with Robert Mesta as the first coordinator. Robert, as the only SJV employee at the time, set out to build the partnership.

The SJV soon grew to include southern AZ and CA and in Mexico, Baja California, Baja California Sur, Sonora, Sinaloa, and Nayarit, as well as Western Chihuahua, Western Durango, and the Gulf of California and its islands. This biodiverse area included coastal waterfowl habitats and four binational watersheds. It also featured 41 Ramsar wetlands of international significance, 18 WHSRN sites, and 67 Audubon-designated Important Bird Areas, parts of 10 NABCI Bird Conservation Regions, and 25 indigenous nations (SJV 2023).

The 2023 Strategic Plan reflected significant growth in the scientific underpinnings of the JV, which strives to "step down" appropriate objectives from each of the international bird plans to SJV geography and develop conservation initiatives accordingly. Focal habitats are arid borderlands, grasslands, forests, islands, coastal wetlands, riparian systems, and fresh watersheds.

The SJV now boasts more than 80 partners, and an eclectic group of persons from universities, state, and federal agencies, First Nations, private industry, private landowners, nonprofit organizations, and other partners from both the United States and Mexico that collaborate to address habitat conservation needs. Its management board is especially diverse with more scientists, university members, and nongovernment organizations than most such boards. Its Science Working Group is equally diverse and uniquely open to anyone with a working interest in birds or habitat stewardship in the region.

Following the 2007 NAWMP Continental Assessment (Chapter 15), the SJV produced a waterfowl-specific supplemental plan that addressed many of the recommendations in that report (Beardmore 2007). Therein, the JV reinforced its focus on coasts of Baja, Sinaloa, Nayarit, and the Gulf of California for Pacific brant, northern pintail, lesser scaup, and mallard, and secondarily surf, black, and white-winged scoters, American wigeon, redhead, blue-winged teal, cinnamon teal, and canvasback.

For other birds, the JV maintains several databases and planning guides to support land managers in making bird-friendly decisions. For example, there is a unique "open pipe" initiative to discourage uncovered open pipes used in gates and fences, chimneys, and flues that pose fatal traps to cavity-prospecting birds. Most of the open pipes were PVC tubes used to mark mining claims on BLM lands. BLM had been seeking to get rid of these but there were thousands of them and many far from roads or trails.

Jennie Duberstein led the SJV science partnership initially and since 2016 has been the overall JV Coordinator. She also has a strong interest in youth education that has served the JV well.

Central Hardwoods Joint Venture

Formed in 2001, the Central Hardwoods Joint Venture (CHJV) partnership is based on common interests in BCR 24, a varied ecosystem covering central KY and TN, northern AL and AR, southwestern MO and IL, southern IN, and northeastern OK. Partners include state, federal, and not-for-profit wildlife agencies and organizations working toward the long-term viability of native bird populations across the dumbbell-shaped region.

The Joint Venture boasts a strong science base and growing experience in multi-objective habitat delivery. "Doing the right things in the right places" is their operating mantra. Working in this largely forested region, but with various land-type associations (https://www.chjv.org/geography/ecological-hierarchies/) that merit customized conservation approaches, the JV has developed sophisticated planning tools for an array of mostly landbird species. They note that, "Species in the greatest need of conservation attention typically have some combination of relatively small ranges, small population sizes, declining trends, and reliance on threatened or already-degraded habitats" and these are prioritized over species with stable or increasing populations.

The Joint Venture has relied on PIF, NABCI, and NAWMP to identify priority species in their jurisdiction. Then, with habitat suitability models for each species, and linked to their GIS database, they have targeted places and actions best suited to helping those populations. Jane Fitzgerald and her team used these GIS tools early and extensively. Conserving intact systems is a priority but so is working with the private forest industry to enhance habitat conditions. Their 2020–2025 strategic plan (https://www.chjv.org/wp-content/uploads/CHJV-Strategic-Plan-2021–2025-final.pdf) includes a commitment to improving their already considerable biological foundation and adapting their conservation strategies accordingly.

The JV also has emphasized "working grasslands", partnering with the National Grasslands Conservation Initiative (formerly the National Bobwhite Conservation Initiative) to encourage cattle producers to incorporate native grasses into grazing operations across the fescue belt of the east central US. Dissertation research by the current CHJV coordinator, Kyle Brazil has shown why that practice is economically viable (Brazil et al. 2020).

Although there is a relatively minor focus on waterfowl in the CHJV, they have stepped-down goals for waterfowl and as of 2024 achieved most of their waterfowl objectives. Jane Fitzgerald ably led this partnership from 2001 until retiring recently and deserves much credit for the impressive progress they have made.

Northern Great Plains Joint Venture

The Northern Great Plains of the United States includes some of the most intact sagebrush-steppe and grassland landscapes in North America. Over 70% of the region is in perennial cover and consequently it is a stronghold for grassland-nesting birds. The grasslands also sustain an important grazing industry, provide carbon storage, soil stabilization, and water retention in a system where productivity is strongly limited by availability of water.

JV Planning began for the area in 2002, with a focus on grassland birds and grazing. Andy Schollett of the FWS led that early work. The JV was formally constituted in 2006 with Ken Sambor from ND as the Coordinator followed by Dan Casey in 2015. The JV includes southeastern MT, southwestern ND, Western SD, and northeastern WY. Grassland health, water management, and grazing management emerged as primary foci. The management board presently includes federal (four), state (four), NGO (nine), landowner, corporate, and First Nations' members. The partners are dedicated to supporting resilient grasslands to ensure that grassland birds, local communities, and ecosystem services are sustained over time. To quote from their strategic plan, "Resilient Lands and Communities – Our conservation projects help recharge aquifers, improve water quality, stabilize soils, sustain bird and pollinator populations, and improve sustainable forage production for ranching". "Resilient lands support resilient rural and indigenous communities".

Their 2022 Strategic Plan (Wrightman et al. 2022) deals at length with serving and expanding their partnership. Their profile is highly collaborative, supporting conservation practices that work for birds and the people who depend on the Northern Great Plains for their livelihoods. Elements include developing decision support tools and facilitating information transfer through workshops on conservation design like their popular wet meadow program. Other efforts include sharing new science products with partners, monitoring project effectiveness, improving understanding of economic tools that can incentivize conservation, and incorporating different cultural perspectives. Related to cultural traditions, bison restorations have been used in some areas to restore historical dynamics in the grasslands.

The JV continues to seek new people and organizations as collaborators. They are also organizing subregional activities such as listening and learning from local tribal members, agricultural producers, and energy partners. The JV is also seeking ways to leverage carbon markets to better serve habitat conservation and offers advice for ranchers interested in succession planning for conservation values. It's a diverse set of initiatives.

The JV has been actively engaged with larger scale planning for grassland conservation including the Central Grasslands Roadmap, a collaborative guide to increase conservation of North America's Central Grasslands which span some 500 million acres across Indigenous Lands, Canada, the United States, and Mexico. The complementary JV8 Initiative is another collaborative planning process involving eight migratory bird joint ventures unified for conservation of the Central Grasslands. Finally, the NRCS Working Lands for Wildlife program also seeks to conserve resilient rangelands for people and wildlife and has been another partner contributing to the JV. Catherine Wightman is the present JV Coordinator.

East Gulf Coastal Plain Joint Venture

This JV encompasses diverse habitats from the northeastern coast of the Gulf of Mexico and its associated coastal plain reaching north to Western Kentucky, the parts of BCR 27 not already served by the Atlantic Coast JV. Like all the JVs, this is a self-directed partnership comprising 14 state, federal, private, and academic organizations coordinating bird conservation efforts in the coastal plain of six states – Eastern LA, MS, Western TN and KY, and central AL and FL. Assembled in 2006, with coordination by Allison Vogt, they published a first detailed implementation plan in 2008 (EGCPJV 2008).

The JV footprint is about 63 million acres of longleaf pine forests, eastern interior grasslands, freshwater mostly forested wetlands, as well as coastal rivers, bays, and shorelines. For waterfowl, some of the greatest interest is in breeding habitat for Florida mottled ducks and wood ducks, along with migration and wintering habitat for other Mississippi and Atlantic Flyway species.

The JV's first implementation plan reflected the partners' objective of ensuring the sustainability of bird populations in their region and the ecological communities of which they are a part. Embracing the philosophy of strategic habitat conservation and a partnership approach, much of the first plan was focused on organizing the JV and launching more detailed planning work. Five major habitat types were identified for implementation. Open pine habitats were chosen as a first major focus. Avian species priorities were identified by cross-referencing a combination of state Wildlife Action Plans and national bird conservation plans.

Their most recent update is a detailed landbird plan (Greene et al. 2021). It includes a list of high-priority landbird species, numerical population objectives for those species, and habitat objectives to meet those population objectives. The plan includes ten-year and long-term habitat objectives by state and major habitat types. High-priority species (29) were chosen based on inclusion in 16 conservation plans, lists, and datasets including those of the PIF Landbird Conservation Plan, State Wildlife Action Plans, FWS's Birds of Conservation Concern, and plans of adjacent Joint Ventures. Rare species were identified but common species in decline were also included.

For coastal systems important to waterfowl, shorebirds, and other waterbirds, actions include restorations of areas lost to development and plans for adaptation to sea level rise in wetlands, aquatic, estuarine, and shoreline habitats. Catherine Rideout presently serves as the JV Coordinator with Rob Holbrook as Assistant Coordinator.

Rio Grande Joint Venture

This second cross-border JV was established in 2006. Focused mainly on the expansive watershed of the Rio Grande, it covers a segment of Gulf Coast habitats for wintering waterfowl, shorebirds, and waterbirds but mainly features habitats for interior landbirds. The JV spans most of southern Texas and in Mexico the Chihuahuan Desert, Tamaulipan Brushlands, and Gulf Coast Prairie.

Currently, the JV is operating under a 2023–2027 Strategic Plan with priorities designated for BCRs 34 and 35 including interior grasslands, riparian corridors, freshwater wetlands, seagrass meadows, barrier islands, and coastal marshes. The partnership's ambitious vision is for the Rio Grande Joint Venture (RGJV) landscapes to support thriving and diverse communities of birds, other wildlife, and people far into the future. Of most immediate concern are grassland birds. Over 90% of migratory grassland bird species breeding in Western North America winter in the grasslands of the Chihuahuan Desert. Many of those populations are in trouble, having dropped by 1/3 since 1970. Specific initiatives are underway focused on reddish egrets, Sprague's Pipit, chestnut-collared and McCown's Longspurs, and Baird's Sparrow.

The JV is led by an International Management Board of about 18 people including Mexican and US members, nongovernment organizations, businesses, and private individuals. The American Bird Conservancy has supported people to coordinate and execute much of the JV staff work, led by coordinator Karen Chapman. The JV's current strategic plan is organized around four detailed themes: conservation planning; conservation delivery; monitoring and research based on the US Joint Ventures' strategic habitat conservation model (https://mbjv.org/science/); and growing and strengthening the RGJV partnership.

Oaks and Prairies Joint Venture

The Oaks and Prairies (OPJV) partnership is heavily focused on grasslands and grassland birds. It includes two BCRs, the Oaks and Prairies grasslands and the Edwards Plateau. Comprising only parts of two states, OK and central TX, and with only two federal agencies (FWS, USDA), the management board is populated with many NGOs, research entities, and landowners. The partnership came together in 2009, in response to the 50-year decline in grassland birds. One species of special concern is the Northern Bobwhite. TX breeding Bobwhite have dropped by some 90% over that time frame as has the annual harvest of OK quail (https://www.opjv.org/) due at least partially to the loss of suitable habitat.

The JV's general approach is to link management actions with national population goals for breeding and wintering grassland birds. The mission of the JV is to plan for and facilitate bird habitat conservation, research, and outreach to ensure sustainable populations of these focal species and others. With >95% of grasslands in the region in private hands, and some 53,000 farms in the JV's focus counties, the partnership necessarily has a major focus on working lands.

One federal program has been particularly important to their work. The Grassland Reserve Incentive Program (GRIP) provides direct payments to qualifying landowners as an incentive for conducting certain grassland bird habitat improvement practices on their property. The most common practices to maintain grassland health are prescribed grazing, brush management, and prescribed burning. Since 2013, more than 120K acres of grassland habitat has been restored or enhanced through GRIP (https://www.opjv.org/). USDA's Natural Resource Conservation Service has provided the lead funding for this with additional support from the states and several NGOs

A grassland bird conservation business plan for 2015–2025 currently guides the work of the partnership, informed by ongoing evaluations and delivery experience. It is available to review at (https://www.opjv.org/about_us). It is built upon spatially explicit biological objectives and relies largely on GRIP delivery, communications, and conservation efforts by supportive local landowners. Bird monitoring and evaluations of program effectiveness provide feedback for program improvements in the future. Led through theearly years by Jim Giocomo of the American Bird Conservancy, the current Coordinator is Robert Perez, also supported by ABC.

Appalachian Mountains Joint Venture

The Appalachian Mountain Joint Venture (AMJV) was formed in 2008. Preliminary planning for the partnership began in 2003 with support and advice from the neighboring Atlantic Coast JV and FWS Region Four but it was a complex partnership that took time to develop. ABC boosted the partners' efforts with financial support beginning in 2007.

In contrast to the two-state OPJV, the Appalachian Mountain JV stretches from the southwestern Appalachians in Alabama to the northeastern highlands in southern New York and includes portions of TN, KY, OH, AL, GA, NC, VA, MD, PA, NJ, NY, and all of WV…overall some 103 million acres in extent. The AMJV is a massive partnership of over 55 state and federal agencies, private conservation organizations, and universities united in the broad mission of conserving habitat for the benefit of birds, wildlife, and people in the core of the Appalachian Region (https://amjv.org/).

The JV contains some of the largest expanses of forest remaining in the Eastern United States. Their objectives are derived from NABCI BCR 28 including objectives from NAWMP, PIF, and all the other international bird plans, plus regional plans for woodcock, ruffed grouse, Bobwhite, and the Wildlife Action Plans of all the states engaged in the JV. The AMJV's first implementation plan was completed in 2008 (Appalachian Mountains Joint Venture Management Board 2008) with the stated intention of updating these at regular intervals. By 2025, however, they do not appear to have developed any waterfowl-specific objectives.

About 230 species of birds occur regularly in the region and 70 of those, mostly landbirds, were chosen as priorities and the needs for these are well illustrated in the 2008 plan. The JV also chose six focal landscapes for which they are developing specific initiatives: (1) Allegheny Highlands (PA/NY); (2) Greenbrier (WV); (3) Virginia Highlands; (4) Southeastern Ohio; (5) Cumberlands (TN/KY); and (6) the Southern Appalachian High Country (TN/NC/VA). Partners work with public and private forests, mainly striving to maintain early-, mid-, and late-successional stages of forests across a range of altitudes weighted by where the species of greatest conservation concerns are found.

The JV is committed conceptually to the adaptive process of strategic habitat conservation (Chapter 13), but that work remains in early stages. A management board of 18 partners at the outset included four federal agencies, ten states, and four NGOs. Additional NGO and university affiliations have been added since then to support the work of the JV. The staff are led by

Todd Fearer an employee of ABC, who began work for the JV as science coordinator in 2010 before taking on the overall Coordinator role in 2012. Brian Smith guided the JV in the first years of its existence.

California Central Coast Joint Venture

The California Central Coast Joint Venture (C3JV) was just recently formalized. Based on a broad and ambitious implementation plan first drawn up in 2022 (https://www.c3jv.org/_files/ugd/f8877e_42be7509c0ed462e8b7910a4fce75f84.pdf). The new JV intends to operate under that plan's guidance from 2025 to 2035.

Covering the mid-coastal region of California from west of the Central Valley to the Pacific it includes eight counties. Boundaries are Santa Cruz on the north southward to the Santa Barbara Channel and encompass parts of BCRs 32 and 17.

Federal, State, tribal (six territories), NGO, academic and industry partners are engaged. The mission of the new C3JV is to work through inclusive partnerships to steward healthy and resilient habitats for birds, other wildlife and people. Their initial plans include identification of high-priority species and explicit conservation targets for six ecosystems including: (1) riparian and freshwater wetlands, (2) scrub and chaparral systems, (3) oaks and prairies, (4) conifer and mixed hardwood forests, (5) coastal and marine systems, and (6) urban and intensive agricultural zones. Waterfowl will be a minor focus in the C3JVs broad conservation program that also plans to include major social and education components. It will be interesting to follow their progress and note what lessons might obtain for other JVs.

The Puzzle of Reporting Structures

The first habitat Joint Ventures assembled under NAWMP focused almost exclusively on waterfowl. Although the coordinators reported to various federal supervisors in Canada or the United States, the JVs had a natural reporting relationship to the international Plan Committee (PC). Each JV developed a prospectus for its operations and some form of agreement or charter, signed by the partners, that described its mode of operation. These documents were subsequently reviewed and endorsed by the PC. This endorsement grew in importance when weighting criteria for NAWCA grants were developed; Joint Ventures endorsed by the PC gained a scoring advantage that persists today.

Because the JV partnerships were voluntary, and members still followed their individual federal, state, or NGO governance procedures, the relationship to the PC was always a little vague. After 1999, when all the habitat JVs became all-bird initiatives to some degree, some JVs began to look for additional technical guidance from PIF, NABCI, and other bird plans. For some newer JVs, the guidance from, for example, NABCI had greater impact on their business than NAWMP although the JVs lacked formal reporting relationships with the other bird plans.

Attempting to clarify this, Steve Williams (FWS) issued a Director's Order (#146) on Joint Venture implementation in September 2002. It specified that to be recognized as a US migratory bird Joint Venture, partnerships must accept responsibility for delivery of national or international bird conservation plans, and they should do so for all migratory bird habitat priorities in their geographic area. Partnerships should be directed by a diverse management board and guided by an implementation plan and an evaluation strategy. In turn, the Service would strive to provide a coordinator for each recognized US JV and help establish base capacity for biological planning, conservation, implementation, and evaluation. Supporting JVs that addressed the full spectrum of bird conservation needs would be prioritized, but assisting new Joint Ventures with planning and organization would be eligible too if funds were available.

Fast forward 20+ years, the JVs still take broad guidance for waterfowl conservation from the PC and its technical support bodies. This occurs mainly through regular updates of the Plan and by periodic meetings with each JV when the partnerships report their progress and current plans. Those reviews were one outcome of the 2007 Continental Progress Assessment (Chapter 15).

As Joint Ventures broadened to include all birds, the role of the PC became more that of reviewing and endorsing the waterfowl component of JV plans and not the entirety of their activities. Today, JVs that are endorsed by NAWMP report to PC regularly. It is not a legal requirement, but it is understood this is what JVs do. JVs focused mainly on other birds but are NAWMP endorsed (Sonoran, Northern Great Plains, and Central Hardwoods) only report once every six years, while others report about every three years. Four JVs (Appalachian Mountains, East Gulf Coast, Oaks and Prairies, and Rio Grande) that are not endorsed don't report to the PC at all. It remains to be seen whether the new California Central Coast JV will do so.

In communicating with the PC, the JVs report primarily, but not exclusively, on their waterfowl work. PC staff provide guidance on reporting agendas. In my experience as both a PC member and JV member, these conversations are usually helpful to both parties. Somewhat surprisingly, no JVs formally report to other bird plans. This may be due to the link NAWMP has to federal JV funding. But perhaps as important, NAWMP remains the only international bird accord signed at a Ministerial/Secretarial level by the three federal governments with a commitment that has been renewed with each Plan Update.

References

Appalachian Mountains Joint Venture Management Board. 2008. *Implementation Plan for the Appalachian Mountains Joint Venture: A Foundation for All-bird Conservation in the Region.* Brian W. Smith (Editor). Appalachian Mountains Joint Venture. Frankfort, KY. 92pp.

Beardmore, Carol J. 2007. *The Sonoran Joint Venture: Bird Conservation Plan –Waterfowl Management Supplement. Version 1.0.* Sonoran Joint Venture. Tucson, AZ.

Brazil, Kyle. A., Patrick D. Keyser, Gary E. Bates, Arnold M. Saxton, and Elizabeth D. Holcomb. 2020. "Continuous grazing of mixed native warm-season grass in the fescue belt". *Agronomy Journal* 112: 5067–80.

East Gulf Coastal Plain Joint Venture. 2008. *Implementation Plan, Version 1.* East Gulf Coastal Plain Joint Venture, Auburn, AL. 106pp.

Greene, Rachel, Kristine Evans, Andrew Cox, Jeffrey Gleason, David Jones-Farrand, and Mark Woodrey. 2021. *Landbird Conservation Plan Summary, Version 1.0.* East Gulf Coastal Plain Joint Venture, Atlanta, GA.

Ralph, C. John and Terrell D. Rich. (Editors). 2005. "Bird conservation implementation and integration in the Americas: Proceedings of the Third International Partners in Flight Conference, 2002 March 20-24. Asilomar, CA. Volumes 1 and 2." in *USDA Forest Service General Technical Report PSW-GTR-191.* Albany, CA. 1296 pp.

Wightman, Catherine, Steve Riley, and Lucinda Morris (Editors). 2022. *Northern Great Plains Joint Venture Strategic Plan.* Northern Great Plains Joint Venture Management Board, Clancy, MT. (https://ngpjv.org/files/2022_Strategic_Plan.pdf).

11 Expanding the Vision – The 1994 Update

Getting Real about the Job Ahead

A key provision that has enabled NAWMP to stay relevant for 40 years has been the commitment made in 1986 to periodic updates of the Plan. Charles Baxter, reflecting in 2007, called this provision, "one of the geniuses of the Plan". The updates have varied in focus, in depth and breadth of engagement by the Plan community, but they have all altered the course of the enterprise to some degree. Prescribed to occur at five-year intervals, they haven't always hit that calendar mark but have always delivered value. The first was completed in 1994. It was delayed, understandably, by the focus of partners on start-up challenges; especially organizing the first joint ventures, developing first-step funding and soon afterward the North American Wetlands Conservation Act (NAWCA). There was also setting up the national Plan offices in Canada and the United States to accomplish, establishing the international Plan Committee and its roles, and launching the Continental Evaluation Team (Chapter 13). All these things demanded attention of Plan leaders.

When the Plan Committee solicited thoughts on the issues and consultative process that should be followed in this first update, Canadian partners quickly formed a task force comprised of Sandy Macaulay (PHJV), Keith McAloney (EHJV), Rick McKelvey (PCJV), Jim Patterson (DUC), Ken Cox and Clayton Rubec (NAWCC Secretariat), and Greg Thompson and Steve Wendt (CWS). Their report (NAWCC Secretariat 1992) recommended that while NAWMP should remain a waterfowl plan, it should also encourage achievement of other benefits to biodiversity as Plan partners had already begun to do. They argued that an update was needed mainly because each JV had developed implementation plans that materially increased cumulative habitat objectives. A focus on the most important areas for waterfowl remained important. They also opined that while a 15-year planning horizon made sense, Plan implementation should be recognized as ongoing and not fixed on a specific sunset year. They found no need to revisit population goals but urged developing knowledge to allow objectives to be set for more species. Finally, they urged more thought about the criteria and pace of establishing additional JVs or expanding JV boundaries.

The Canadian team also acknowledged an ongoing debate about value of leases vs. purchases. Leasing habitat was popular with some Canadian partners because it was much easier to negotiate with landowners; however, US funders preferred longer-term security. The crux was whether these leases could lead to a longer-term relationship with landowners. Are leases a gateway to more than a lease? With only a few years of experience, the answer was unclear.

A parallel US task group was also formed, and the two groups reconciled an "Issues Paper" in December 1992 led by Bob Streeter (FWS) and Greg Thompson (CWS), thus providing guidance to the update writing team. The binational group agreed that the 1986 population objectives

DOI: 10.4324/9781003732310-11

should be reaffirmed. Items for debate included: Should JVs beyond the present 12 be approved? What policy should exist about expanding boundaries of the existing joint ventures? All JVs have expanded their quantitative habitat objectives – should these be accepted or challenged somehow? How might partners account for less than permanent protection when reporting accomplishments, and under what circumstances should the Plan embrace short-term habitat protection? The 1986 Plan envisioned 66–75% of dollars being spent in breeding areas. Over the first several years, it had been about 32%. What new funding strategies might help address this difference? Should there be more joint ventures created for species of concern (e.g., northern pintail, canvasback, and sea ducks). How might the Plan be restructured to include Mexico? And how might the Plan better respond to opportunities to influence agricultural policy?

As broader public consultations began, national offices released a review of accomplishments during the Plan's first five years (USFWS and CWS 1993). Highlights included $500 million invested, and more than 2 million acres of wetlands and associated habitats protected, restored, or enhanced by the JVs. The 1986 objectives had been refined by the JVs, and acreages and projected costs had grown. The Plan's initial cost estimate of $1.5 billion had grown to about $6 billion! To date, most of the funding had been devoted to migratory and wintering habitats in the United States and less to breeding habitats in either country. The Plan had attracted broader partnerships and greater interest than foreseen, offering both more opportunities and challenges.

After further drafts, debates, and reviews, two main things changed because of this initial update. First, based on planning by the joint ventures, the estimated scale of work needed to attain Plan objectives grew dramatically and, second, Canada, the United States, and Mexico negotiated a way for Mexico to formally join the NAWMP partnership.

The 1994 update was subtitled, "expanding the commitment". It might as well have been, "getting real about the job ahead". Table 5 of the Update (Environment Canada et al. 1994) listed acreage objectives for protection, restoration, and enhancement of habitats for the original joint ventures and five more that appeared in the years after 1986. In fairness, the 1986 authors acknowledged that their preliminary estimates of habitat needs (1986, page 13) were a shot in the dark, informed by professional opinion, but not the kind of detailed planning and decision-support models that the joint ventures were expected to assemble. A growth in objectives was expected. Still, the differences between the two plans were dramatic: for the US Prairie Pothole Region, 6.1 million acres vs. 1.1 million; for Prairie Canada, 7.2 million vs. 3.6 million; for California's Central Valley, 935K vs. 80K, for the Gulf Coast and Lower Mississippi combined, 4.2 million vs. 0.7 million, and for Eastern Canada 542K acres vs. 70K. Big numbers, but with the enthusiasm of joint venture partners and the advent of new funding sources this did not seem impossible. My own reflection at the time, which I believe was widely held, was something like, "OK, there's a massive job ahead. Let's get to it!"

Mexico Formally Joins the Partnership

Although Mexico was included in discussions about the North American Plan in the early 1980s, the federal government there was not ready to commit, due in part it seems to concern about allowing foreign involvement with internal Mexican affairs. Through discussions in 1987–1988, they did, however, agree to form with Canada and the United States a Tripartite Committee for the Conservation of Migratory Birds and their Habitat. That endured through a transition to the Trilateral Committee for Wildlife and Ecosystem Conservation and Management in 1992. These committees began to address Mexico's priorities and eventually helped pave the way for Mexico to sign onto NAWMP (Steve Wendt, pers. comm. March 2025). US interests at the onset were more focused on wintering waterfowl in Mexico than the broader suite of species and ecological matters of greater interest in Mexico.

One unique thing about the Mexican federal system is that at six-year intervals, whenever a new party assumes leadership of the government, the wildlife agencies like SEMARNAT (Secretariat of Environment and Natural Resources) experience nearly a complete turnover of staff. Though some professionals might be hired back by the new government, there can be a long pause in doing business, so continuity in planning can be elusive.

Interest in the funding opportunities attached to NAWCA seems to have been one key to Mexico's eventual engagement (Bob Streeter, pers. comm., July 2024). As noted previously, in 1990, Streeter had been assigned to coordinate assembly and function of the new NAWCA Council. Because NAWCA legislation mandated that a portion of annual funding be directed to Canada and Mexico, Bob contacted various federal officials in Mexico to alert them to the new funding opportunity for wetland projects. Because of ongoing shifts of the Mexican federal agencies responsible for different aspects of wildlife management (pollution control, land ownership, species status, habitat protection, etc.), a pathway for project development could not be determined before the first round of NAWCA proposals in 1990. Streeter subsequently contacted several US nongovernment organizations with counterparts in Mexico to encourage them to develop wetland conservation projects for the Council to consider. Conservation International, Ducks Unlimited, and The Nature Conservancy responded and coordinated submissions in time for Council consideration in that first window. Officials in the Secretaria de Desarrollo Urbana y Ecology (SEDUE) did require that their organization approve any Mexican projects that were to be submitted to the NAWCA Council but allowed these to proceed.

After the first Mexican projects were funded, interest picked up in developing additional submissions. With the approval of SEDUE staff, Streeter's office supported workshops in three regions of Mexico, where US and Mexican NGO's brought in dozens of Mexican state people, universities, and other regional and local NGO's to inform them of the opportunity for funding projects and the process for developing and submitting proposals to the Council. More Mexican project proposals followed but the breakthrough occurred in 1992.

Streeter recalled that in about 1991 a new federal wildlife director, Dr. Graciela De la Garza, was appointed in Mexico. She was subsequently invited to a meeting of the US NAWCA Council in the Ace Basin, SC in 1992. For both NAWCA business and ahead of the first Plan update on which work was just beginning, there was a desire to conclude a formal agreement with Mexico. After the meeting, Streeter gave De la Garza a ride to the Savannah airport where they had some time to talk while waiting for their flights. They discussed Mexico and NAWMP, and on a scrap of paper began outlining the conditions under which De la Garza would be willing to promote formal Mexican membership. Time spent with the Wetlands Council, seeing how it was operating, visiting some of the projects on the ground, the collegiality of partnerships had impressed her, so she was inclined to see Mexico participate more fully under certain conditions. Bob recalls (pers. comm. August 2024) that two of those conditions were that for Mexico, the focus had to be broader than waterfowl, rather on conserving biodiversity associated with wetlands, and that local people living in or near these wetland systems would have to be involved in any matters that affected their livelihoods. Lastly, they wanted assurance that the United States and Canada through the Plan would not dictate the management of hunting in Mexico. Those proved to be conditions that Canada and the United States were prepared to work within.

De la Garza subsequently assigned people to work toward bringing Mexico into the Plan. The FWS provided resources for them to travel north or to send US staff south. Bob remembers that "those things we agreed to on the napkin were incorporated into the update" and that allowed her to go up the chain to obtain the Secretary's commitment to sign it. A draft was developed through a few iterations, approved and readied for signatures by the Minister of Environment Canada, the Secretary of the Interior United States, and the Secretario de Desarrollo Social Mexico and Mexico was in. The NAWMP Committee was subsequently expanded to include Mexican members as well.

In the early 1990s, Mexican federal and state governments invested few resources in wildlife conservation and expressed more interest in wetlands for a variety of values other than waterfowl. The NGO and academic communities in Mexico, however, had already stepped forward to support initial waterfowl conservation planning and project delivery, as well as offering valuable technical connections with various international NAWMP committees and individuals. De la Garza had the trust and support of the Mexican NGO community which helped facilitate this early partnership (Eduardo Carrera, pers.com. September 2024).

Evolution and Status of NAWMP in Mexico

Today, Mexican partners are fully engaged in two substantive cross-border habitat joint ventures, the Sonoran JV and the Rio Grande JV, which have brought new resources and staff expertise to conservation of Mexican habitats. Mexico has embraced NAWMP as one of its several priorities for bird conservation designed by regional partnerships and delivered in a cultural context that fits their situation. Usually, this has meant focusing on wetland ecosystems broadly rather than waterfowl uniquely, and by engaging local communities in project development, often with a sustainable use element. Conservation education is often an integral part of their projects as well. With its unique political structure, history, diverse ecosystems, economic context, and different information base, it is unsurprising that this has taken a different form than the JVs in Canada and the United States.

Wetland and Waterfowl Resources in Mexico

During discussions about Mexico joining the NAWMP partnership, Environment Canada assembled a helpful report (Canadian Wildlife Service 1992) on wintering waterfowl populations in the country based on survey data when consistent surveys were flown in 1955–1991. These were broken down by species and regions and the report offered an assessment of the main habitat stressors and highest priority wetlands ($n = 30$) for conservation. It painted a compelling picture of the importance of Mexican habitats for multiple species. The western and eastern coastal systems, and the central grasslands, are all vital areas for waterfowl in Mexico. New triennial aerial surveys now help track the abundance and distribution of ducks and geese in the country (Otto and Frechtel 2001; Mcllink et al. 2009).

The Pacific Northwest region, especially coastal wetlands from the Colorado Delta south along the Baja California coast and southeast along the coasts of Sinaloa and Nayarit support some 60–80% of Pacific Black Brant where they winter on extensive eelgrass beds. The coast of Sinaloa and Nayarit including areas of rice culture near the coasts support large numbers of wintering northern pintails, northern shovelers, green-winged teal, and American wigeon. Main threats to habitats in the region include tourism development, shrimp farming in the shallow lagoons, and water quality impacts of agriculture and industrial/urban development. Eastward, the interior grasslands support upward of 20% of mid-continent white-fronted geese and numerous ducks dependent on annually variable water abundance. The Laguna Madre on the Gulf Coast supports redheads and other ducks, plus many waders and shorebirds. The mangrove coast of the Yucatan Peninsula provides important stopover and wintering habitat for blue-winged teal and many other species including American flamingos. Based on the federal mid-winter surveys and the extensive wetland inventories accomplished by Ducks Unlimited de Mexico, A.C. (DUMAC), it has been estimated that in total, Mexican wetlands winter about 20% of North America's waterfowl with species combined.

DUMAC and its partners, with NAWCA support, identified 31 major wetland systems of highest priority in the country (https://dumac.org/humedales-prioritarios-para-aves-acuaticas/). In sum, these systems support about 80% of the wintering waterfowl in Mexico. Six of these wetland complexes are near the Gulf of Mexico, 15 in the North Pacific region, and ten in the Northern and Central Altiplano regions. Beginning with 1994 and subsequent NAWMP updates, these areas have been used to designate vital waterfowl habitat regions in the country. Conservation efforts in these places have included wetland mapping and protection, freshwater impoundments, mangrove restorations, restoration and management of coastal wetlands, and technical training.

A very successful capacity-building program focused on early-career Latin American and Caribbean professionals, DUMAC's RESERVA, with support from the USFWS, offers a two-month intensive course leading to a Diploma in Natural Resource Management and Conservation. Begun in 1989, by June 2024 it completed its 55th class and had helped train 692 young conservation biologists in the region – truly an outstanding initiative.

Box 11.1 Reflections on NAWMP in Mexico – 2025

By Eduardo Carrera Gonzales, CEO, Ducks Unlimited de Mexico

The history of wetland conservation in Mexico has always been linked to the environmental policy of federal programs, which is the responsibility of the Secretariat in charge, currently the SEMARNAT. This agency directs the environmental policy for bodies of water through the National Water Commission (CONAGUA), with support from the National Forestry Commission (CONAFOR), the General Directorate of Wildlife (DGVS), and the National Commission of Protected Natural Areas (CONANP).

Within this national policy, there are legal instruments framed by the General Law of Ecological Balance and Environmental Protection (LGEEPA) and the National Water Law (LNA), which support the conservation of natural resources, including wetlands, but framed in the broad water agenda. While the government structure is there, unfortunately the budgets in these agencies are limited and they have not had the resources required to efficiently address the conservation problems faced in these ecosystems and thus guarantee quality habitat for the species that depend on them.

With Mexico's inclusion in the North American Waterfowl Management Plan in 1994, a new era of work on the conservation of wetlands and the management of waterfowl began. This initiative, complemented by the NAWCA, established a vision of conservation integrated across the three countries. This continental initiative, accompanied by availability of funds, has been used by many organizations in Mexico, supporting the fulfillment of the needs identified in each Plan update.

Through NAWMP and NAWCA, wetland conservation in Mexico has benefited the habitats of migratory and resident waterfowl and the local communities that live around them. In this period (1991–2024), $81,691,297 have been received for 373 projects in the country. These resources have generated significant economic impact when added to the complementary funding from private Mexican institutions and government funds in support of these conservation projects, for a total of $176,422,354 dollars in a ratio of about two to one.

This funding opportunity encouraged the development of national conservation plans in Mexico, such as the Strategy for the Management, Conservation, and Sustainable Use of Waterfowl. That document has served as a public policy instrument to direct conservation efforts and the application of funds. This has supported the development of conservation initiatives, which try to address the threats faced by wetlands and the wildlife species that depend on them, but considering in each of these, the social component that in Mexico cannot be separated from these conservation processes.

Plan projects in Mexico have been carried out aimed at the acquisition of land for conservation; land and wetland conservation agreements; concessions for the conservation of federal properties; wetland restorations and management projects (e.g., mangrove restoration); designation of protected natural areas at the federal or state level that coincide with priority wetlands; and ecological easements aimed at land conservation. Combined, these actions have achieved almost 4.2 million acres under conservation, restored, or under some management scheme.

Interactions that the Plan has generated among professionals from the three countries have created important learning processes among the parties. In Mexico, it has encouraged the adoption of practices and application of techniques for addressing problems faced by wetlands in our country, adapting these to local socioeconomic conditions, always seeking to give the best response to the challenges in full compliance with the responsibilities of the country as a signatory member of the Plan. At the same time, Canada and the United States have learned from Mexico's traditions of always including local people in its conservation projects.

We recently completed the 2024 update of the North American Waterfowl Management Plan, where once again the NAWMP Mexican Committee made up of DGVS staff and civil society organizations participated and made important contributions to the plan, especially regarding the component of both migratory and resident species, as well as the implementation of our own concept of Human Dimensions.

There is still much to do. In some ways, it seems that the challenges are becoming larger, more complex, and more difficult to address. Initiatives such as the Plan have made it possible for the three countries in North America to combine their efforts to achieve common goals of conserving wetlands of importance for migratory birds. Each country carries out its initiatives, based on its legal framework and government programs that guide the environmental policies of each country, and through which they face their unique economic and social challenges as they fulfill their responsibilities as signatories of the Plan.

In my opinion and from many other partners in Mexico, we can affirm that there has not been an initiative like this, which has managed to combine the efforts of the governments of the three countries, landowners, and nongovernmental organizations, in achieving a continental conservation goal as has been the case with the North American Waterfowl Management Plan.

Congratulations on 40 years of continuous work and I toast to 40 more years!

Box 11.2 Jim Patterson: Farmers and Agricultural Policy in Canada

Someone else who was always thinking about people broadly as part of the solution to habitat conservation was Jim Patterson (CWS). Jim was invited to succinctly describe at an international conference (Patterson 1994) his hopes for NAWMP and the Wetlands for the Americas program. He observed that up until the late 1970s waterfowl management in North America had essentially two components – population management through regulation of harvest and a network of refuges and other protected areas. But because privately-owned farmland provided the majority of breeding waterfowl habitat, the traditional approach of habitat protection through nature reserves was not adequate.

Many NAWMP habitat programs, therefore, were designed to work with private landowners by providing incentives for sustainable land use practices that benefited waterfowl as well as soil and water resources. In theory, with the adoption of sustainable agricultural practices, the national treasury, the environment, waterfowl, and the farmer could all win. Then, Patterson's big leap of faith was that incentive payments to teach these lessons would not need to be provided in perpetuity for these ecological and economic benefits. Rather, he believed that those benefits should serve as an incentive to adjust farming practices to a more self-sustaining basis. Jim also saw needs and opportunities in the realm of government policies, especially trade policies like the North American Free Trade Agreement (NAFTA) and the World Trade Organization's General Agreement on Tariffs and Trade (GATT) to address even more environmental concerns while removing trade-distorting subsidies. The approach, insisted on by Mexico of considering local communities in wetland conservation projects, was a different take on this idea, but both presaged the now growing emphasis of expanding support for NAWMP through the multiple benefits that wetland and grassland conservation offer to people.

References

Canadian Wildlife Service. 1992. *The Importance of Mexico to Migratory Waterfowl in North America.* A report to the NAWMP Committee. Ottawa, ON. 24pp.

Environment Canada, U.S. Department of the Interior and Secretario de Desarrollo Social Mexico. 1994. *1994 Update to the North American Waterfowl Management Plan: Expanding the Commitment.* Ottawa, ON. 47pp.

Mellink, Eric, Monica Riojas-Lopez, Jaime Luevano Esparza, and Jennifer A. Wheeler. 2009. "Historic changes in mid-winter use of Laguna Cuyutlan, Colima, Mexico by waterfowl." *Ornitologia Neotropical* 20: 171–79.

North American Wetlands Council Canada Secretariat. 1992. *Report of the NAWMP Update Task Force.* Canadian Wildlife Service. Ottawa, ON. 12pp.

Otto, Mark C., and P.A. Frechtel. 2001. *Temporal and Spatial Changes in the Mexican Mid-Winter Inventory 1937–1998.* U.S. Fish and Wildlife Service. Laurel, Maryland. 8pp.

Patterson, James H. 1994. "Thc North American Waterfowl Management Plan and Wetlands for the Americas programmes: A summary." *Ibis Supplement* 137: S215–18.

U.S. Fish and Wildlife Service and Canadian Wildlife Service. 1992. *North American Waterfowl Management Plan: Potential Issues Identified for the Plan Update.* U.S. Fish and Wildlife Service, Arlington, VA. 6pp.

U.S. Fish and Wildlife Service and Canadian Wildlife Service. 1993. *North American Waterfowl Management Plan. A Review of the First Five Years.* Report to NAWMP Committee. U.S, Fish and Wildlife Service. Arlington, VA. 31pp.

12 Major Decisions Ahead – The 1998 Update

By the late 1990s, the NAWMP enterprise faced potentially transformative decisions. The original term of the Plan through the year 2000 was approaching, yet it was clear that the job of securing the future for waterfowl habitat on the continent was far from done. The habitat objectives, expanded by fourfold in the 1994 Update, made that clear. The work needed to continue but several events in the early '90s left the partnership questioning its initial course.

Conservation at Landscape Scales

Throughout most of the 1980s, precipitation had been below normal across the prairie pothole country leaving few small wetlands holding water. In the United States, however, a conservation-friendly USDA Farm Bill (1985 and subsequent versions) created the Conservation Reserve Program (CRP) that incentivized farmers to plant perennial grasses instead of annual crops on millions of acres of land including much of the "duck factory" of the Dakotas, southwestern Minnesota, Iowa, and northeastern Montana. Finally, in the summer of 1993, a wetter period ensued which only deepened over the next few years. Ducks returned to the area in the millions, the expansive CRP grasslands provided much-improved nesting cover (Reynolds et al. 2001), and duck populations rebounded year after year to unprecedented numbers of total ducks in 1997–2000 (USFWS 2001).

This reinforced the notion that landscape-scale changes were indeed part of the "essential formula" for bountiful duck production. Achieving success at such a scale was highly desirable, but perhaps not attainable by buying small patches of good cover here and there across the 300,000 square miles of the Prairie Pothole Region. More recent large-scale, long-term analyses (Weegman et al. 2024) reinforce the notion that landscape-scale variation in conditions on both breeding and wintering areas can materially affect the productivity of prairie-breeding dabbling ducks.

While the work of the Continental Evaluation Team (Chapter 13) and evaluation studies in several joint ventures had been gradually improving the biological foundations of the Plan, much progress was still needed to confidently connect continental population objectives to JV population and habitat objectives, and to understand how best to design and execute projects to affect waterfowl populations. Clearly, more research and management trials were needed to improve the cost-effectiveness of habitat interventions (Anderson et al. 1996; Continental Evaluation Team 1996; Johnson et al. 1997). Congressional reauthorization of NAWCA was approaching and any new input from the Plan community had to be transmitted by the end of 1997, adding pressure for taking stock of Plan progress promptly.

Lastly, bird conservation had changed dramatically from the earliest days of NAWMP. Partly because of the NAWMP model, major initiatives for landbirds (PIF) and more (NABCI) had

DOI: 10.4324/9781003732310-12

coalesced (Chapter 9). Some innovative JVs like the LMVJV had already broadened their partnerships to begin projects for birds other than waterfowl. So too, had the PPJV, the ACJV, the EHJV, and others. Some conservation leaders thought the obvious next step would be to subsume NAWMP under a new larger governance umbrella like NABCI. Others cautioned that in the process, we might lose the support of hunters, the Flyway Councils, and some NGOs who had collectively led the way for duck habitat conservation in the first decade of the Plan.

NAWMP leaders decided to tackle these issues by accelerating the next update. The debate began, frankly, before the ink was dry on the 1994 Update. The urgency is evidenced by the shortest update cycle (four years) in the 40-year history of the Plan. As David Smith, then Executive Director of the US NAWMP office recalled (pers. comm. April 2025), it was a "delicate time that presented several possible and divergent futures". The bird initiatives could compete, struggling over inadequate funds. They could replicate organizational structures, or they could try to advance landscape conservation together. The "big question" Dave remembers was how NAWMP and all the other interests could keep focused on their respective priorities without getting lost in a greater whole?

The 1986 Plan recognized that land-use practices and policies affecting extensive areas of the continent would have to be altered for NAWMP goals to be met. Conservation efforts would have to extend far beyond the limits of protected public lands and deal with whole landscapes, including private lands. Partners would need to venture beyond the comfort of long-established wildlife programs and relationships to embrace programs and policies that directly affected the ecological health of large landscapes – to benefit not only wildlife but people as well (Patterson 1994). The early effects of CRP in the Dakotas served to reinforce these ideas. Nevertheless, little such progress had been made in the first decade of NAWMP in other places. Robert Bailey, a member of the NAWMP negotiating team and later a member of the Continental Evaluation Team wrote to Bob McLean, Canadian Director of the NAWMP Office (February 23, 1997), urging "a re-focus on policy instruments for achieving NAWMP goals", something his mentor Jim Patterson had long championed. Bob raised the idea of a "Landscape Policy Institute" as a new coordinating mechanism for policy advancements in Canada, but that structure did not materialize.

A Broadening Policy Context

Since 1986, Canada, the United States, and Mexico had concluded or strengthened other alliances in conservation and trade that created additional obligations, opportunities, and challenges. These included the Convention on Wetlands of International Importance signed in Ramsar, Iran in 1971 (https://www.ramsar.org/), the 1992 North American Free Trade Agreement and the parallel North American Agreement on Environmental Cooperation, and the Tri-Lateral Committee for the Conservation and Management of Wildlife and Ecosystems. While each of these initiatives reflected an increasing awareness of the economic and environmental benefits of international cooperation, together they formed an increasingly complex federal context within which NAWMP was being pursued. Yet, it was difficult in the late 1990s to pinpoint many incremental accomplishments for waterfowl conservation flowing from these additional policy-level forums. One exception seems to have been the value of these other alliances in the engagement of Mexico, which was developing its socioeconomic and conservation policies more in tandem than were Canada or the United States (Chapter 11).

Another element vital to the success of NAWMP broadly is the role of the continent's indigenous peoples for which migratory birds have cultural and, in some cases, dietary importance. The Parksville Protocol agreed to in 1995, enabled amendment of Article IV of the Migratory

Bird Convention to include language about the commitment of both countries to preserve and enhance migratory birds, but also allow governments to establish laws, decrees, or regulations to allow the taking of migratory birds at any time of the year for specific purposes consistent with the conservation principles of the Convention. Through subsequent amendments, traditional subsistence harvest of waterfowl was recognized as legitimate. The roles of indigenous communities in the management of migratory birds and the stewardship of vast areas of migratory bird habitat continue to evolve. In Mexico too, where almost all the land is either common (ejido) or private, it is especially important that residents play an active role for conservation to be effective. There was a growing consensus that success of the Plan would depend on effective partnerships among all sectors of society that have a role in conservation.

Stakeholder Consultations

With all these issues ascendant, the Plan community launched its largest consultation process since 1985. A three-country working group was appointed by the PC to explore a constellation of issues and return to the PC in 1996 with recommendations for how to proceed. Charles Baxter and David Smith (US), Richard Goulden and Gregory Thompson (CAN), Humberto Berlanga (MX), and Mark Konef (CET liaison) comprised the team. Their report, "Preparing for the 1998 Update: a Bridge to the Future", was delivered to the Plan Committee at a meeting in Quebec City in October. The group had considered 74 issues identified by the PC related to the Plan's context, biological foundations, structure, and future. They distilled these to 15 and prepared a set of five options to guide the 1998 Update. Their summary table of five options would guide public consultations. Options ranged from continuing the current path of the Plan (1) to development of a new integrated continental conservation plan and delivery structure for ALL migratory and nonmigratory birds (5). All five options included recommendations for enhancing the scientific foundations of the Plan; the differences were mainly related to the elements and degrees of integration with other bird initiatives. What degree of integration might make the most sense? Biological planning? Monitoring? Prioritization? Funding? Administration? Or more? They posed two summary questions, both assuming that a formal framework for the conservation of North American waterfowl would exist beyond 2000:

1 What refinements are needed in the biological foundations for waterfowl conservation upon which the 1986/1994 Plan rests?
2 How might it (the Plan) be linked biologically and institutionally, with other continental bird conservation and habitat strategies that have emerged since 1986?

As you might expect, reactions to these wide-ranging ideas were varied and strong. I was at the two-day meeting in Quebec City speaking to the biological priorities on behalf of the Continental Evaluation Team (Chapter 13). Our recommendations were adopted except there was little support from the PC members for more work on geographic prioritization of habitat investments. For other bird integration, Options one and five were nonstarters so most of the debate centered on options two, three, or four but agreement was elusive.

The PC decided on a two-step consultation process: first, present all five options to the conservation community in all three countries for their reviews; then after analyzing the responses, go back to the community for feedback on a recommended path forward. The first consultation kits were distributed at the end of December 1996 with a two-month window to respond. Responses (n = 166) (Berlanga et al. 1997) were received from federal and state government agencies, nongovernment waterfowl organizations, and other bird conservation groups. Five major themes emerged from Canadian responses: (a) maintain the waterfowl and wetlands focus of the Plan while doing more for other wetland-dependent species; (b) increase recognition of Plan benefits to waterfowl and wetlands, biodiversity conservation, sustainable development, and other ecosystem benefits; (c) coordinate with other bird plans but this seemed best done at the joint venture level where integrated planning at a project level might be most efficient, (d) preserve existing partnerships and encourage new ones; and (e) strengthen and enhance funding for all birds. Ten common themes emerged from US responses, several of which overlapped with Canada's results. More variable responses were received on matters of the formality and geographic scale of desirable coordination with the other bird initiatives and cooperative funding. There was broad agreement that maintaining flexibility to act on local opportunities was important. Preferences from Mexico varied but generally favored greater integration with all birds. For most organizations in Mexico, waterfowl are no more important than any other birds. Habitat is the common ground for coordination and cooperation, and a watershed approach with local community engagement was particularly attractive there.

Thus, the second consultation proposed three themes to be viewed in a new 15-year time horizon, with continuing five-year updates. The desired scale and formality of coordination with other bird initiatives remained unresolved. The PC did agree that at all organizational levels, flexibility should be maintained to enter into any agreements promoting the Plan's goals, and new partnerships were encouraged.

After digesting the second set of consultation responses in December 1997, the Plan Committee charged the writing team to express three "visions" under three headings for the years ahead. In so doing, they reiterated the focus of NAWMP on waterfowl but in the broader context of conservation efforts for all birds, other wildlife, and resources such as clean water. They also drew attention to expanding NAWMP's influence to entire landscapes, often working landscapes, where the hope for substantial and sustainable gains seemed most promising. The final 1998 Update (NAWMP Committee 1998) described these priorities as follows:

Strengthening the Biological Foundation

> *Plan partners enhance the capability of landscapes to support waterfowl and other wetland-associated species by ensuring that Plan implementation is guided by biologically based planning, which in turn is refined through ongoing evaluation.*

- Develop measurable objectives for planning and evaluation.
- Expand monitoring and assessment capabilities.
- Tighten the focus of interventions to limiting factors and places.
- Evaluate conservation strategies.

Toward Landscape Conservation

Plan partners define the landscape conditions needed to sustain waterfowl and benefit other wetland-associated species, and participate in the development of conservation, economic, management, and social policies and programs that most affect the ecological health of these landscapes.

- Implement waterfowl conservation in a sustainable landscape context.
- Expand conservation coordination with other wildlife initiatives…making wildlife conservation relevant to more people.

- Seek landscape solutions that benefit waterfowl and other needs.... especially agriculture, forestry, water, and trade policies.
- Implement community-based projects within a landscape context...helping to sustain both economic progress and ecological stewardship.

Broadening the Scope of Partnerships

Plan partners collaborate with other conservation efforts, particularly migratory bird initiatives, and reach out to other sectors and communities to forge broader alliances in a collective search for sustainable uses of landscapes.

- Broaden partnerships with other migratory bird conservation initiatives.... Working together should yield benefits for all.
- Seek partnerships with other economic sectors to meet common goals.... Private landowners, the soil and water conservation community, and water resource interests seem promising.
- Support and encourage conservation partnerships with communities.

Mexico's federal response in this final consultation was not fully supportive of the US/Canada consensus on a waterfowl focus (Bob Streeter, pers. comm. July 2024). PC members Humberto Berlanga and Eric Gustafson explained that while working for waterfowl and wetlands is a good way to develop conservation strategies in Mexico, too much specialization was unrealistic. Mexican delegates did agree that the Plan was an acceptable mechanism to achieve broader conservation outcomes since funds targeted for the benefit of waterfowl also protected other species. Mexico would continue to view the Plan as part of a larger strategy for protecting wildlife habitat in Mexico, and Mexican partners would continue to seek complementary funds for protecting other species and habitats. Berlanga asked that the Plan Committee endorse Mexico departing from the traditional NAWMP focus to diversify the opportunities that the Plan provides in Mexico, and the PC acceded to Mexico's request.

Looking back from 2025, it's clear that several joint ventures found traction with these broadened visions of the Plan. Progress in partnerships with landowners and rural communities is increasingly common as discussed above in the Playa Lakes, Intermountain West, Central Valley, Rainwater Basin, and elsewhere.

The 1998 Update recognized that policy work to affect large acreages had much potential for helping to achieve Plan goals. There have been notable "wins" in this respect such as conserving wetlands for flood abatement in the upper Midwest, affecting wetland drainage policy in Manitoba, and USDA programs like WRP and GRIP. But there have been significant setbacks too such as the loss of most of the CRP acreage in the Prairie Pothole Region, the loss of protection for isolated wetlands formerly protected under the US Clean Water Act, and little success in developing a strong wetland policy in Saskatchewan, among other places. Policy work requires constant vigilance. The 2024 Update (Chapter 23) again called for the Plan community to enlarge our conservation footprint on the land in places vital to waterfowl, by recognizing and marketing the multiple values of wetlands and grasslands to society.

Where to Next?

In sum, the 1998 Update offered strong guidance and substantive challenges to the Plan community. The question of whether NAWMP would remain as the "keeper of a waterfowl conservation vision" was answered affirmatively. NAWMP would remain. It would not be absorbed under NABCI. It would continue its quest to sustain waterfowl populations but do so in the spirit of broader cooperation with the other bird initiatives and other users of the land.

Affecting whole landscapes at a continental scale was a daunting ambition. The joint venture partnerships that were just consolidating their strategies, programs, and administrative functions were now challenged to seek even broader partnerships both within and beyond their accustomed conservation communities. Given this broader context, it was also not surprising that the attention of Plan leadership and its technical support bodies turned strongly to the challenge of enhancing the biological foundations of the enterprise. Given the challenges of materially affecting duck populations and the uncertainty about just what efforts, in what places were most likely to achieve that, were vital questions to answer.

There had been, through late 1996 into the fall of 1997, a draft technical companion document, developed mainly by the Continental Evaluation Team (Chapter 13), which it hoped would be published as a supplement to the 1998 Update. But time ran out to develop the strategy fully at both national and regional levels, so the work was deferred to 2000 and served as the starting point for the next update. Much of the capacity to do this biological foundation work lay in the federal agencies and waterfowl NGOs and there were important new ideas emerging around the concept of adaptive management that seemed highly relevant for NAWMP. Thus, that work accelerated at both the international and joint venture levels and led to the next major stock taking for NAWMP – the 2004 Update on Strengthening the Biological Foundations.

Box 12.1 Kenneth Cox – Wetlands' Champion (1944–2004)

Ken Cox, an astute and tireless crusader for wetland conservation, led the Secretariat of the North American Wetland Conservation Council Canada for many years, later expanded to include the NABCI Canada Council. He focused national attention on wetland policy through numerous policy papers, a wetland mitigation policy, wetland conferences, and more. Ken was also a key representative from Canada with the United States, and the projects funded through the Wetlands Council had a major impact on the success of NAWMP in Canada. Ken had a knack for bringing people together and a passion for wetland advocacy with government decision-makers, business leaders, and resource sector associations. He saw the "big picture". He also helped facilitate updates of the Plan from 1994 to 2004. A recognized leader in international efforts such as RAMSAR, Cox was also a recipient of the NAWMP International Canvasback Award. His legacy lives on in thousands of wetlands that he helped protect.

Box 12.2 Rod Fowler – DU Canada's Thoughtful Prairie Leader (1944–2002)

As the PHJV was coming together, Stew Morrison asked DUC's Prairie Region Manager, Rod Fowler, to take the lead in positioning the company for a major role in the delivery of NAWMP. Rod, a chartered accountant by training, had earlier served as Provincial Manager in Manitoba so he understood prairie conservation and the company's strengths and vulnerabilities. Importantly, people trusted Rod. He had good instincts and a constructive manner of engaging others. He was a good listener, a mentor to many, and adept at solving problems. He was the right person for Ducks Unlimited in the right role, and for me, he became a valued advisor and friend.

Rod led DUC's engagement in the PHJV through a formative decade before being promoted to Director of Conservation nationally and eventually to Executive Vice-President, the top staff role in the company. He helped launch Canada's other habitat joint ventures too and was on the international steering committee for the 1998 Update of the Plan. Rod became the most prominent supporter within DU for embracing the potential advantages of broadening JVs to include other birds, a movement he supported in Canada along with Gerry McKeating (PHJV), Art Martell (PCJV), and George Finney (EHJV). Regrettably, Rods years as DUC's leader were shortened by illness but his work in shaping and facilitating the growth of the Plan in Canada was profound.

References

Anderson, Michael G., Michael W. Tome, Robert O. Bailey, Richard K. Baydack, Mark D. Koneff, Thomas E. Martin, Jeffrey W. Nelson, James K. Ringelman, Clayton Rubec, and Robert E. Trost. 1996. "NAWMP evaluations: How can we generate the feedback that plan partners need?" *Proceedings of the International Waterfowl Symposium* 7: 250–57.

Berlanga, Humberto, Robert S. McLean, and Byron K. Williams Co-Chairs. NAWMP Committee. 1997. *Phase II Consultation Kit, the 1998 Update to the North American Waterfowl Management Plan.* June, 10, 1997.

Continental Evaluation Team. 1996. *Annual Report of the Continental Evaluation Team to the NAWMP Committee*. USFWS. Arlington, VA.

Johnson, Fred A., Michael G. Anderson, Richard K. Baydack, Jeffrey W. Nelson, James K. Ringelman, Mark D. Koneff, Robert O. Bailey, Thomas E. Martin, and Clayton Rubec. 1997. "Enhancing biological performance of the North American Waterfowl Management Plan." *Transactions North American Wildlife and Natural Resources Conference* 62: 377–85.

NAWMP Committee. 1998. *Expanding the Vision:1998 Update North American Waterfowl Management Plan.* Canadian Wildlife Service, U.S. Fish and Wildlife Service, Secretaria de Medio Ambiente y Recursos Naturales. Ottawa, ON. 43pp.

Patterson, James H. 1994. "The NAWMP and Wetlands for the Americas: A summary." *Ibis Supplement* 137: S215–18.

Reynolds, Ronald E., Terry L. Shaffer, Randy W. Renner, Wesley E. Newton, and Bruce D. J. Batt. 2001. "Impact of the conservation reserve program on Duck recruitment in the U.S. Prairie Pothole Region." *Journal of Wildlife Management* 65: 765–80.

U.S. Fish and Wildlife Service. 2001. *Waterfowl Population Status, 2002*. U.S. Department of the Interior. Washington, DC. 51pp.

Weegman, Mitch D., James H. Devries, Robert G. Clark, David W. Howerter, Daniel Gibson, J. Patrick Donnelly, and Todd W. Arnold. 2024. "Ecological and anthropogenic drivers of waterfowl productivity are synchronous across species, space, and time." *Ecological Applications* 34, no. 4: e2679.

13 Strengthening the Biological Foundations of NAWMP

Evaluations Begin on the Prairies

For those planning the first habitat actions under NAWMP in the 1980s, the biological values of the work envisioned were not well known. The strength of the scientific basis for program choices varied from region to region and program to program, but everywhere there remained material uncertainty. Research on prairie breeding waterfowl had been underway for decades, led by the Delta Waterfowl and Wetlands Research Station in Manitoba, the Northern Prairie Wildlife Research Center in North Dakota, and the Prairie Migratory Bird Research Centre in Saskatoon. The biological background for planning was strong, yet the surest approach to enhancing waterfowl populations was unclear. Perhaps because these prairie scientists had a sober sense of that uncertainty, it was there that biologists first invested in developing a comprehensive JV biological evaluation plan.

Implementation planning for the PHJV began in the summer of 1986, just weeks after the Plan was signed (Chapter 4). By early 1987, work on a companion evaluation plan got underway and was concluded in 1988 (Delta Waterfowl and Wetlands Research Station 1988). Those of us developing the evaluation plan[1] were also engaged in implementation planning which helped connect the important dots between habitat plans and biological evaluations. The scale of proposed habitat expenditures and the imperative of using those funds wisely drove that commitment by the joint venture.

A multi-partner Evaluation Coordinating Committee was set up in 1989, reporting to the JV Advisory Board, and I was chosen as the first chairperson. Programs comprised supplementing the annual May waterfowl surveys with additional ground counts to enhance precision of population estimates. Also included were periodically repeating historical CWS wetland habitat surveys, creating an open competition for short-term directed studies with management implications, and designing a comprehensive assessment of JV program effectiveness. That assessment entailed surmounting design challenges, developing unprecedented coordination with conservation delivery teams, obtaining major funding, and getting commitments to transparency.

By 1990, the evaluation work began with survey augmentation and the first round of directed studies funded by CWS. Funding for the multimillion-dollar PHJV Assessment was finally assembled in 1993 and the project got underway after a pilot season in 1991 to work out logistical and methodological challenges.

The US PPJV had two major assets close at hand as they began plotting an approach to evaluations. First, the NPWRC in Jamestown, founded in 1965, had a cadre of world-class waterfowl and wetland scientists and an ongoing program of research on prairie ducks. Second, USFWS regional offices in Denver and Minneapolis had established two habitat and population

DOI: 10.4324/9781003732310-13

evaluation team (HAPET) offices in Bismarck, ND and Fergus Falls, MN. Their original functions were to coordinate a waterfowl population and production survey (the four-square-mile surveys) and develop spatial production models using GIS tools and duck productivity models being developed by Lew Cowardin and others at NPWRC. Digital wetland inventory data had just become available and that made spatially explicit planning possible where inventory coverage existed, including the wetland management districts within the PPJV. Beginning in 1990, the HAPET offices were further tasked with acquiring needed biological information to "feed" this model building and enhance conservation planning capability. Ron Reynolds, Dan Cohan, and others from HAPET conducted workshops at every Wetland Management District demonstrating predicted impacts of alternative management actions, which got managers thinking about options. A decade later, HAPET employed similar technology to integrate planning for birds other than waterfowl.

With these technical innovations just emerging, the first PPJV Implementation Plan in 1989 featured only two pages about evaluations, noting in principle the need to evaluate conservation strategies and create a database for accomplishments. It acknowledged the need to guide expenditures to the most cost-effective practices, but the initial focus would be on habitat delivery. Evaluations would become operational later. A fulsome evaluation program for the PPJV was developed in 1993/1994 (Arnold et al. 1994). Unsurprisingly, PPJV scientists proposed pursuing elements much like the PHJV had chosen; namely, monitoring landscape changes, population estimation, and tracking recruitment rates in IA, MN, ND, SD, and MT. Monitoring nest success by HAPET was also part of their plan, along with estimating hen survival, predator populations, and abundances of other wetland birds. Their assessment plans included testing the mallard productivity model, rather like the parallel PHJV study, by counting breeding pairs, assessing cover attractiveness, nest success, breeding effort, duckling survival, summer hen survival, and more. The planners also believed, based on studies in hand, that reducing rates of predation on duck nests would have to be accomplished if PPJV goals were to be met. In time, their ambitious assessment plan was set aside as efforts shifted to an evaluation of the USDA Conservation Reserve Program on waterfowl production (Reynolds et al. 2001).

Box 13.1 Evidence of Declining Nest Survival

Concurrent with the beginnings of NAWMP, scientists at Northern Prairie, Saskatoon, Delta, and elsewhere had been studying the population ecology of prairie predators and their relationship to waterfowl nest success and hen survival. Evidence emerged that nest survival of prairie ducks had been declining for some time. Cowardin and colleagues (1985) had surprised many when their large study of radio-tagged breeding mallards in ND found rates of recruitment insufficient to maintain the breeding population. Nest success seemed to be declining over time which made improving nest success one primary goal for prairie conservation planners. Using data from the 1930s to the 1980s, Wendy Beauchamp and others (Beauchamp et al. 1996a, 1996b) reviewed 143 nesting studies and found that the likelihood of hatching, though highly variable among places and years, averaged >30% in the 1930s and 1940s, around 20% in the 1960s and 1970s, but <15% in the 1980s. Interestingly though, the regression slopes of the declines did not differ between studies from unmanaged and variously predator-managed study sites. Later, nesters such as gadwall, blue-winged teal, and northern shovelers were somewhat more likely to hatch than early nesters like mallards and northern pintails.

A Continental Evaluation Team

Aware of the uncertainties about how to enhance waterfowl populations, Harvey Nelson and his small NAWMP staff (Fig. 2.2) encouraged developments of JV evaluations, although how such work would be funded and by whom was unknown. Nelson also anticipated that quantitative information on NAWMP program performance would be important to ensure continued support for habitat conservation. In the fall of 1989, Nelson and biologist Dave Sharp invited a group of us to Chicago to begin imagining what a continental-scale evaluation program might look like. We met at the Ducks Unlimited office in Long Grove, IL – Nelson, Sharp, Bob Blohm (FWS), Bob Bailey (CWS), Frank Dunstan (Audubon), Jeff Nelson and Bruce Batt (DU), Danielle Thibeault (Canadian NAWMP Office), and me (Delta).

In his invitation, Nelson astutely noted that "an improved understanding of the linkage of habitat and population parameters will be essential". Encouraged by Harvey to think broadly, we concluded that tracking, monitoring, and evaluation are interrelated processes and would all be essential for successful implementation of NAWMP. We agreed that there was a critical need to establish a more definitive relationship between specific acreages of habitat conserved and numbers of waterfowl and other wildlife affected. This was vital for setting priorities, developing implementation strategies, and choosing management options. We agreed that at the outset, a Continental Evaluation Team (CET) should only concern itself with biological evaluations and not the more usual government "programmatic" evaluations. I left the meeting tasked with drafting a guidance document for joint venture evaluations before the next meeting five months later! That was a naïvely ambitious schedule, but we were underway!

Over the next couple of years, the Team developed a continental evaluation strategy for NAWMP, consulting extensively with Plan partners along the way. The final evaluation plan (Sharp et al. 1992) was designed to address five basic questions at both the continental and joint venture scales:

1. Are we meeting the waterfowl habitat and population objectives identified in the NAWMP and joint venture plans?
2. Are we utilizing the most appropriate management techniques available?
3. Are there changes that need to be made?
4. Are there information/research gaps that prevent adequate evaluation? If so, how can these be addressed?
5. Are there ongoing evaluation or oversight activities necessary for successful implementation of NAWMP?

Initially, the scope of responsibilities included tracking accomplishments, monitoring populations and landscape changes, testing program effectiveness, and recommending program adjustments. We did not include policy analyses, socioeconomic studies, or political considerations that would be left to others.

With planning concluded, the Team was repopulated.[2] I was appointed Canadian cochair and Mike Tome (USFWS) the US cochair. The CET continued to meet twice annually to promote and monitor progress in addressing the recommendations of the strategy. We also began producing substantive annual reports on the status of NAWMP evaluations and related biological knowledge for the international Plan Committee.

The first useful thing the team did was encouraging the federal agencies to develop and implement accomplishment tracking in each country. Although the national systems were not fully congruent and were soon complicated by the advent of the North American Wetlands

Conservation Act and its reporting requirements, a serviceable system was created that has served NAWMP partners well. On the Canadian side, Danielle Thibeault of the Ottawa office deserves much credit for accomplishing that task with the help of Len Warren and others at DU Canada.

Population tracking remained the responsibility of the federal governments, and several steps were taken to enhance the precision of those surveys early during the NAWMP era (Graham Smith 1995). Use of the surveys continues to evolve including in the 2024 Update of the Plan (Chapter 23).

Habitat databases differ among the countries, so joint ventures have needed to cobble together various data sources to track landscape change. National agricultural land-use reporting systems are one such category of divergent data. Habitat monitoring options have also evolved rapidly with different satellite platforms and GIS tools and will likely continue to differ among nations for the foreseeable future. Still, most JVs appear to have what they need to track trends in their most important landscape metrics at scales that are useful for conservation planning.

The CET also strongly encouraged progress in joint venture evaluations. The CET's "Guidelines for JV Evaluations" unfolded through four drafts (December 1989 through August 1991) and multiple consultation meetings with joint venture technical teams. It was finally issued in January 1992 with the imprimatur of the Plan Committee, as part of the overall evaluation strategy for NAWMP (Sharp et al. 1992). Along with it was a request that the JVs report on the status of their evaluation activities later that summer. In August, the CET reviewed those reports and provided feedback, initiating a process of periodic review and coaching of JV evaluations that continued and evolved over two decades.

Early Evaluations Elsewhere

The Lower Mississippi Valley team, led by Charles Baxter, Seth Mott, Charles Loesch, and Ken Reinecke, began thinking about biological evaluations early on and emerged with a science plan shortly after the PHJV did (Loesch et al. 1994). Their strategy, as per advice from the CET, was to monitor progress in meeting the JV's goals and objectives and testing the validity of the assumption about foraging habitat limitations that underpinned those objectives. Their initial plan included 14 specific issues to be evaluated with a focus on wetland habitats, waterfowl, and other migratory birds. Led by Randy Wilson, the JV's science coordinator, the JV's science team comprehensively updated their plan in 2002.

The Central Valley JV's first implementation plan (1990) included a brief evaluation and monitoring section that laid out seven questions requiring research investments at the Valley-wide level, the river-basin level, or the local level. Their technical committee was charged with advancing that work but by the summer of 1991 they were simply tracking acre accomplishments. In 1996, this effort was augmented with a more ambitious plan (Miller et al. 1996). It proposed hiring an evaluation coordinator and undertaking studies testing the biological assumptions underlying JV plans, plus pursuing multiple directed studies.

By 1991, the Gulf Coast JV had begun short-term studies to identify waterfowl habitat needs and a few tests of whether projects were doing what was intended in their designs. A similar report was offered by the Atlantic Coast JV, while the Eastern Habitat JV in Canada added investments in policy reviews and socioeconomic analyses.

By the summer of 1992, 12 of the 13 habitat JVs had completed a concept plan. Of those 12, eight had completed their first implementation plan, four (PHJV, GCJV, PPJV, and CVJV) also submitted an evaluation plan to the CET. Three others (LMVJV, EHJV, and PCJV) had evaluation plans in preparation. About then, the CET began moving its meetings around the continent to meet with JV implementation and evaluation teams, thereby gaining greater insight into the challenges they faced and providing the opportunity to discuss more fully what the CET was advocating. By January 1994, we reviewed the Lower Mississippi plan and revised submissions from the US Prairie Pothole JV and the EHJV. By August 1994, we had a first plan from the Pacific Coast and an update from the PHJV. So, a pattern was established that at minimum encouraged progress by the JVs in developing their own evaluation endeavors and provided an opportunity to learn from what each other were doing.

To enhance learning among JVs, the CET organized occasional workshops of multiple joint ventures. One of the first was for wintering waterfowl joint ventures held at Five Lakes Outing Lodge in Arkansas in 1995. Alan Afton, Mike Conroy, Mickey Heitmeyer, Jim Kelley, Ken Reinecke, and Loren Smtih contributed their scientific expertise, and the JVs involved were the Lower Mississippi Valley, Central Valley, Gulf Coast, Playas Lakes, and Rainwater Basin. The workshop, led by Jim Ringelman, posed three overarching questions: (1) Are the strategies of the nonbreeding JVs reasonable given current knowledge? (2) What is the strength of evidence for effects of nonbreeding JVs on waterfowl demography?, and (3) What major assumptions need to be tested and how might we do that? An emerging conclusion was there needed to be renewed emphasis on testing key assumptions underlying nonbreeding JV programs. It was a valuable exchange of plans, perspectives, and progress and an opportunity for cross-pollination among JV evaluation leaders. In later years, multiple PPJV / PHJV science and policy forums provided similar opportunities for common learning between these adjacent JVs.

Adaptive Management Thinking

The CET became an important crucible for the growth of adaptive management thinking across the NAWMP partnership. The concept of adaptive management, a more formalized way to think about learning as a vital and linked component of management actions, had been developing for some time (Walters 1986; Hilborn and Walters 1992) led by people engaged mostly in fisheries or forest management but also spearheaded by Fred Johnson, Ken Williams, Jim Nichols, and others as an approach to waterfowl harvest management (Nichols et al. 1995; Johnson et al. 1996).

I'm not sure if Fred Johnson ever formally became a member of the CET but he thoroughly engaged with us over several years helping us understand the potential of a more formal adaptive approach to improving NAWMP performance (Johnson et al. 1997). It became evident to the CET members that this philosophy might be very helpful to the habitat management community as we considered how to gain more reliable knowledge about the effectiveness of JV actions.

The notion of management as experiment wasn't entirely new to waterfowl habitat managers. The Small Unit Management Project at NPWRC in the early 1980s was an experiment to assess how small units of predator-free nesting cover might affect regional duck production (Cowardin et al. 1998; Krapu et al. 2000). The PHJV Assessment (Howerter et al. 2014) was a massive test of the impact of various scales and elements of habitat delivery on mallard biology and dabbling duck production. The Assessment design had pushed delivery staff to double their planned intensity of conservation impact on experimental landscapes to enhance the opportunity to learn. The waterfowl habitat management fraternity had rarely thought about management as experiment in such a comprehensive and costly way.

Cross-pollination between the CET and people like Johnson, Williams, and Nichols who were working on these ideas for waterfowl harvest management was inspiring and moved us to consider how broader application of adaptive management principles might advance our understanding of NAWMP effectiveness and do so in an efficient manner. (Anderson et al. 1996; Johnson et al. 1997). These ideas were expressed most inclusively in the CET's 1996 annual report to the Plan Committee (Continental Evaluation Team 1996). That report forcefully introduced the notion of a rigorously adaptive approach to NAWMP evaluations and spurred the idea of a larger technical companion document to accompany the 1998 NAWMP Update. Ultimately, for two reasons, I think, that supplement did not come to fruition. The CET was probably pushing too hard too fast and suggesting a more top-down direction for JV evaluations than Plan partners were ready to accept in 1997. Also, work was still needed to flesh out the plans we were pursuing so most of the work was deferred until the 2004 Update focused on biological foundations again. At that point, we all concluded that a pair of documents was needed – the "usual" high-level policy agreement and a supplementary Implementation Framework.

An important augmentation to the initial concepts of cyclical planning, delivery, evaluation, and replanning – the heart of the ARM concept – was initially described by Pohl-Wostl (2009) as "double-loop learning". This is the notion that normal program assessment and improvement through ARM cycles might occasionally lead to the need to step back, usually to a higher perspective, and consider if the system being pursued is still the right one. This might also mean reconsidering the essential objectives on which the AM cycle is based. Double looping thus can become triple looping. Sequentially, then the questions become: Are we doing things right? Are we doing the right things? And finally, do we have the governance or system right? This concept describes well what happened to the NAWMP partnership in 2008–2012 when we took a multistep path to a new and enlarged set of fundamental objectives. Our growing commitment to adaptive management as the way to conduct NAWMP business most certainly helped us when confronted with these transformative ideas at the end of the decade....but more on that later. First, the CET concluded that it was time for it to evolve too.

The NAWMP Science Support Team

The FWS in response to the growth of adaptive management approaches within the agency reorganized part of its technical capacity into an Adaptive Management and Assessment Team (AMAT) in 1998. One of its purposes was to address large-scale biological planning and evaluation needs of both NAWMP and the US Adaptive Harvest Management initiative that had emerged in 1995 (Williams et al. 1999). Based at Patuxent Wildlife Research Center, but with strategic nodes elsewhere, this team was thinking broadly about how the concepts of adaptive management could help guide multiple aspects of conservation, but its focus was limited to the United States. The AMAT assembled a strong technical staff including Fred Johnson, Rex Johnson, Jim Dubovsky, Jim Wortham, and Mark Koneff with statistical support from Andy Royle, mapping expertise, and more.

Another vital development was a decision to seek additional funding from Congress to support US joint venture activities. Among other things, this enabled joint ventures to retain science coordinators to support conservation planning and evaluation work. Not all JVs chose to do this, but many did. A September 2000 Congressional Conference report recommended approval of \$5.49 million to joint ventures in FY2001, with a projection of \$10.4 million by FY2004. These totals included administrative support for the national North American Waterfowl and Wetlands office. Among the JVs, the largest shares of funds were earmarked for the US Prairie Pothole, Lower Mississippi Valley, Atlantic Coast, and Gulf Coast joint ventures. The Canadian JVs were not eligible. Before FY2004, the FWS proposed an increase in joint venture base funding to $29.9 million by FY2010 rationalized in part by the transition of migratory bird joint ventures to an all-bird focus. Eligible activities would include monitoring, evaluation, and applied research. The funding continued (sub-activity *1234/NAWMP Joint Ventures*) but the proposed tripling of funds did not happen. By FY09, funding sat at $13.5 million, and about $16.8 million in FY2024. Nevertheless, this annual funding has been vital to advancing JV evaluation work.

As the Joint Ventures began to recruit more science staff, it was obvious to the CET that we needed to find a mechanism for engaging those new folks in NAWMP evaluation work. In mid-1999, CET leaders along with some of the AMAT team began thinking about how we might create a more effective evaluation structure. Fred Johnson (FWS), Areal Rojo (SEMARNAT), and I wrote a proposal for creating a new NAWMP Science Support Team (NSST) to succeed the CET (Anderson et al 1999). The NSST would engage fully the expected new cadre of JV science coordinators, existing science support groups like the HAPET offices in the PPJV, the evaluation team in the PHJV, and others along with some national science leaders, in a less top-down approach to evaluation design – one of the fair criticisms that had been leveled at the 1996 CET adaptive management proposal. Operating by consensus and reporting to the Plan Committee, the NSST's mission would be: *to help strengthen the biological foundations of the North American Waterfowl Management Plan and facilitate continuous improvement of Plan conservation programs.*

Taking Stock in 1999 – the First 15 Years

At the Plan Committee's 1999 fall meeting in Patzcuaro, MX, the CET offered an overview of the state of the Plan's biological underpinnings at the end of the millennium encompassing the first 15 years of the Plan. We argued that strengthening the biological foundations of NAWMP was vital for improving the basis for management decisions, and that as NABCI unfolded and the joint ventures broadened their conservation mandates, the Plan Committee must be the "keeper of the waterfowl flame".

Vitally, the Plan's biological foundations link continental population goals with regional conservation objectives and strategies, and therefore, depend upon knowledge of landscape conditions that affect waterfowl abundance and vital rates (Anderson et al. 1996; Williams et al. 1999). Biological foundations in our lexicon encompassed population objectives, geographic priorities, habitat objectives, and the linkages between habitat variation and population processes. These linkages in turn should give rise to effective conservation strategies that maximize returns on conservation investments.

Unfortunately, a period of unexpected turmoil had intervened in the 1990s with a decision by the US government to excise science capacity from the USFWS and vest it in the US Geological use Survey instead (USGS), widening the gap between habitat managers and scientists (Cohn 2005). This was bridged in time, but momentum was lost for a while. On the positive side, during the 1990s, numerous joint venture programs had been improved by evaluation feedback. Evaluation planning had progressed and many of the first regional studies had been effectively focused.

At large scales, however, scientists still could do little to associate plan actions with population responses. NAWMP population objectives did not consider uncontrolled environmental variation, and habitat trends in most JVs were poorly quantified, so understanding the true effect of NAWMP actions was elusive. Plan objectives also remained only weakly linked from the continental scale to regional scales, despite initial efforts to do so by Mark Koneff (2003) and the CET. At least three problems obtained: (1) What portions of the total continental populations were tallied in annual surveys? and (2) How do these populations distribute themselves in the nonbreeding season? In other words, how many birds, of what species, occur in migration and wintering JVs and how might that pattern vary annually and seasonally? Finally, (3) what habitat conditions are needed to support these birds through the nonbreeding season and in preparation for spring migration?

Continentally, plan programs remained only weakly prioritized regarding likely impacts on waterfowl demography. Planning assumptions for habitat interventions were only rarely made explicit or rigorously tested. Evaluations were just getting underway in several important JV regions (e.g., Gulf Coast, Central Valley). And perhaps most seriously, in many parts of the NAWMP enterprise, management and research remained poorly integrated with little formal replanning accomplished despite knowledge gained.

Why hadn't we done better? For one thing, partners were anxious to deliver programs and achieve acreage results to encourage more funding. Mike Tome, former CET Chair, reflected (pers. comm. April 2025) that some management agencies believed they "knew what they were doing" and "didn't need costly monitoring and evaluation". The mantra in some places was "move dirt and move on". Some partners may have also lacked confidence in a truly long-term view of the NAWMP enterprise. Why focus on costly assessments unless you expect to keep at it for a long time? Collectively, we had become very good at project planning but less accomplished at biological planning, and we underestimated the complexity of the challenges we faced and the levels of uncertainty in our ecological and human systems. Limited funds and human resources of course posed constraints as well.

It was our hope in the 1990s that adaptive management – the continuous improvement in management performance over time could become a shared vision for how to proceed. But how were we to achieve that? First, improving management performance starts with the specification of clear and measurable objectives. To some, JVs this still meant acres of habitat and not a population response, and step-down objectives were still unsettled, meaning linking continental Plan objectives down to regional conservation strategies. It also implied identifying key planning assumptions and uncertainties that could undermine management results and subjecting these to critical scrutiny. Important too was accepting that when evaluation evidence accumulates, something needs to change to achieve shared goals. That is, are we willing to change? Where the potential value of reducing uncertainty is cost-justified, implementing habitat programs with dual objectives of immediate conservation results and learning to manage better in the future seemed wise. This implied enhancing monitoring capabilities and building supporting institutional capabilities at both the JV and National levels, which is where we proposed that the new NSST would serve.

The proposal was received with enthusiasm by the Plan Committee. After our presentation and a thorough discussion, the PC unanimously approved the NSST mission and objectives and gave us a conceptual "go!", but with the caveat of finding the right people and funds to support the work, which the PC charged the three national offices to do. Johnson, Rojo, and I were named interim leaders until more permanent solutions could be found, and the national offices undertook to share the new concept with the JV coordinators at a meeting the following month. The PC members agreed to present the concept to the JV management boards and the Flyway Councils in their jurisdictions. They also recommended a follow-up scoping workshop with partners to further elucidate the Team's priorities and procedures.

Specifically, the NSST proposal recommended a shift to "viewing planning, implementation and evaluation as integral components of everyday management" – in other words, an adaptive management approach. The NSST would report annually to the PC and Plan partners on the status of the biological foundations of the plan, evaluation results, and implications for future conservation activity. We proposed that the Team would include national technical leaders and one member from each NAWMP joint venture.

Thus, we left the pivotal meeting in Mexico at the end of 1999, on the cusp of concluding the original 15-year planning horizon of NAWMP, with approval in concept for the NSST, and consensus that enhancing the waterfowl plan's biological foundations would be the focus of the next Update. While the NSST was established, and the CET dissolved, we were charged by the Plan Committee to undertake a final round of scoping meetings with the partners who would have to make the new Team work.

Box 13.2 Adapting Communications Too

While increased attention was being paid to evaluation and adaptation of conservation programs, people involved in communications and marketing of NAWMP were thinking adaptively too. Dave Case had been thoroughly schooled in adaptive management by Fred Johnson and others in the process of assisting the Adaptive Harvest Management Working Group with their communication challenges. Case had long encouraged applying that same analytical rigor to evaluating communication strategies. An IAFWA project was designed to formalize that approach. The project was conducted jointly by Southwick and Associates, Responsive Management, and DJ Case. The voluminous project reports included all the survey results and 15 recommendations to improve the effectiveness of those investments (IAFWA 1997).

Willows, California December 1999

Less than a month after the PC endorsed the NSST, several NAWMP joint ventures held a pivotal workshop in California with a focus on biological planning and evaluations. Organized mainly by Mike Eichholz (CVJV Monitoring and Evaluation Coordinator) status reports from the Central Valley, Pacific Coast, San Francisco Bay, Playa Lakes, Intermountain West, Gulf Coast, Upper Mississippi River and Great Lakes, and Lower Mississippi Valley JVs displayed widely varying progress in testing planning models, assessing limiting factors, and monitoring outcomes. John Eadie (UC Davis) challenged the group to think about assessing the potential linkages between wintering JVs and waterfowl survival and subsequent recruitment, and the implications of inevitable annual variation for planning and evaluations. Jim Sedinger (UN Reno) challenged everyone to think critically about spatial scales and what questions need to be answered at various nested scales. We also had a first exploratory discussion about linking evaluations of multiple joint ventures in the nonbreeding period. While the questions being posed were difficult and progress slow, limited by funding and time, these kinds of forums became increasingly important for making progress in understanding joint venture biological effectiveness.

Memphis, June 2000

The scoping meeting in Memphis was designed primarily to solicit feedback from the joint ventures on the NSST concept. We hoped to emerge with a consensus regarding the major functions of the NSST and its makeup. Twenty-eight attendees offered a useful mix of perspectives. Seth Mott and I served as facilitators. Points of consensus included: (1) the need for a technical body with a continental-scale perspective to advise the PC; and (2) the mission and proposed major objectives were appropriate with a little more elaboration. Most of the meeting was spent exploring the role of the NSST vs. other JV bodies, flyway technical committees, agency institutions (e.g., AMAT), and the like. Six major roles (with numerous sub roles) were agreed to and refined via email after the meeting.

The NSST would be populated mainly by JV-appointed technical members, although time and funding to support their engagement remained a concern. The need for parallel science teams at the JV level was apparent and strongly supported. We agreed to convene an organizational meeting in Bismarck in October where Ron Reynolds of the HAPET office in Bismarck would be elected as the NSST's first chairperson.

Preparing for the 2004 Update

With the focus of the next update on biological foundations, the Plan Committee tasked Seth Mott and me with organizing a meeting in October 2000 to scope the technical information needed for the update and explore approaches for assembling data. All JVs were invited and represented except two. The Flyway Councils had technical committee representatives attend as well. Steve Wendt (Canada), David Smith (United States), and Ariel Rojo (MX) served as national representatives. The meeting revolved around three topics foremost in the update guidance received from the Plan Committee: population objectives, species and geographic priorities, and the effects of habitat variation on waterfowl demography. We reached reasonable consensus on an initial approach to each topic and a six-month window for reporting progress.

Population objectives: Participants favored continuing use of the 1970s benchmark as NAWMP goals but agreed it would be useful to express these with confidence intervals linked to environmental variability. Attendees also wanted to explore establishing objectives by discrete populations (e.g., eastern mallards, mid-continent mallards), to try to resolve inconsistencies with Flyway management plans and help improve the linkage of continental population objectives to JV scales.

Species and geographic priorities: The group agreed that continental priorities should reflect population trends, especially any estimated "deficit" between expected and observed population size under varying environmental conditions. The proportion of the population harvested should also be considered. Thus, they hoped to account for both uncontrolled environmental changes and importance to people as elements in setting priorities.

Effects of habitat variation on waterfowl demography: Featured here were testing assumptions of food availability as the primary limiting factor outside breeding areas and contrasting winter survival of ducks over time with variable wetland conditions. It also seemed desirable to develop some "common currency" that might provide broad insights about geographic priorities, and to establish informative habitat monitoring over broad geographic areas.

With work of the NSST getting started and the next update close at hand, Plan partners were set for the next stage in the evolution of the Plan as the second 15-year window was upon us.

Box 13.3 Charles Baxter – Tireless Advocate for Science and All Birds (1948–2009)

Charles Baxter's voice is found throughout this book. Charles, true to his Baptist roots, was indeed an apostle for NAWMP, for science-based conservation, and an early advocate for JVs to embrace all-bird conservation. A compelling speaker in his unassuming drawl, but he led with the passion of an evangelist.

Baxter was born in Corsicana, TX and graduated from Texas A&M University. He moved to Vicksburg, MS in 1980 and retired in 2007. His meritorious career with the USFWS earned him numerous Service awards. Initially, a biologist with the Division of Ecological Services, Charles became one of the first joint venture coordinators and grew to be a mentor for us all.

Charles was smart, relentless, and unafraid to speak to power with conviction. He cared deeply about conservation and people. I think he would have been a supporter of the expanded vision of the 2012 Revision had he lived to see it. Matched by few others, I missed his clarity of thought, especially when we confronted complex, difficult questions. His leadership through the first two decades of NAWMP was matched by few!

Notes

1 Bruce Batt and Michael Anderson (Delta), Richard Wishart and Jeffrey Nelson (DUC), and Kent Brace (CWS) with a long list of consultants.
2 Members were Rick Baydack (U. Manitoba.), Clayton Rubec, Steve Wendt, and Bob Bailey (all CWS). Michael Tome (FWS, Cochair), Lew Cowardin (USGS), Jeff Nelson (DU), Jim Ringelman (CO) Mike Anderson (DUC), and Robert Trost (USFWS). In 1995, Tom Martin (Montana CWRU) replaced Lew Cowardin. Later, Graham Smith (FWS) joined the team.

References

Anderson, Michael G., Fred A. Johnson, and Ariel Rojo. 1999. *Strengthening NAWMP by Improving the Foundations for Management Decisions – A Proposal for a New NAWMP Science Support Team.* North American Waterfowl Management Plan Committee. Unpublished Report. USFWS. Arlington, VA. 8pp.

Anderson, Michael G., Michael W. Tome, Robert O. Bailey, Richard K. Baydack, Mark Koneff, Thomas E. Martin, Jeffrey W. Nelson James K Ringelman, Clayton Rubec, and Robert E. Trost. 1996. "NAWMP Evaluations: How can we generate the feedback that Plan partners needs?" *Internation Waterfowl Symposium* 7: 250–57.

Arnold, Phillip, Keith Harmon, Jeffrey Nelson, and Ronald Reynolds. 1994. *An Evaluation Program for the United States Prairie Pothole Joint Venture.* NAWMP. Bismarck ND and Fergus Falls, MN. 24pp.

Beauchamp, Wendy D., Rolf R. Koford, Thomas D. Nudds, Robert G. Clark, and Douglas H. Johnson. 1996a. "Long-term declines in nest success of Prairie Ducks." *The Journal of Wildlife Management* 60, no. 2: 247–57.

Beauchamp, Wendy D., Thomas D. Nudds, and Robert G. Clark. 1996b. "Duck nest success declines with and without predator management." *The Journal of Wildlife Management* 60, no. 2: 258–64.

Cohn, Jeffrey P. 2005. "After the divorce: Improving science at federal wildlife agencies." *BioScience* 55, no. 1: 10–14.

Continental Evaluation Team. 1996. *Annual Report of the Continental Evaluation Team to the NAWMP Committee.* USFWS. Arlington, VA. 11pp. plus appendices.

Cowardin, Lewis M., David S. Gilmer, and Charles W. Shaiffer. 1985. "Mallard recruitment in the agricultural environment of North Dakota." *Wildlife Monograph* 92: 35pp.

Cowardin, Lewis M., Pamela J. Pietz, John T. Lokemoen, H. Thomas Sklebar, and George A. Sargeant. 1998. "Response of nesting ducks to predator exclosures and water conditions during drought." *Journal of Wildlife Management* 62, no. 1: 152–63.

Delta Waterfowl and Wetlands Research Station. 1988. *Evaluation Program. Prairie Habitat Joint Venture. North American Waterfowl Management Plan.* Portage la Prairie, Manitoba. Unpublished Report. 27pp.

Hilborn Raymond and Carl J. Walters. 1992. *Quantitative Fisheries Stock Assessment: Choice, Dynamics and Uncertainty.* Routledge, Chapman and Hall, Inc. New York. 570pp.

Howerter, David W., Michael G. Anderson, James H. Devries, Brian L. Joynt, Llwellyn M. Armstrong, Robert B. Emery, and Todd W. Arnold. 2014. "Variation in mallard vital rates in Canadian Aspen Parklands: The Prairie Habitat Joint Venture Assessment." *Wildlife Monographs* 188: 1–37.

International Association of Fish and Wildlife Agencies. 1997. *Expanding the Reach and Resources of the North American Waterfowl Management Plan.* Public Outreach Project Final Report and Recommendations. Washington, DC. 19pp. (https://nawmp.org/sites/default/files/2025-02/iafwa-nawmp-report.pdf).

Johnson, Fred A., Byron K. Williams, and Paul R. Schmidt. 1996. "Adaptive decision-making in waterfowl harvest and habitat management." *Internation Waterfowl Symposium* 7: 26–33.

Johnson, Fred A., Michael G. Anderson, Richard K. Baydack, Jeffrey W. Nelson, James K. Ringelman, Mark D. Koneff, Robert O. Bailey, Thomas E. Martin, and Clayton Rubec. 1997. "Enhancing biological performance of the North American Waterfowl Management Plan." *Transactions North American Wildlife and Natural Resources Conference* 62: 377–85.

Koneff, Mark D. 2003. *Derivation of Regional Waterfowl Planning Objectives from NAWMP Continental Population Objectives.* USFWS Unpublished Report.

Krapu, Gary L., Pamela J. Pietz, David A. Brandt, and Robert R. Cox. 2000. "Factors limiting mallard brood survival in prairie pothole landscapes." *Journal of Wildlife Management* 64: 553–61.

Loesch, Charles R., Kenneth I. Reinecke, and Charles K. Baxter. 1994. *Lower Mississippi Valley Joint Venture Evaluation Plan.* NAWMP. Vicksburg, MS. 34pp.

Miller, Michael R., M. Robert McLandress, and Frederick A. Reid. 1996. *Central Valley Habitat Joint Venture Evaluation Plan.* NAWMP. Sacramento, CA. 67pp.

Nichols, James D., Fred A. Johnson, and Byron K. Williams. 1995. "Managing north American waterfowl in the face of uncertainty." *Annual Review of Ecology and Systematics* 26: 177–99.

Pahl-Wostl, Claudia. 2009. "A conceptual framework for analysing adaptive capacity and multi-level learning processes in resource governance regimes." *Global Environmental Change* 19, no. 3: 354–65.

Reynolds, Ron E., Terry L. Shaffer, Randy W. Renner, Wayne E. Newton, and Bruce D. J. Batt. 2001. "Impact of the conservation reserve program on Duck recruitment in the U.S. Prairie pothole region." *Journal of Wildlife Management* 65: 765–80.

Sharp, David, Michael Anderson, Robert Blohm, Jeffrey Nelson, Frank Dunstan, Robert Bailey, and Michael Tome. 1992. *Evaluation Strategy North American Waterfowl Management Plan.* North American Waterfowl Management Plan Committee. Unpublished Report. 14pp.

Smith, Graham W. 1995. "A critical review of the aerial and ground surveys of breeding waterfowl in North America." *Biological Science Report* 5. U.S. Department of the Interior. National Biological Service. Washington, DC. 252pp.

Walters, Carl J. 1986. *Adaptive Management of Renewable Resources.* Macmillan Pub. Co. New York. 374pp.

Williams, Byron K., Mark D. Koneff, and David A. Smith. 1999. "Evaluation of waterfowl conservation under the North American Waterfowl Management Plan." *Journal of Wildlife Management* 63: 417–40.

14 2004 Update – Biological Foundations Take Center Stage

With the NSST formed and preliminary inquiries requested by the Plan Committee underway, an *ad hoc* group met ahead of the June 2001 PC meeting to consider the content and thrust of what we optimistically hoped would be a 2003 Update. PC cochairs David Smith and Bob McLean were there along with Steve Wendt, Ken Cox, Seth Mott, Rex Johnson, and me. We pondered the merits of a new 15-year Plan vs. another five-year update. We kicked around content ideas and tag lines. We agreed with Seth that most of the feedback we had been hearing was a desire for the Plan to provide more leadership on priorities for continental waterfowl conservation. We agreed that another key message should be "NAWMP is here to stay!" to settle the question of Plan's duration after the first 15 years. We also floated the idea of writing two documents: a summary for policy makers and a broad audience, and a more technical document to provide the advice we hoped to deliver to the technical community.

Soon thereafter, the Plan Committee formalized the Update Steering Committee (USC) and Writing Team.[1] Our first task was to prepare an outline for the PC in July and by November, an expanded outline to be circulated broadly to Plan partners – celebrating accomplishments but describing the work ahead. We received a couple of dozen useful responses from individual states, joint ventures, and NGOs that were generally supportive of our plans. The idea of paired documents was sustained too so our team prepared for a deeper dive into the chosen themes.

What Are These "Biological Foundations"?

After the first meeting of the Update Steering Committee, Seth Mott, Rex Johnson, Steve Wendt, and I were asked to describe more explicitly what we had been talking about! What did we mean by NAWMP's biological foundations and what did we mean by adaptive management for NAWMP? Because this was central to the approach to conservation under the Plan, I quote at length what we offered:

> The <u>biological foundations</u> of NAWMP comprise the ecological understanding of factors that affect the distribution and abundance of waterfowl, and especially the linkages between landscape elements (e.g., water abundance, land use, habitat quality) and waterfowl birth rates, death rates and population change.
>
> NAWMP <u>population objectives</u> are set based on historic abundances of species and consensus among waterfowl stakeholders about desired populations levels that: 1) ensure population viability; and 2) provide for subsistence harvest, sport harvest, and other forms of public enjoyment. Once established, population objectives in turn determine what habitat resources are needed and what harvest levels are permissible to sustain each population over the long term, recognizing inevitable and uncontrollable climatic variation.

DOI: 10.4324/9781003732310-14

> Knowledge of the linkages between landscape features, habitat variability and population change is central to science-based wildlife management. The cost-effectiveness of conservation actions depends crucially on providing the right resources, in the right places, for each species, thus sustaining populations throughout their annual cycle. The elements that allow managers to improve performance therefore include the models (hypotheses) about factors affecting waterfowl population change, hypotheses about ecological factors limiting population growth, and other biological assumptions made during conservation planning.

Because of a rich scientific history and extensive practical management experience, the Plan had a broad foundation on which to build conservation plans. This foundation varied greatly, however, among species and geographic regions. For instance, we knew a great deal more about mid-continent mallards than we did about King Eiders in the central Arctic or masked ducks in Mexico. Rex Johnson recalled (pers. comm. May 2025) that while it was understandable that we focused on building out the biological foundation for mallards, we underestimated how long it would take to lay the foundational fundamentals for birds nesting outside the prairie pothole region.

Despite the Plan's strong foundation, waterfowl live in a complex and ever-changing world. Their habitats are under unrelenting pressures from human development and impacts like climate change. Managers must make conservation investments in the face of uncertainty about the impacts of their actions on populations. It is vital, therefore, to continually strengthen the scientific basis for management actions to achieve the objectives of the Plan. Mainly, this means integrating closely the traditionally separate work of biological planning, implementation, and evaluation in cycles that enable managers to adapt to new knowledge and changing circumstances.

The term *adaptive management* is used in many ways. The Steering Committee used it in a broad sense to describe the process of employing iterative cycles of planning, implementation, and evaluation to improve management performance. This was intended to be a more inclusive concept than adaptive resource management which typically involves formal optimization of alternate management actions based on competing mathematical models, as in annually updated Adaptive Harvest Management for waterfowl, allowable fish catch, etc. In contrast to traditional isolated research, we regarded adaptive management as the explicit use of conservation program delivery to facilitate learning and subsequent improvement of program performance over time. To enable progress, such conservation programs must have clear quantifiable objectives, predictions about the outcomes of alternative management actions, monitoring procedures to measure the outcomes, an evaluation process to compare outcomes with the original predictions, and finally, recommendations for future delivery of the conservation program. Predictions may comprise quantitative models or just simple conceptual models. The key is to view planning, delivery, monitoring, evaluation, and replanning as an integrated whole.

NAWMP's broad use of adaptive management is perhaps closest to Kai Lee (1993) who described it as,

> An adaptive policy is one that is designed from the outset to test clearly formulated hypotheses about the behaviour of an ecosystem being changed by human use. In most cases, these hypotheses are predictions about how one or more important species will respond to management actions.

More traditional approaches to waterfowl research (e.g., asking what factors limit duckling survival of northern pintails?) or field-testing new conservation ideas before they are widely applied remain useful for generating insights, some of which will eventually be subjected to adaptive management cycles.

A NAWMP Science Forum

As work on the new update began, the NSST and the Plan Committee organized a major NAWMP Science Forum in January 2002 at the National Wetlands Research Center in Lafayette, Louisiana. The purposes were to share knowledge of how evaluation and replanning have improved JV effectiveness already, to enhance the dialog among Plan Committee, NSST, and joint venture members about biological foundations, and to identify important knowledge gaps that should be featured in the 2004 Update.

Ron Reynolds, the NSST Chairperson, reported on how evaluations in the prairie pothole region have improved program delivery there. Speakers from other breeding areas, migration areas, and wintering joint ventures followed. Then reports from Mexico and the species JVs were shared. Rex Johnson pulled together the common themes. Karla Guyn and Stuart Slattery promoted new species action groups for northern pintails and scaup, respectively, and I talked about climate change and waterfowl conservation. Monitoring needs (Dale Caswell) and progress in refining objectives (Koneff) rounded out the presentations. This was followed by breakout sessions to identify key action themes for 2003 and beyond. The three-day event was the broadest and most in-depth workshop on the biological foundations of NAWMP ever undertaken and certainly the deepest dive into technical issues for the Plan Committee in its nearly 16-year history.

We emerged with a sober but helpful sense of the many operating uncertainties in the NAWMP enterprise, the importance of continuing to share information and ideas across the community, and the value of enhancing the biological foundation knowledge proposed for the coming Update.

After digesting the results, the NSST produced its first substantive set of recommendations for the Plan Committee. It recommended strengthening the role of the Plan Committee itself as stewards of the Plan and as active advocates for informed conservation actions. These included actively supporting and guiding the NSST, representing Plan interests with NAWCA, NAWCC, NABCI, and the Flyway Councils, organizing more forums for creative thinking like the Lafayette meeting, and encouraging ongoing progress in five-year updates. Fostering iterative cycles of planning, implementation, and evaluation within JVs and at the continental level was strongly recommended too. All this seemed important while also paying heed to species and places of special concern (e.g., sea ducks) and the overarching need for enhancing our abilities to measure landscape change. Collectively, this was a tall order indeed! Rex Johnson vividly recalled (pers. comm. May 2025), "feeling anxiety back then about the magnitude of the challenge we laid out and how few of us were available to work on these tasks".

Species and Geographic Prioritization

The Lafayette forum and partner consultations thereafter affirmed the desire to consider species and geographic prioritization in conservation investments. Rex Johnson (FWS) and his team had produced in December 2000 a compendium of waterfowl harvest in the conterminous United States from 1980 to 1999. With county-specific harvest data for each species, and tables aggregating the data by BCRs, it provided a rich pattern of recent harvest distribution. Following the Lafayette forum, Rex's team, the USC, and the Plan Committee worked through several options for determining and communicating conservation priorities. It proved to be complicated!

In the end (NAWMP Plan Committee 2004b), two approaches were used. First, continental species priorities were created using a combination of population trends and US harvest data. Population trends were assessed by WBPHS trends from 1970 to 2002 and species classified as increasing, stable, unknown, or decreasing. Importance in the harvest was categorized as high, medium, or low. Goose and Swan populations were assessed relative to Plan objectives. From this analysis, mallards, black ducks, lesser scaup, northern pintails, common eiders, and several Canada goose stocks emerged as the highest taxonomic priorities.

Second, suggested priorities for breeding birds within regions were developed using WBPHS data with Breeding Bird Survey data to supplement for species poorly surveyed by the WBPHS. The NSST then reviewed habitat threats in the regions and used expert opinion to rank these in five categories from high to very low risk. For nonbreeding waterfowl, abundance by region was estimated using mid-winter survey data for the United States and Mexico and expert opinion for Canada. For mid-latitude migration areas, geographic distributions of harvest estimates were used to provide relative importance rankings. Arranged by BCRs and individual species, this analysis resulted in an unwieldy collection of 16 dense tables for breeding and nonbreeding ducks and five more tables for geese and swans. The authors urged that planners treat these results as a starting point and not a rigid prescription for regional planning.

For reasons that are not altogether clear today, an attempt to aggregate geographic priorities for species by BCR was begun but not carried through to the final 2004 Update. Granted, there were many assumptions needed to feel confident about the results, but I suspect mainly it wasn't politically comfortable to promote what emerged as a very heavy weighting toward breeding-area BCRs. Geographical prioritization was at odds with the contemporary desire of many partners to cover the continent with JVs, a goal boosted by the emergence of "all-bird" JVs. Rex Johnson recalled (pers. comm. May 2025) that the Plan Committee and some partners "disliked our approach that suggested some JVs were less important than others, a view much at odds with the prioritization investigation that the Update team had been asked to undertake". It became a contentious process. Thus, the product in the 2004 Implementation Plan proved useful within the joint ventures for waterfowl conservation planning, if not very impactful at the continental scale.

More recently, Roberts et al. (2023) revised this prioritization schema by considering newly available biological and social data along with the broadened goals of the 2012 Plan, which I will return to in Chapter 23.

Drafting and Consulting

In April 2002, the PC charged the writing team with beginning the technical document first, a product targeting mostly JV technical practitioners. The marketing-focused strategic framework would follow. The PC urged the writing team to be bold about rigorous prioritization of investments within joint ventures and seek to strengthen regular dialog and reporting of JVs to the PC. The writing team sent a first draft to the Plan Committee on July 12th. After a quick turnaround by the PC, a draft for public comment was circulated in August and extensive comments were received in the fall.

That first loop once again reinforced the value of such consultations for developing consensus. Feedback was generally positive. People applauded the intention for the PC and NSST to enhance communications with the JVs and agreed with the value of more JV-JV contact as well. The idea grew of NAWMP as a "learning community" (Bennis and Biedermann 1997) sharing diverse experiences to help move the partnership forward. Several reviews offered that in talking about adaptive management we needed to recognize that cyclic replanning needed to apply to policy and socioeconomic efforts too, not just habitat programs. In fact, there was general concern that we underplayed the importance of agricultural policy and other policies in our quest. The recent CRP experience in the Dakotas seemed instructive, for example. Others noted the potential value of touting more loudly the ecosystem benefits flowing from waterfowl projects. With this feedback, the writing team set about revising the technical document, now called the "Implementation Framework", and drafting the policy document, now described as "Strategic Guidance". After running these by the Plan Committee in May, both were released for a final round of partner consultations in July.

A Broader Active Role for the Plan Committee

Meanwhile, two other teams were at work updating the terms of reference for the Plan Committee itself and building a plan for actioning the main recommendations of the Update. Over a series of meetings, the cochairs, now David Smith and George Finney along with Scott Yaich, Seth Mott, Danielle Bridgett, and I reviewed the original Terms of Reference for PC and worked through several drafts trying to capture the desire of the PC for a more proactive role and one responsive to the needs laid out in the developing Update. Smith and Finney also stayed in touch with the evolving leadership in SEMARNAT where Felipe Ramirez Ruiz emerged as the third PC Cochair.

We landed with the statement included in the update (see pp. 84–85), which reaffirmed that the Plan Committee, an international body, exists to provide leadership and oversight for the activities undertaken in support of the Plan. It posited two broad functions described as leadership (with seven explicit roles) and Plan management (with 12 explicit roles). This guidance remained unchanged until lightly amended in the 2024 Plan Update.

We recommended that membership remain at 18, 6 each from Canada, the United States, and Mexico, selected from agencies responsible for waterfowl management in those countries and appointed by the directors of the federal wildlife agencies. The Plan Committee, in turn reports through the NABCI Canada Council, the Director of the USFWS, and secondarily the four Flyway Councils in the United States, and through the Director General de Vida Silvestre of SEMARNAT.

The review team also discussed desired skill sets and linkages of PC members to achieve these wide-ranging goals but chose to share those thoughts with the three national cochairs rather than include recommendations for changes in the Update. Soon afterward, several of those recommendations for diversifying member skill sets were acted upon, first in Canada and Mexico.

A Comprehensive NAWMP Progress Assessment

A series of stimuli including the 1999 Joint Venture symposium in California, early work of the NSST, the 2002 NAWMP Science Forum in Louisiana, and critiques of the first draft of the 2004 Update (especially Bob Bailey's letter from the Delta Waterfowl Foundation), set the wheels in motion to design a first comprehensive continental assessment of NAWMP's work. I recall vividly a meeting in Ottawa in April 2003 with George Finney, David Smith, Danielle Thiebeault, Scott Yaich, and Seth Mott that began with reviewing PC roles but switched into a creative flip-chart session trying to give form and function to a continental progress assessment. We beat around purpose, scope, content, logistics, an "International Assessment Team", a plausible schedule, and vitally, how might the PC use the results of such an assessment to advance waterfowl conservation? We publicly vetted the idea at the November 2003 Duck Symposium in Sacramento. The framework was developed further in Chicago in May 2004 when the Plan Committee, several joint ventures, and the NSST met together. The assessment that evolved was vital to the evolution of NAWMP and is the topic of Chapter 15. But first, the conclusion of the 2004 Update.

The First Twin NAWMP Updates

In final form, the 2004 Update comprised two documents: The "Strategic Guidance" document (NAWMP Plan Committee 2004a) was intended for policymakers, a more general audience, and for the required political signatures from Canada, Mexico, and the United States. The "Implementation Framework" (NAWMP Plan Committee 2004b) was intended for active Plan partners. The three so-called vision statements from the 1998 Update were retained with a little tweaking:

1 Plan partners define and attain the landscape conditions needed to sustain waterfowl populations,
2 Plan partners forge broad alliances with other conservation efforts and communities to achieve Plan objectives, and
3 Plan partners continually improve the biological foundations of waterfowl conservation.

These were wrapped up under a new, succinct purpose statement: *The purpose of the Plan is to sustain abundant waterfowl populations by conserving landscapes, through partnerships, guided by sound science.*

Plan population goals remained unchanged from 1986, but with added information on recent and long-term trends for each species. Joint Venture habitat objectives were mostly carried over from 1998, although some were adjusted to reflect recent JV planning. Mostly, however, this section carried a strong caveat (p. 45) that these habitat objectives were not uniformly compelling, and in need of further study to link population and habitat objectives with greater confidence.

Encouragingly, it laid out a vision for the Plan partnership as a "learning community". It envisioned the Plan's adaptive approach to management succeeding if joint ventures continued to expand their capacities for regional planning, implementation, and evaluation and then reported what they learned throughout the Plan community ensuring that partners learn from one another. A "learning community" also requires freedom to innovate, accompanied by a disciplined approach to monitoring and assessment. Leadership that values learning is essential as is sharing what each team or individual learns. Everyone thus has a role in leading the partnership forward.

Box 14.1 Learning Together

None of us is as smart as all of us.

(Bennis and Biederman 1997. Organizing Genius)

Rex Johnson remembers this as a period of rapid evolution among JVs, and as with any period of change, there were challenges, most acutely between JVs with biological planning and evaluation capacity and those without. He recalls that some of the "have nots" were unwilling to spend the resources to build science capacity and responded by expanding their boundaries and more actively embracing NABCI as some felt they were viewed as of secondary importance to NAWMP. The PC and NSST had to hold this diverse group of JVs together.

The update also included the announcement of the new Pintail Action Group, stimulated by a workshop in 2001 and approved by the Plan Committee in May 2003 as a focus group, different from a species joint venture but still meriting NAWMP support (Chapter 8). The revised plan also signaled the Plan Committee's intention to launch a first-ever comprehensive NAWMP Progress Assessment in 2005 in partnership with the joint ventures and the NSST.

Near the end of production, the draft continental map, shaded to indicate relative importance to waterfowl based on the BCR-level population trends and threat analysis by Rex Johnson and his NSST team, was replaced with an updated version of the original 1986 map simply depicting, "areas of continental significance to North American ducks, geese, and swans". This was a disappointment to many of the authors of the 2004 Implementation Plan but was consistent with the PC decision to back away from the geographic prioritization listing discussed above. The 1986/1994/1998 map based on expert opinion had some value but lacked consistency and eschewed data-driven prioritization. Some places like the prairie pothole region were depicted with BCR boundaries. For others, like the US Great Basin, only large wetland complexes were identified. This change came after overwhelming objections from some Flyway Councils and joint ventures to using BCRs for this purpose. Renaming BCRs as WCRs (waterfowl conservation regions) accommodated the grumbling, but spatially they were essentially identical.

Mexico remained committed to signing the Update despite a change in leadership at SEMARNAT that limited their participation in the process. Unfortunately, however, there was not yet a finished and approved Mexican waterfowl conservation plan to append to the Update as all had hoped earlier in the year.

Changes to the revised Implementation Framework followed affirmation of its intended technical audience, and it expanded to 120pp with several useful appendices including:

- a A recommended approach to regional-scale model-based strategic planning and evaluation written by Rex Johnson et al., a precursor to their comprehensive guide to strategic habitat conservation (USFWS 2008; Johnson et al. 2009) that was later widely adopted by the USFWS and still serves today as guidance for many US Joint Venture implementation planners.
- b The continental and regional species prioritization analyses described above.
- c Revised terms of reference for the Plan Committee.
- d A discussion of continental monitoring needs from the NSST.

With a dash of bravado, both documents concluded with this:

> With this 2004 Plan, the Plan community reinforces its unwavering commitment to waterfowl conservation and particularly to the central role of science in guiding Plan actions. The adaptive processes advocated here offer a clear path to success, even in the face of ecological and sociological uncertainties. Our continent's spectacular waterfowl have a bright future if we continue to strive on their behalf. We have a solid Plan. We have a history of achievement. A future of waterfowl in abundance is now ours to secure.

Lastly and appropriately, the Update noted that the Plan partnership had lost two champions during the work on the 2004 Update, Dr. James H. Patterson and Mr. Kenneth W. Cox, and the Update was dedicated to their memories.

Box 14.2 NAWMP across My Career

By Seth Mott, USFWS Retired

Ed. Note: Seth's 30-year journey from JV Assistant Coordinator to international Plan Committee member offers a unique perspective on NAWMP.

In the mid-1980s, I was the Chief of the Game Division in the Mississippi Department of Wildlife Conservation. It was at a meeting of the Mississippi Flyway Council that I first heard about the North American Waterfowl Management Plan, and I was excited to help our agency engage in implementation.

The Lower Mississippi Valley Joint Venture (LMVJV) launched in 1987 and program planning accelerated. Charles Baxter was selected as JV Coordinator and stationed in Vicksburg, MS. I had worked with Charles in the past and in 1989 left my state job to work as his Assistant Coordinator. In my opinion, the next decade was the "golden age" of NAWMP. It was very rewarding to meet and work with partners who shared my enthusiasm for Plan goals. As a high priority for many conservation organizations, the NAWMP enterprise expanded across the continent. During this period, three features evolved that I view as hallmarks of the Plan.

The first was recognizing the importance of private lands. The 1986 Plan acknowledged that long-term solutions for waterfowl conservation "will require the coordinated action of governments, private organizations, landowners and other citizens". In the United States, the 1985 Farm Bill, with its conservation provisions, helped foster a desire in the agricultural community for involvement with habitat projects. For example, Ducks Unlimited Inc. solicited donations of used steel pipes from oil and gas companies. The pipe was fabricated into water control structures with grant funds from NAWCA and JV partners, then distributed to landowners by state and federal wildlife agencies. In return, landowners signed ten-year agreements to winter-flood harvested croplands or natural wetlands. Since those days, every JV has engaged in innovative projects and partnerships that advanced NAWMP goals while addressing local conservation priorities as diverse as aquifer recharge, sea-level rise, salmon restoration, and urban open space. Today, these values joint ventures provide to local communities help ensure that NAWMP endures.

The second feature was a focus on biological foundations. All JVs were charged to produce a strategic plan. The LMVJV plan, among the most rigorous, included quantified objectives for how much, what kinds, and where habitat for wintering waterfowl was needed. Fortunately, existing research was available for building that framework but more work on associated uncertainties was needed to turn our initial framework into a solid foundation for conservation. The LMVJV was among the first to include research, monitoring, and evaluation in our strategic plans.

I call my third feature accountability and assessment. From the beginning, NAWMP leadership stressed the need to report Plan accomplishments to agency executives, legislators, and funding partners. Through the 1990s and beyond, such reporting consisted mostly of tabulations of dollars expended and acres impacted. Only later were some JVs able to assess their effects on waterfowl populations directly.

Due to the "partnership" approach of NAWMP, almost every project had multiple participants contributing cash, in-kind support, or staff time. As Assistant JV Coordinator,

I had responsibility for soliciting and compiling accomplishment information from the ten states of the FWS Southeast Region, then forwarding results to the national office in Washington. Initially, I used pencil and paper forms, a calculator, and a fax machine to do this. The NAWMP Office soon recognized the need for an automated system and thus was born the International Tracking System (ITS). ITS was a custom-programmed database that had the capability of compiling, sorting, storing, and searching accomplishment and funding data from across the country. It was a big step forward, but initially, I still had to do the data entry from paper forms!

Following the 1994 Update, the US NAWMP Office directed JVs to include monitoring and evaluation as part of their plans. Some JVs, including the LMVJV, had already done so, but others preferred to focus attention and funding solely on habitat projects. Almost every meeting of JV coordinators included discussions on this topic. Reaching consensus on the value of monitoring and assessment was relatively easy; finding funds and people to do the work was a real problem.

In 1999, I left the LMVJV for the national NAWMP Office (later renamed the Division of Bird Habitat Conservation), first as a JV program specialist and later as Branch Chief responsible for NAWMP and joint ventures. There, the scope of my responsibilities expanded to national and international scales and I began investing more of my time in biological foundation and assessment matters along with added administrative duties.

As the century closed, so did the 15-year planning horizon of the original NAWMP. It was time for the new Update described in this chapter. Coordinating and writing this update became a large part of my responsibilities. With its completion, my attention turned to organizing and coordinating a comprehensive continental assessment of progress toward achieving Plan goals (Chapter 15). Most importantly, we sought a critical assessment of Plan effectiveness to inform adaptive program adjustments. The Plan Committee also wanted an evaluation of the effectiveness of NAWMP institutions and processes. To boost credibility of the assessment, we empaneled waterfowl experts from outside the federal government as an independent Assessment Steering Committee.

In large part, the final report affirmed that NAWMP and the Joint Ventures were serving as a cohesive force for conservation, impacting more than 13 million acres across North America. However, there was only scant evidence about how those acres, apart from any effects of hunting and natural environmental variation, had improved continental waterfowl populations. This was no surprise as the continental Plan never had an experimental design intended to test its biological success. Soon after release of the report, I was interviewed by a national outdoor writer who asked if the assessment "proved" NAWMP has increased duck numbers; yes or no. I said "no" and he had his headline! Nonetheless, in the face of continuing loss of wetlands and grasslands to agriculture and other developments, there was broad acceptance that NAWMP was a vital endeavor and should continue. Recommendations in the Assessment Report (Chapter 15) became focal points for future NAWMP actions.

All the work described above was a prelude to my most exciting NAWMP experience, the Joint Task Group (JTG) (Chapter 16). The impetus for creating the JTG was a desire by managers for greater certainty about the effects of both harvest and habitat management actions. The NAWMP Assessment and Adaptive Harvest Management reviews had exposed ambiguity, and potential conflict between harvest and habitat objectives for waterfowl. The JTG comprised 13 eminent waterfowl scientists and managers from across the United States and Canada, plus me. We were charged with exploring options and recommending solutions for reconciling the use of waterfowl population objectives for both harvest and habitat management. I'd characterize our discussions in the JTG as an effort to create a "unified theory" for waterfowl management by creating models that could explicitly link changing habitat conditions with waterfowl vital rates and hunter satisfaction. We used the word "coherence" for this desired future state of waterfowl conservation. I found our debates and discussions to be mentally challenging and enlightening, and we were satisfied that our final report was a significant contribution to our profession. As described later in this volume, the JTG report and ensuing workshops and surveys laid a foundation for the 2012 Plan Revision and subsequent NAWMP efforts for the next decade.

In 2009, I was asked to serve as the acting Science Advisor to the FWS Director while a search was conducted to fill the position permanently. Once the new Science Advisor was in place, she asked me to stay, and so in 2010, I became Deputy Science Advisor. A new President and Congress had recently made climate change science, mitigation, and adaptation national priorities and as a result federal agencies received substantial funding to pursue these concerns. The bulk of this new funding was funneled through our office and directed to creating Landscape Conservation Cooperatives or LCCs. The LCCs were modeled on the NAWMP joint ventures but had broader goals of preserving biodiversity in the face of accelerating climate change. Not surprisingly, the creation of these new landscape partnerships overlaying the existing network of joint ventures resulted in some confusion and conflict. Ultimately, the JVs kept their traditional identities, working with established partners to address conservation opportunities through the lens of the NAWMP and other bird plans, while the LCCs funded mainly climate-related research. Though a few LCCs shared funding and data with the JVs, the two groups were often perceived as distinct initiatives. I had hoped to encourage more collaboration between the two programs, but my efforts were largely limited to maintaining friendly relations in both groups. In 2016, a new administration changed federal priorities again, with LCC funding redirected to non-climate science needs. The LCC network eventually crumbled but the JVs endured. Why? The JVs were partnerships built first on shared goals and were cooperatively funded. Conversely, the LCCs were built around a funding source seeking partners.

Soon thereafter, the US Cochair of the Plan Committee asked me to accept the vacant position on the Plan Committee reserved for a second FWS administrator. That gave me a platform to stay involved with NAWMP following the JTG report and 2012 Plan Revision. Many of the recommendations I helped write in the JTG report and Plan Updates were directed to the Plan Committee for action and now I was positioned to help address them. The waterfowl management community had responded favorably to the JTG's call for "coherence" among the goals for waterfowl numbers, habitat, and hunter satisfaction and several meetings, surveys, and working groups ensued. Unfortunately, concurrent recommendations for adding technical capacity to support new approaches were mostly fruitless. The US JVs had received some budget increases that helped expand the NSST, but other funding for waterfowl work was flat or declining. The Science Office had some funds available for unmet science needs, and I suggested it as potential match for funds from Canada or the states, but to no avail.

This was one example of waning influence of the Plan Committee. While my fellow PC members remained enthusiastic about NAWMP, membership on the Plan Committee had been delegated downward in agency hierarchies and NAWMP funding needs competed with a host of other issues. Bluntly, duck numbers and duck hunting were crises of the 1980s. Climate change, endangered species, and wildlife diseases gained the conservation spotlight in the 21st century. A personal example of this trend played out later when a new Science Advisor decided that I should spend all my time working on Science Office issues, resulting in my resignation from the Plan Committee.

I retired from FWS in 2018, more than 30 years after I first heard of the NAWMP. The best thing about NAWMP for me was the hundreds of people I met and worked with in some capacity. For most of those years, Mike Anderson was a prominent colleague, so I readily agreed when he asked for help in preparing this book. In doing so, I reviewed Plan and JV documents since my retirement and was heartened to see that NAWMP and the Joint Ventures remain vigorous and innovative today.

Note

1 Seth Mott (FWS), Steve Moran (Rainwater Basin JV), Jon Andrew (FWS), Rex Johnson (FWS), Scott Yaich (DUI), Eduardo Carrera (DUMAC), Ken Cox, George Finney, and Steve Wendt (all CWS), and me.

References

Bennis, Warren G. and Patricia W. Biederman. 1997. *Organizing Genius: The Secrets of Creative Collaboration.* Perseus Books. New York. 239pp.

Johnson, Rex R., Charles K. Baxter, and Michael E. Estey. 2009. "An emerging agency-based approach to conserving populations through strategic habitat conservation." Pages 201–24 in Joshua J. Millspaugh and Frank R. Thompson (Editors). *Models for Planning Wildlife Conservation in Large Landscapes.* Elsevier Press. Burlington, MA. 624pp.

Lee, Kai N. 1993. *Compass and Gyroscope – Integrating Science and Politics for the Environment.* Island Press. Washington, DC. 255pp.

North American Waterfowl Management Plan, Plan Committee. 2004a. *North American Waterfowl Management Plan 2004. Strategic Guidance: Strengthening the Biological Foundation.* Canadian Wildlife Service, U.S. Fish and Wildlife Service, Secretaria de Medio Ambiente y Recursos Naturales. Ottawa. 22pp.

North American Waterfowl Management Plan, Plan Committee. 2004b. *North American Waterfowl Management Plan 2004. Implementation Framework: Strengthening the Biological Foundation.* Canadian Wildlife Service, U.S. Fish and Wildlife Service, Secretaria de Medio Ambiente y Recursos Naturales. Ottawa. 106pp.

Roberts, Tony, Joshua Dooley, Alan Hanson, Kate Martin, Kyle Spragens, and Greg Yarris. 2023. "North American Waterfowl management plan species prioritization – 2023 revision." *North American Waterfowl Management Plan Science Support Team Technical Report* 2023-01. 13pp.

United States Fish and Wildlife Service. 2008. *Strategic Habitat Conservation Handbook A Guide to Implementing the Technical Elements of Strategic Habitat Conservation (Version 1.0).* Report from the National Technical Assistance Team February 11, 2008. Washington, DC. 24pp.

15 A Continental Progress Assessment

Work on the 2004 Update (Chapter 14) convinced Plan partners to pursue a comprehensive assessment of NAWMP accomplishments and information needs before beginning the next update. Such a continental assessment would build on the biological foundation elements highlighted in 2004. The project, one of the most ambitious, time-consuming, but ultimately valuable efforts ever sanctioned by the Plan Committee, occupied a group of experienced waterfowl scientists and managers from 2005 to early 2007. It also engaged the technical teams in every Joint Venture (JV) and national waterfowl managers.

Previous reports on Plan progress consisted mostly of tallies of acres conserved and dollars expended, comparisons of recent waterfowl survey results with Plan objectives, or regional assessments of JV accomplishments. More critical examinations of the Plan included a status assessment by the JVs and the Continental Evaluation Team (1996–1997) in an unpublished Technical Companion Document drafted along with the 1998 Update, and reports of iterative planning cycles in a few of the habitat JVs. A review published by Ken Williams et al. (1999) identified gaps in Plan evaluations to date. Critically, they noted weakness in our ability to relate changes in habitat abundance with effects on waterfowl populations. Thus, as of the late 1990s, tracking of Plan performance had been primarily an accounting of partnerships, funding, and the geographic distributions of activities. When compared to the summed habitat goals of the 1986 Plan, <30% of the Plan's habitat goals had been attained at a cost of about 90% of the funding target, but the partners could not assess with confidence what those changes meant to waterfowl populations.

Most JV evaluations had studied impacts of conservation at local-regional levels and most of those had occurred in breeding areas. Other analyses sought correlations between duck abundance and habitat conditions. Those assessments were helpful but limited. Given NAWMP's scale, conventional statistical design approaches were problematic leaving partners with little ability to predict population responses to conservation actions. Williams et al. (1999) concluded that the Plan partners needed: (a) broad-scale habitat monitoring that focuses on biologically relevant landscape patterns; (b) improved population monitoring for some species in some areas; and (c) a much more refined understanding of the functional associations between populations and habitats. Understanding such associations seemed essential for informed conservation decisions in the future.

Plan leaders recognized that a comprehensive, continent-wide assessment of progress toward achieving NAWMP biological goals must be an undertaking of the entire NAWMP partnership and could be accomplished only by the JVs working with the Plan Committee and the NSST. The endeavor needed to encompass the full scope of NAWMP activities, including the institutional relationships among the JVs, the NSST, and the Plan Committee itself. As advocated in the 2004 Update, partners also needed to assess the status of adaptive processes necessary to ensure regular improvements over time.

DOI: 10.4324/9781003732310-15

Purpose and Desired Outcomes

Results of the continental assessment were expected to inform the next update by clarifying the most vital needs for management actions; identifying additional resources needed for implementing conservation solutions; and importantly, allowing the Plan Committee to share with financial stakeholders a set of compelling science-based recommendations for future investments. Because the JVs delivered nearly all Plan programs, it was vital for this assessment to also stimulate critical thinking and self-evaluation by the JVs about how they might enhance their effectiveness for waterfowl.

Desired outcomes therefore included:

1 Regional and continental accounting of progress toward achieving the biological goals of NAWMP.
2 Renewed regional and continental population objectives and estimates of the landscape conditions necessary to achieve those objectives.
3 Adaptive processes of planning, implementation, and evaluation in place and advancing throughout the Plan partnership.
4 Improved geographic and programmatic targeting of conservation investments.
5 Relationships among the key institutional components of NAWMP (Plan Committee, NSST, JVs, NAWCA Councils, and Flyway Councils [FWC]) are renewed, strengthened, and clarified to help achieve Plan goals.

To begin engaging the JVs, a standard matrix of questions was developed by the PC and NSST focused mainly on biological foundation matters, including accounting for progress toward biological goals. A comparison of current habitat conditions with those future conditions deemed necessary for attaining Plan goals was also requested. The PC also wanted to know what adaptive processes were in place presently at various spatial scales, and what the current working relationships were among the Plan's institutions. Lastly, we sought thoughts on whether these elements need to be renewed, strengthened, or clarified for partners to achieve Plan objectives.

The Assessment Team Comes Together

A time-limited International Assessment Steering Committee (hereafter ASC) was recruited by the Plan Committee to lead the effort. It included members from the Plan Committee, the JVs, and the NSST, as well as other individuals with strong scientific or institutional links to Plan activities.[1] A full-time Coordinator (Dave Paullin, USFWS and former Central Valley Coordinator) was hired to help herd the process including leading efforts to gather information needed for the assessment, orchestrating sessions with all the JVs, and coordinating the compiling, analyzing, and reporting of results. Dave was an inspired choice. His credibility within the JV community helped greatly with engaging that key audience, and he had a knack for helping us work through diverse opinions and building consensus, even if we drove him a little crazy for 23 months!

The ASC was first tasked with fleshing out an operating plan and settled on a four-step process beginning with (1) submission of a questionnaire to the JVs and other stakeholders, followed by an internal ASC review of the responses, then (2) a face-to-face meeting with Paullin, a subset of ASC members, and JV members (usually biologists, planners, and the JV Coordinators). The results of those written and in-person inquiries were then aggregated, subjected to peer review by other ASC members who were not part of the live meetings, thus allowing (3) the development of a draft set of conclusions and recommendations for each JV, and (4) a synthesis report for the Plan Committee and NAWMP partners broadly.

The ASC met for the first time at the National Conservation Training Center (NCTC) in Shepherdstown, West Virginia in April 2005.

It felt like a great team – we had Plan veterans, new faces, academics, people with JV and Flyway experience, all three nations, and everyone motivated to attain the goals of the project. Our charge came from both the PC and the NSST, and with Paullin's leadership the team worked together splendidly.

At our inaugural meeting, Dave Smith provided historical context for our work. Seth Mott reviewed the path taken by evaluation ideas since 1986. Then, I provided an overview of antecedents to the Assessment and the Assessment framework as developed to date. I also shared a related presentation and discussion on "NAWMP Population Goals and Harvest Management: Conflict or Coherence?", a matter we deferred to a different process (Chapter 16). To provide sideboards for our work, we agreed that it would not be ASC's task to revise the NAWMP goals but to focus on our collective success in achieving goals to date.

We debated how to divide the workload and settled on Dick Bishop's suggestion that two to three members of the ASC should attend each of the live interviews, including someone not familiar with the JV being interviewed. Bob Clark urged us to undertake a test case first that could help us refine our approach to the rest. We discussed at length who might serve on which interviews, striving for two or three ASC members plus Paullin for participation in each, plus Seth Mott whose task was connecting the ASC to the FWS national office attending as many as possible. We also drafted processes for partner feedback and an approach to synthesis when the consultations were finished.

The Interviews Begin

After the NCTC meeting, a questionnaire was refined along with a plan for the first in-person interviews. I attended the pilot interviews with the Central Valley, US Pacific Coast, and San Francisco Bay JVs along with ASC members John Eadie and Mark Petrie. We met in Novato, CA on June 15th with the Pacific Coast and San Francisco Bay folks, and in Sacramento the following day with the Central Valley team. We had the distinct advantage that Dave Paullin had been the longtime coordinator of the CVJV, so he understood the issues from both "sides" which was very helpful in navigating those two days. We were also joined by Red Hunt, former Director of California Fish and Game and a member of the original 1986 Plan Committee who added valuable historical perspectives to the discussions. We were grateful for the patience and constructive engagement of our JV colleagues as we bumbled through our first two days of interviews: to Rob Holbrook, Sandy Scoggin, Susan De La Cruz, Fritz Reid, John Takekawa, Carey Smith, Gary Page, and Beth Huning for discussions on Day One; and to Bob Schaffer, Ruth Ostroff, Joe Fleskes, Fritz Reid, Dennis Orthmeyer, and Rob Holbrook on Day Two.

Our ASC team learned much from that experience, both about the three JVs and about the process of enabling an effective learning experience for both the ASC and the JV teams. These three JVs contributed more than they will ever know to the useful outcomes of the continental assessment. Happily, while every JV meeting was unique, the constructive attitudes of the sessions prevailed. In the end, I sincerely believe that we learned from one another, which was an outcome we all hoped for, and I give much credit to Coordinator Paullin for guiding us along that path.

I agree with John Eadie (pers. comm. November 2025) that the ASC process of visiting with every JV moved the bar in ways we hadn't fully anticipated. Because we were serious about the need to move forward but also came to understand the circumstances of each JV, the review process reinforced the community-wide, continental feel of the whole NAWMP effort. We asked, they gave honest feedback, we listened, and we all learned. The process itself was as impactful as the final report and led me to strongly recommend (Chapter 24) investing in the process again.

After 18 months of surveys and interviews and more months of drafts, debates, and synthesis, the final assessment report (Assessment Steering Committee 2007) was delivered to the Plan Community in February 2007, 23 months after the ASC met for the first time. During the Assessment, we reviewed materials acquired *via* written questionnaires, interviews, and supporting materials provided by all the habitat and species JVs, FWC, Mexico, and the Pintail Action Group.

NAWMP programs in Mexico were reviewed in February 2006. At the time, México did not have a delivery structure like JVs in Canada and the US. Mexican Wildlife Conservation Units or "UMAs" are based on individual landowners or local social communities known as *ejidos* and were concentrated in the six northern states. Their important conservation principles included providing long-term economic benefits as well as training and capacity building to local people. Conservation success in México had been measured mostly by the number of UMAs engaged, but measures of habitat quality were lacking. The ASC acknowledged that México was capacity-limited for NAWMP planning and delivery. Nor did the country have well-developed science infrastructure, such as training in monitoring or evaluation protocols. The team subsequently offered several recommendations for consideration in capacity building. The ASC also heard that linkages with other JVs, particularly the Sea Duck and Arctic Goose JVs would be helpful and welcomed by México – so too with the Gulf Coast JV, and the Sonoran JV, which have since developed. Strengthening ties with the NSST seemed desirable too.

The ASC Report

General Impressions

The ASC's report (2007) described our quest as challenging, informative, and rewarding. We learned that the Plan community had affected millions of acres since 1986, but much work remained to be done before the goals for sustaining waterfowl populations would be realized. We were also deeply impressed with the professionalism, expertise, and dedication of the many Plan partners we met and so expressed confidence in our collective ability to move the entire Plan effort closer to attaining its continental waterfowl population objectives. The partners, frankly, made it an inspiring experience.

Paullin mused (pers. comm. April 2025) that the biggest value for him was getting to see the NAWMP enterprise from such varied perspectives. Part of that came from answers to the questions we asked, but mostly, "there was so much to learn from face-to-face interactions between people at all levels and from all places" that the sum of those interactions was deeply revealing. Paullin continued, "NAWMP brought value added to all our enterprises…more energy and more focus. The whole became greater than the sum of its parts". The ASC concluded that the Plan had been a cohesive force, bringing focus to waterfowl and wetland conservation and management efforts in North America. People in the waterfowl conservation partnership remain committed, were better organized, were more coordinated, and getting more done than ever before. These were great strengths to build upon.

The JV organizational model had stimulated an impressive amount of habitat conservation and innovative ways of delivering programs. The Plan by then had influenced over 13 million acres of breeding, migration, and wintering habitat in North America while investing some $3 billion. Notwithstanding these accomplishments, the ASC believed that Plan partners had an opportunity to improve our collective performance. On behalf of the team, I gave the waterfowl scientific community a first look at our conclusions and recommendations at the Fourth International Duck Symposium in Bismarck in August 2006. By then, we had interviewed 22 JVs, spoken in person with 115 JV leaders. We had received from them >500 pages of written responses to our 40 initial questions, and >100 pages of responses to a different set of questions asked of the four Flyways. In September, we shared a preliminary report with Plan partners and received 25 substantive reviews. Most of the report was well received although we had some pushback on the general notion of the PC becoming more active in managing the JVs, and predictably some objection to our recommendation for greater emphasis on the prairie breeding regions. Entrepreneurial JVs have a natural resistance to herding; the trick is finding a productive balance between freedom to innovate and sufficient consistency.

Conclusions and Recommendations

The Plan Committee met in Savannah, GA in January 2007 to review the ASC's many recommendations and produce a response that would be released along with the final report of the ASC. Much of the creative conversation with the PC centered on prioritization and timing of their response to the many recommendations. I was in an odd position at that meeting having been deeply engaged with the ASC but also having been appointed to the international Plan Committee in 2005 just as the ASC began its work, and the minutes reflect my comically awkward mix of roles and inputs!

Table 15.1 lists the main recommendations of the ASC quoted from the report with minor editing for length and deletion of suggested individual assignments. Some are elaborated further below the Table. These were organized in four groups: (1) plan impacts on populations; (2) adaptive processes; (3) targeting of conservation investments; and (4) institutions working better together.

Table 15.1 Major Recommendations of the ASC (2007)

Plan impacts on populations

1 The Plan Committee should ensure the development of a clearer and more robust accountability framework for the achievement of NAWMP biological objectives involving all organizational levels in the Plan Community.
2 A comprehensive review of Plan objectives should be a high priority leading up to the next Plan Update.

Adaptive processes

1 Adaptive management, as the way of approaching NAWMP delivery, needs to be embraced and employed more widely.
2 While technical structures might vary, it is essential that all JVs develop the ability to address basic biological foundation issues. at www.routledge.com/9781041224389.
3 Plan progress requires a fundamental commitment to support monitoring and evaluation activities, within and among the JVs and through the NSST.
4 Implementation of the National Strategy for the Management of Waterfowl and their Habitats in Mexico, along with research to help inform further conservation planning, should be important priorities for Plan partners.

(*Continued*)

Table 15.1 (Continued)

Targeting of conservation investments

1 Greater progress, and therefore substantially more diverse and leveraged funding, is needed in breeding areas where many duck populations seem to be most limited.
2 JV actions should be guided by explicit biological models. This should be the standard planning method in all JVs.
3 In biological planning, diving ducks, sea ducks, over-abundant goose species, and species of special concern (e.g., lesser scaup, northern pintails) deserve greater attention.
4 Global climate change must be given more consideration in JV regional targeting, program emphasis, and project design.
5 Wetland conservation in the western Canadian and US boreal forest regions should be more explicitly connected to the Plan.
6 Increasing engagement in public policy will require development of new funding sources, new expertise, new measures of accomplishment, and new approaches to evaluation. Policy initiatives should be subjected to the same sort of evaluation rigor as for traditional wildlife programs.
7 Programs that lead to long-term protection or restoration of natural wetland systems should continue to play a major role in meeting waterfowl needs.
8 North American Waterfowl Management Plan partners should continue to promote farming practices that are economically appealing to producers and benefit waterfowl.
9 Maintaining or strengthening conservation provisions of the US Farm Bill is crucial for both breeding and nonbreeding JVs. Of similar importance should be expansion and further development of wildlife-friendly farm policies in Canada.
10 Enhancing policy and legislative protection for wetlands should be a major priority in Canada, the United States, and Mexico.
11 Plan partners should promote policy disincentives for converting native grassland to cropland in the United States and Canadian Prairies.
12 Plan partners should be actively involved in policy decisions that affect wetland water supplies for wildlife.

Institutions working together better

1 The Plan Committee needs to exercise greater and more visible leadership within and beyond the Plan Community, and more active management of Plan functions.
2 The Plan Committee should advocate that waterfowl harvest and habitat managers develop a coherent and coordinated approach for setting and achieving Plan objectives.
3 The NSST should be revitalized to tackle several follow-on assignments from this assessment and from the parallel JTG on NAWMP population objectives. The NSST also needs greater human and financial resources to advance its work. Plan partners collectively must accomplish this.
4 The 2005–2006 Assessment experience has been extremely positive for the Plan Community and ought to be repeated periodically.
5 The species and habitat JVs should communicate more and better integrate their work.
6 To revitalize communications and strengthen accountability within the Plan Community, the federal secretariats in our three countries should become more thoroughly and continuously engaged in NAWMP business.
7 The Plan Committee and JVs must maintain a strong dialog with the NAWCA Councils around the needs of NAWMP.
8 Simultaneously, it is vital for Plan partners to find new, substantive, and complementary sources of financial support for programs that are critical for achieving NAWMP objectives.
9 The Plan Community should strive continually to engage more stakeholders in NAWMP.

Plan Impacts on Populations

In a few of the oldest JVs with the longest history of science support, partners had estimated the effects of NAWMP programs and other landscape changes on waterfowl vital rates (e.g., PHJV) or regional foraging capacity (e.g., CVJV, LMVJV), which were assumed to be linked to waterfowl survival or subsequent reproductive rates. At the continental scale and for most individual

JVs, however, we were unable to critically evaluate progress toward attaining Plan population objectives, and we were unable to describe with confidence the landscape conditions needed to achieve those objectives. This is attributable to two factors: (1) limited ability by most JVs to assess ongoing landscape change; and (2) limited knowledge to link landscape composition and variation to waterfowl population vital rates.

The ASC urged attention to better monitoring of key habitat trends such as the extent of wetlands (all JVs), nesting habitat (breeding JVs), or foraging habitat (wintering JVs). Estimates of habitat gains and losses are essential for estimating true conservation progress and setting appropriate habitat objectives. Along with that, we urged the development of improved biological understanding of how landscape variation and habitat interventions influence waterfowl vital rates. While inclusion of such metrics in JV annual evaluations would be impractical, research in selected circumstances (e.g., several breeding ground JVs and perhaps one or two important wintering JVs) would help develop more informed investment strategies for the Plan. Such knowledge should also allow informed reviews of regional habitat goals and more valuable performance metrics.

Adaptive Processes

Monitoring and assessment should be treated as integral components of management. It is not sufficient to view these as optional "add-on" components to "regular" conservation business. Adaptive management is how the NAWMP enterprise should function. Some JVs had strong evaluation programs and used them to assess program performance and guide shifts in management actions. However, commitments to science support varied markedly among JVs.

To strengthen management decisions, JVs should explicitly state and test their most critical biological assumptions. JVs especially need to reduce uncertainty about the linkages between continental population goals and local/regional habitat goals, and between landscape change and waterfowl vital rates. Because of these uncertainties, most JVs had not yet defined the broad landscape conditions necessary to reach and sustain waterfowl populations at objective levels.

For Mexico, we urged the completion of their draft National Strategy for management of waterfowl and their habitats, along with incremental research and monitoring to help inform conservation planning.

Targeting of Conservation Investments

Assessing conservation needs at a continental scale, the ASC concluded that greater progress and therefore more funding were needed in breeding areas where current evidence suggested many duck populations were most limited. While there remained uncertainty about the seasons and regions of greatest limitation for many species, bioenergetic analyses suggested that sufficient food resources were currently available for waterfowl in several of the major wintering JVs. In contrast, for much of the PPR, in many years, reproductive success remained below levels that can maintain stable duck populations.

For all JVs, the ASC concurred that explicit biological models should be the standard planning method and that these should be updated as new evidence emerges. We also noted the need for better population estimation of several sea duck species, factors limiting species of present concern (e.g., scaup, northern pintails), and habitats vulnerable to changing climate.

The ASC agreed with many respondents that work on public policies, as well as programs for conservation on private lands were essential for achieving the large-scale impacts needed to sustain waterfowl over the long term. These initiatives will require a similar level of evaluation scrutiny we expect from traditional habitat projects. Despite their presumed importance, only a few policy initiatives have been rigorously evaluated regarding biological impacts. Focus on policy instruments also creates new demands for social, economic, and varied ecological information and most JVs were just beginning to address such needs.

The ASC surmised that agricultural habitats will remain important for meeting the nutritional needs of nonbreeding waterfowl in many JVs. On the breeding grounds, the need to affect the productive capacity of large landscapes makes it essential that Plan partners work effectively with other users of the land. Programs that positively affect both duck production and farm income are sorely needed. Programs that advocate winter cereals in an annual crop rotation, incentivize perennial grass cover, or reward ecological values provided by cooperating landowners are examples.

Maintaining or strengthening conservation provisions of the US Farm Bill seemed advantageous for both breeding and nonbreeding JVs. Further development of wildlife-friendly farm policies in Canada should be helpful too. Enhancing policy and legislative protection for wetlands was viewed as a major priority in all three countries. Multiple levels of governments promulgate policies that affect wetlands and all need to be lobbied for waterfowl-friendly actions. Decisions that govern the distribution and pricing of water were becoming increasingly important to wetland managers in many nonbreeding areas, but particularly in the semiarid West. In the future, freshwater seemed likely to become even more limiting to wildlife management options.

Institutions Working Together Better

The ASC urged that the Plan Committee exercise greater and more visible leadership across the Plan Community. The PC should build upon its functions described in the 2004 Update and strengthen these through renewed dialog with the JVs and other stakeholders. The JVs expressed a desire for more interaction and integration between themselves and the Plan Committee, NSST, and the FWCs. More sharing of experiences among JVs seemed desirable too.

We urged that the NSST be revitalized to tackle several follow-on assignments from the assessment and from the parallel JTG on NAWMP population objectives (Chapter 16). The NSST needed greater human and financial resources to advance its work. There was nearly universal support among Plan partners for enhancing the mandate and the capacity of the NSST and a desire that the NSST be a functioning technical arm of the Plan Committee.

The 2005–2006 Assessment experience was seen by many as extremely positive for the Plan partnership and ought to be repeated periodically. The ASC urged the Plan Community to adopt a two-stage process wherein NSST and JV representatives report annually on biological progress to the PC and the JVs, and that every five to ten years a more formal review panel be commissioned to complete a broader comprehensive assessment of progress – much like this current assessment. The ASC urged also that JVs undertake periodic independent peer review of their biological foundations and conservation strategies in the normal course of program management.

The ASC also urged the PC to maintain a strong dialog with the NAWCA Councils around the needs of NAWMP. US federal support through NAWCA and nonfederal matching funds remain critical to conservation progress under NAWMP. A strong dialog among Plan stakeholders is needed to ensure continuing complementarity of these essential continental efforts. The Plan partnership also should strive to engage more stakeholders in NAWMP. Strong and expanding partnerships have been a key factor in NAWMP's success and should continue to propel

NAWMP forward. It is also vital that the JV partners develop additional sources of funding to help achieve their all-bird goals while not compromising support for waterfowl conservation.

The Plan Committee Responds

The meeting in Savannah concluded with the PC endorsing essentially all the ASC recommendations. The top priority chosen was setting up a stronger accountability framework for Plan progress with the PC, NSST, and the JV partners including a regular reporting cycle. The PC also felt it was urgent to improve the ability to estimate net habitat changes in JV regions and develop a stronger linkage between the continental population objectives and regional habitat objectives.

The PC endorsed the ASC's conclusion that improving reproductive success in the mid-continent prairies remained the primary challenge in achieving Plan goals, and that achieving this would likely require not only more traditional conservation projects but also advances in wildlife-friendly agricultural and wetland policies in both Canada and the United States. They further endorsed enhancing conservation partnerships on working lands throughout the JVs. The PC supported adopting an adaptive management framework but in a way that encouraged innovation in approaches by individual JVs. They encouraged more attention to species of current concern (e.g., northern pintails, scaup, and several sea ducks) and closer coordination between the species and habitat JVs.

The federal secretariats of the three countries agreed to lead in revitalizing communication and outreach to NAWMP stakeholders, and the PC established a Funding Task Force to develop a "NAWMP value proposition" highlighting opportunities for additional stakeholder investments.

Closing Thoughts

By 2007, there was little doubt that the NAWMP had been an important addition to the long history of waterfowl conservation. Yet, partners recognized that improvements were needed to realize the long-term vision of abundant and sustainable waterfowl populations. The Continental Progress Assessment was a courageous undertaking of self-examination. At the conclusion of the Assessment, many of us felt that we owed it to our stakeholders, our predecessors, our successors, and the birds to do our very best to implement these recommendations to achieve the NAWMP vision.

Given the value obtained from this first comprehensive assessment, the PC declared its intention to support subsequent assessments at about ten-year intervals in concert with Plan update cycles. In hindsight, some of the recommendations in 2007 might have been beyond the reach of some JVs at the time. Progress in achieving these recommendations now should be revealing and offer clues as to how the partners might do more to inform their future decision-making processes.

Some Reflections in 2025 on the JV Experience

Conceiving the habitat JVs in 1985 was a stroke of genius by the NAWMP negotiating team. They realized correctly that no single agency could possibly accomplish the grand continental program envisioned in the Plan. Moreover, a willing coalition of regional partners with the knowledge to plan, and the positioning to deliver, made eminent sense. They would need a compass, of course, objectives to which all could aspire, and external resources to help, but they would also bring unique assets to bear.

Above all, the collective NAWMP JV experience has proven that much can be accomplished by motivated people working together on a grand scale when they share common objectives and are largely free to find their own paths. They bought into shared objectives but focused on local solutions. Common cause, trust, and collaboration were indeed the way to achieve things at this scale!

NAWMP became a proven model for organizing continental conservation. Following the Continental Progress Assessment, the group that interviewed all the JVs offered a description of the characteristics of successful JVs (Online Appendix G see www.routledge.com/9781041224389), but here are my impressions 18 years later of what the most successful JVs do:

- They develop conservation plans based on the best biological and landscape information available, leavened by delivery experience. The Gulf Coast JV 35 years ago described biological planning as a vital "opportunity to think" and periodically renewing plans fosters essential rethinking.
- They monitor the efficacy of the programs they deliver and adapt them to improve returns on investments. They learn whether they are doing enough of the right things in the right places.
- They become learning organizations, sharing new knowledge both within and among related JVs.
- The partners build trust and social connectivity, the "glue" that is necessary to unite people in major efforts over many years.
- Increasingly, they focus on working lands, advocating conservation that works for wildlife but also for the people who live and work in their regions. They listen to their clients and engage them in finding solutions.
- They focus on a sound business approach, leveraging funds from multiple sources and producing adequate returns on investments for partners.
- Where public lands with waterfowl values are available, these are also a valuable part of the conservation estate and need protection.
- They help sustain outdoor traditions by providing access for a wide array of human recreational pursuits.
- They invest in understanding the contributions of their habitats to clean and sustainable water supplies, flood prevention, storm surge protection, pollution control, pollinator habitat, biodiversity conservation, and other values for both rural and connected urban communities.
- They are constantly seeking new partners, individuals, and organizations who want to have an impact. People enjoy playing on winning teams.

Note

1 Members were Ken Abraham (ON Ministry of Natural Resources), Michael Anderson (DU Canada), Richard Bishop (IA), Bob Clark (CWS), Lorne Colpitts (MB), John Eadie (University of California, Davis), Seth Mott (FWS), Dave Paullin (FWS), Mark Petrie (DU Inc.), Eric Reed (CWS), Frank Rohwer (LA State University), Mike Tome (USGS), Ariel Rojo (Direcion General de Vida Silvestre), Humberto Berlanga (SEMARNAT), and Eduardo Carrera (DUMAC). Two PC Cochairs, David Smith (FWS) and Steve Wendt (CWS) were *ex officio* members.

References

Assessment Steering Committee. 2007. *Continental Progress Assessment. North American Waterfowl Management Plan.* Final Report USFWS, CWS, and SEMARNAT. U.S. Fish and Wildlife Service. Arlington, VA. 98pp.

Williams, Byron K., Mark D. Koneff, and David A. Smith. 1999. "Evaluation of waterfowl conservation under the North American Waterfowl Management Plan." *Journal of Wildlife Management* 63: 417–40.

16 Seeking Coherence in Waterfowl Management

As plans for the Continental Progress Assessment were hatching, a concurrent review of the Adaptive Harvest Management (AHM) program in the United States was underway by a task force of the International Association of Fish and Wildlife Agencies (IAFWA), the USFWS, and Flyway Council staff. Their work was focused mostly on the role of population objectives in harvest management. Following some initial exploration about common information needs of NAWMP and AHM at the 2003 Duck Symposium in Sacramento, and again after an AHM meeting in Denver, Fred Johnson (FWS), Mike Runge (USGS), Seth Mott (FWS), and I concluded that a formal explication of the "coherence" challenge that both initiatives faced was warranted. Runge took the lead in developing our thought piece (Runge et al. 2006) but before it was published, we had also engaged the harvest and habitat management communities with the idea of exploring the issue more deeply.

The concern was basically this: harvest managers, specifically those focused on mallard harvest in the United States, and NAWMP habitat managers were both acting to affect the size of the same mallard population. NAWMP partners were striving to support the survival of adult hens and the rate at which they recruit young into the population by investing in habitat over the long term. The other program was striving to maximize harvest opportunity annually while ensuring the long-term sustainability of the hunted population. What each program was doing independently affected the other and so logically their objectives should be set in unison. The details of how these two efforts might interact are complicated and depend on processes like the effects of bird density on survival and reproduction. We wanted to draw attention to this situation to at least ensure that these two major branches of waterfowl management were not working at cross purposes and perhaps find a way to structure both efforts to mutual advantage. The notion of "coherence" spoke to the idea that harvest and habitat management are linked and that the objectives of both programs ought to reflect that linkage.

As Runge et al. (2006) noted, a harvest strategy can affect whether population objectives of the Plan are met, irrespective of the success of the Plan's habitat conservation efforts. Conversely, habitat conservation activities under the Plan can influence harvest potential and, therefore, the amount of hunting opportunity available. At the onset of AHM in 1995, the framers included a simple linear function that devalued harvest if mallard populations fell below the 8.6 million Plan objectives (technically 8.8 million with the addition of pair estimates from MN, WI, and MI). Doing this was a conservative measure designed, to produce a compromise between maximizing harvest and attaining the Plan's mallard objective.

For the mid-continent mallard population and several other species, there is evidence that birds at higher population densities tend to reproduce at a lower rate than do birds in smaller populations. The mechanism for this is not known but could involve some effects of resource limitation on breeding areas or reduced condition of birds in spring from crowded winter

DOI: 10.4324/9781003732310-16

quarters. For our purposes, the only important ramification is that with a Plan objective set higher than the average population size at which maximum productivity, and thus maximum sustainable harvest, might be achieved sets up a conflict. For mid-continent mallards, the goal of 8.8 million birds is contrasted with an estimated maximum sustained yield (MSY) population size of 5.9 million birds. This relationship is readily depicted in a yield curve (Figure16.1) simplified here as a deterministic relationship without the unavoidable noise attendant to annual variation in water abundance or other things that might affect bird numbers.

One implication of these relationships is that if habitat conditions, and thus carrying capacity ("K"), could be improved (Figure 16.2, top curve), the sustainable harvest potential would be increased. Conversely, a loss of habitat (bottom curve) would reduce the harvest potential. But the choices of target points on the yield curve matter to setting goals and thus habitat targets and harvest regulations.

For NAWMP, because of the potential effects of harvest on population size, population objectives are only interpretable if related to a specified harvest strategy. For instance, if we improve habitat conditions, or conditions deteriorate, an observed population size can only be used to evaluate the Plan's success with a specified harvest strategy. Habitat conservation and harvest management are inextricably linked. Habitat conservation can affect the size of the harvestable surplus by improving the potential for population growth. Harvest management can affect the degree to which available habitat is used. We must consider the effects of both to understand population dynamics. Clarification of the Plan population objectives with the concurrence of harvest managers will, in turn, be necessary for consistent specification of regional habitat objectives.

Runge et al. (2006) urged that we use our current understanding of environmental and harvest dynamics of duck populations, derived from AHM and other research, as a basis to help specify the nature of Plan population objectives. Mid-continent mallards were the obvious place to start because we knew a lot about them, and they have long been the most important duck in the hunter's bag. Lesser-known species, especially those exhibiting declining trends, like black

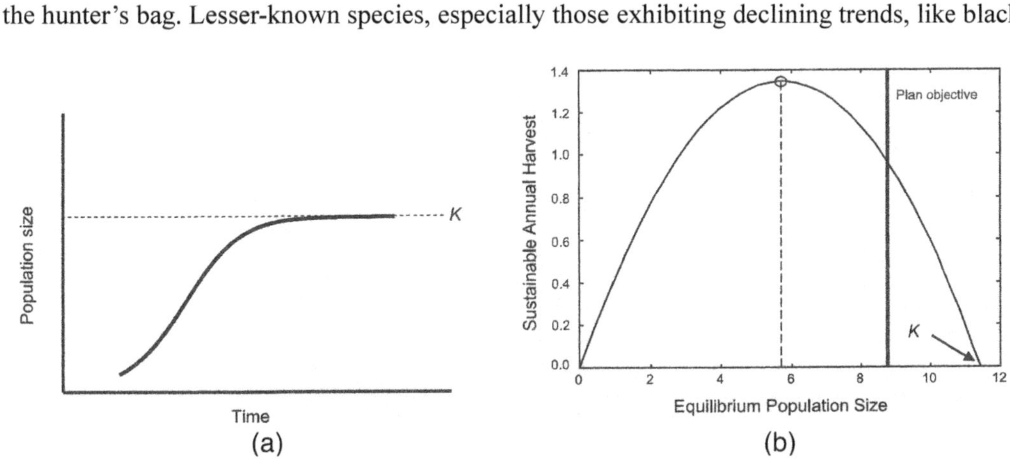

Figure 16.1 Under an assumption of density dependence, as the size of a population grows over time (solid line left figure), it will eventually reach a limit based on the resources available and other limiting factors (carrying capacity or K; the horizontal dashed line). Likewise, if the population is temporarily larger than K, it will decline toward the same limit. Based on this relationship, it is possible to estimate sustainable harvest as a function of population size (right figure). MSY is realized at a population size of K/2 where K is the demographic carrying capacity of the system (Anderson et al. 2007a, Figs 2.2 and 2.3).

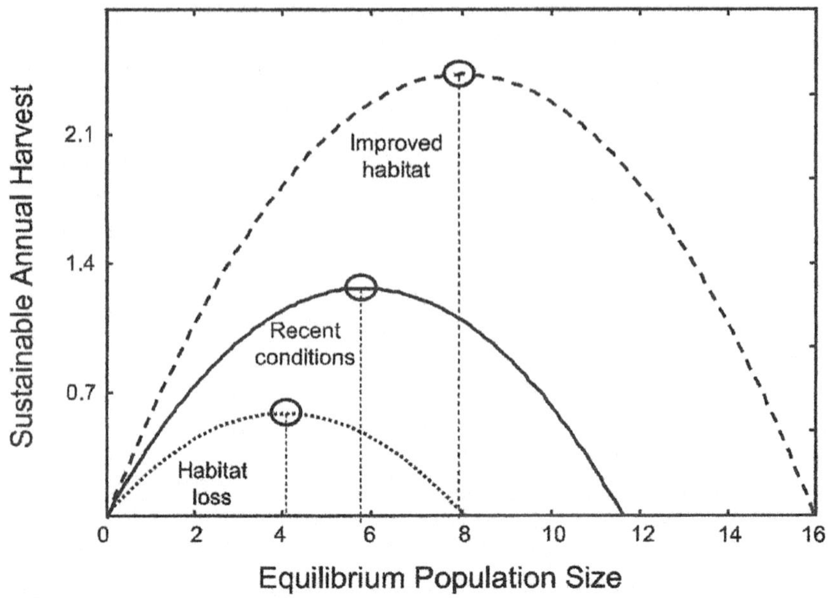

Figure 16.2 Changes in habitat quality can increase or decrease sustainable annual harvest potential (Anderson et al. 2007a, Fig. 2.5).

ducks, northern pintail, lesser scaup, and others seemed to merit attention soon too. The paper concluded that it seemed imperative that the habitat and harvest communities work together to harmonize their program objectives, at least for species important to harvest management.

The Joint Task Group

In November 2004, some of the authors of the Runge et al. paper met in Orlando with the AHM working group to urge advancing this work on coherence. We had challenging and productive discussions and afterward, the group recommended to the AHM Task Force that they include the issues around coherence of NAWMP and AHM population objectives in their ongoing work. In May, we sent a proposal to the leaders of NAWMP and IAFWA to form a crossover team to take this on. Many leaders of harvest and habitat management programs soon agreed and so the Plan Committee Cochairs (Steve Wendt and David Smith) and the IAFWA AHM Task Force Chair (Don Childress) in June 2005 sanctioned the appointment of a time-limited Joint Task Group (JTG), comprised of scientists and managers drawn from both management communities, to explore options and recommend a preferred solution to reconciling the use of Plan population objectives for both harvest and habitat management. An underlying element in this decision was concern that without a common approach to waterfowl management planning, conflicts between harvest and habitat management goals might erode support for both programs.

The charge to the JTG consisted of a purpose statement and four tasks. The purpose was *to develop options and recommendations for clarifying NAWMP population objectives and their use in harvest management.* Specific tasks were to:

1 Clarify the biological meaning of NAWMP population objectives.
2 Consider whether and how those objectives should be incorporated in AHM.

3 Describe the implications for both harvest management and NAWMP of adopting various options.
4 Recommend a preferred option.

The JTG was asked to deliver a draft report within the year for review by the governing bodies, followed by stakeholder review, reconciliation, and a final report. Members appointed to the JTG were Seth Mott, Coordinator FWS, Michael Anderson IWWR/DUC, Dale Caswell CWS, John Eadie UC-Davis, Jeffrey Herbert MT, Min Huang CN, Dale Humburg MO, Fred Johnson FWS, Mark Koneff FWS, Mike Runge USGS, Tom Nudds U of Guelph, Eric Reed CWS, Jim Ringelman DUI, and Barry Willson FWS. Helpfully for NAWMP, four of us (Eadie, Mott, Reed, and Anderson) were simultaneously serving on the NAWMP Assessment Steering Committee. Others were already involved in the AHM reviews (Huang, Runge, Johnson, Herbert, Caswell, and Koneff). Two had been engaged in the earliest planning for AHM (Humburg and Ringelman), and two others were fresh to the process.

Runge and Johnson brought critical knowledge of the technical issues previously described, occasionally drawing on other experts like Scott Boomer (FWS) for additional input. Seth Mott served as our coordinator, as well as offering sage advice on NAMWP matters. We met first in September in Arlington where our conversations were wide-ranging, novel, and portended some of the most stimulating debates of my waterfowl career. We left few stones unturned in our search for a path forward. One deceptively simple question that emerged at our first meeting was, “Where is the fire? What is the compelling case for doing all this coordination?” The question took the better part of two years to answer. But we started kind of bouncing off the walls. Here’s a verbatim example of a recorded exchange from that early meeting:

Anderson: We have a desire to harvest waterfowl. Harvest potential depends on habitat and habitats are under siege. The Plan was developed to address this, to ensure landscape would be able to produce ducks. But 20 years later, we still have only a rudimentary idea how to affect waterfowl population size.

Mott: In some places, we’ve gained ground, in some we’ve lost, in others we don’t know. On the one hand, we have a desire for harvest; on the other, we have this challenge of retaining or increasing the productive capacity of waterfowl habitat. But presently, we don’t have a way to link harvest and habitat conservation.

Nudds: If we’re not using resources effectively, we could be wasting money and not accomplishing anything. We might also be wasting resources on the harvest side by working at too fine a scale. We need to work together to figure out how best to invest resources to accomplish shared goals.

Caswell: The sustainability of waterfowl management is the basic concern. If we don’t get better at how we invest conservation dollars, harvest management will become an exercise in allocating the last couple of ducks.

Group: Growing sense of not really understanding what motivates hunters and other stakeholders. Are we addressing the right issues? What do hunters want? What do landowners want?

Eadie: Crisis at the harvest level: Huge investments. Constraint imposed by Plan goals. Crisis at habitat level: Are we really making progress? There is demographic ambiguity in NAWMP goals so we don’t know if they’re fully useful for planning. Fundamental underpinnings of waterfowl management are at risk because we haven’t knit all these pieces together.

Herbert: The approach must change now. Linkage is the key. We must do a better job of managing that linkage. This is a challenge to the underpinnings of all waterfowl management.

Johnson: It comes down to two fundamental questions: (1) how much habitat do we need, how much is it going to take to support populations at some desirable level; (2) how much harvest does it take to satisfy hunters? These are not independent questions. You can't answer either one unless you do the linkage.

Humburg: NAWMP started out with recognition of harvest, but the pieces have drifted apart. We think they need to be put back together.

Eadie: What's the risk of not doing that? What's the risk of status quo?

Ringelman: We risk becoming irrelevant, working at cross purposes, and wasting resources. You don't know if what you're doing is complementary or not. And you don't know if people like what you're doing. We lack a Grand Unifying Framework for Waterfowl Management – a Duck GUF!

Wilson: This is a multi-scalar issue. The linkages that are needed are not only between harvest and habitat, but between broad and local scales.

You can sense that we had work to do! Tom Nudds remembers (pers. comm. May 2025) the conversation and the day. As the enormity of the task was sinking in, someone broached whether we were talking about a Plan update or new Plan. And who would do it? The silence, a rarity with this group, was deafening. Fundamental replanning would come about six years later.

So, the JTG met, drafted, argued, consulted, and revised. We chose to share our developing ideas in presentations to the joint Flyway Councils and the Plan Committee at the March 2006 NAWNRC in Columbus, OH, and the next day we offered a pair of papers (Johnson 2007; Anderson et al. 2007b) as part of a day devoted to the status and future of waterfowl harvest management. Fred Johnson focused on lessons learned and the how the nature of the debates over AHM had changed since 1995. One growing realization was that specification of regulatory alternatives involved social trade-offs, which are difficult to address with the lack of data regarding hunter preferences, participation, and satisfaction. More social science investments were needed. We had also learned that model discrimination is a slow process given imprecision in model predictions and monitoring data, potential inadequacies in model sets, and passive (rather than active) approaches to learning (see also Nichols et al. 2007). Anderson et al. (2007b) reviewed the numerous threats to waterfowl habitat in North America and the daunting challenges of sustaining the productive capacity of those habitats, despite apparent gains made since the advent of NAWMP. We also highlighted the inescapable linkages between harvest potential and the abundance and quality of waterfowl habitat and uncertainties around increasing or maintaining the carrying capacity of continental habitats. The challenge for the waterfowl management community, we argued, was to focus more squarely on our most fundamental problem – securing the future productive capacity of waterfowl habitats.

The JTG regrouped in April 2006 to review feedback we had received from the initial airing of our ideas. While we encountered little firm resistance, it was also clear that the technical elements were only partially absorbed. The ideas were newer to the Plan community than to the harvest side and that was reflected in the feedback. One vexing issue for the JVs was divining how they might affect "K" or carrying capacity (Figure 16.1, 16.2), or whether this was too simple a construct to be useful in the complex world where we work. What connection is there between what managers do on the land and "K" in the equilibrium population models? We committed to probing that further. Many reactions to our presentations concerned social matters, particularly speculation about what most hunters want in a harvest framework. Second, debate about where we might wish to set our continental objectives on the shoulder of the yield curve (Anderson et al. 2007a, table 4.3) stimulated lively debate.

Dale Humbug urged that we needed a broad forum of stakeholders to make progress on the question of hunter desires. Seth and I argued that we needed to figure out how to maintain momentum after our group broke up with the conclusion of our report. John Eadie noted that ambiguity about what a new decision process would be troubled people. Who would oversee any new "system" that we might recommend? Tom Nudds presciently pointed out that so far, our consideration of human dimensions had only addressed hunters but that others who have competing interests, especially landowners, would need to be addressed in time. These were observations that presaged key elements that would emerge in the 2012 Revision of the Plan and remain only partially resolved in 2024. In short, we began to recognize that harmonizing and linking objectives for harvest and habitat management was complex and resolving them would take time and extensive consultations. These are complex ideas and especially complex institutional matters. Any changes in Plan objectives would necessarily require extensive dialog between harvest and habitat managers and there was no governing body with the mandate to do that. Finally, a better understanding of stakeholders, especially hunters, left us believing that unification of waterfowl management at the continental scale was multidimensional and we urged the waterfowl partners to plan for an extensive review of options in advance of the next NAWMP Update (Chapter 18).

As we were finishing the final report (Anderson et al. 2007a), we added two significant pieces. The first offered some initial ideas about how habitat joint ventures might increase "K" presented as an appendix developed by Jim Ringelman, John Eadie, and me. We considered both breeding and nonbreeding joint ventures. We also found our way back to the question of, "Where's the fire?" from the JTG's very first meeting. We concluded (pp. 41–45), that the waterfowl management community faced an impending crisis. The risk was that the foundations of all our management efforts were not cohesive. We argued that we should establish meaningful, measurable goals that integrate these components and provide a unifying framework for waterfowl conservation. The three-legged stool, as the model came to be called, urged that a revisioning of the Plan was called for that would explicitly link harvest, habitat, and hunters (later people more broadly) with the kind of system thinking that first brought the Plan into existence.

Our final report was delivered in July 2007. We noted that although the original framers of the North American Waterfowl Management Plan recognized the linkages among harvest, habitat, and hunters, the waterfowl management community had not explicitly integrated those elements in a cohesive technical framework. Such integration would require an explicit connection between harvest and habitat management based on basic concepts of population dynamics. Yield curves simultaneously depict carrying capacity, maximum sustainable harvest, equilibrium population sizes, and total harvest over a range of harvest rates and we believed that yield curves provided the appropriate conceptual framework for the integration of harvest and habitat management.

The 1986 Plan established an explicit benchmark for hunter satisfaction – waterfowl populations that existed in the decade of the 1970s that seemed generally to satisfy hunters. Underlying the highly variable moisture conditions of that decade lay a foundation of habitat – wetland basins, grasslands, etc. – that represented the potential of the landscape to produce and sustain waterfowl. We therefore interpreted those original Plan goals as the desire to expand (or at least maintain) the current yield curve, which in turn could be interpreted as increasing (or at least maintaining) the continental carrying capacity – the habitat base.

Harvest objectives that seek less than the maximum sustainable harvest provide a conservative hedge against uncertainty and a measure of protection for species less productive than mallards managed under a common mallard-centric harvest framework. That should also result in higher average population sizes for people to enjoy and require less variable harvest regulations. Thus, we urged that harvest management objectives could be best described as desired points along the yield curve, and we recommended a "right shoulder strategy" to integrate habitat and harvest management and benchmark continental objectives for both programs.

Deliberate policy dialog would be required to balance trade-offs for a continental population objective as a shoulder point that would determine both the harvest objectives and the desired expansion of the yield curve. A desire for greater harvest opportunity would require increased carrying capacity and thus, a greater investment in habitat improvement. Harvest opportunity would be lower if habitat conservation efforts proved ineffective.

The specific point chosen on the yield curve would be influential and require a decision based on human social values. In a sense, the trade-off would be about how much is enough. How much habitat improvement is feasible, affordable, or tolerable by others? How much harvest potential is needed to satisfy stakeholders? Assuming our basic model was correct, we could achieve an average population size of 8.8 million for mid-continent mallards now, simply by decreasing average harvest rates to about 6.1%. Or we could maintain current harvest rates and achieve an average population size of 8.8 million by increasing the continental carrying capacity by about 15–20%. Where is the right balance? If we cannot find an acceptable combination of harvest and habitat management goals that results in an average breeding population size of 8.8 million, that would motivate revisiting the continental population objectives of the Plan.

Regardless of the choice, considerable technical work would be required to integrate uncertainties about habitat and harvest management, including stakeholder desires, clarification of key ecological uncertainties such as the functional form of density dependence, and models that link local habitat management with waterfowl demography.

Recommendations from the JTG were intended to motivate the policy decisions and the technical work necessary to unite all aspects of waterfowl management. Implementation would require communication and coordination among diverse stakeholders, completion of new technical assessments for habitat and harvest management, the addition of technical capacity in human dimensions, and leadership from agencies with mandated responsibility for waterfowl management. Quoting the JTG report, with minor condensing, the six main recommendations were:

1 The FWS and CWS, in consultation with the Flyway Councils, should adopt a "shoulder strategy" for northern pintails and mid-continent mallards. The principles of coherence and integration should be central components of the revised US federal Environmental Impact Statement on waterfowl harvest.
2 The Plan Committee should adopt the same "shoulder strategy" to ensure coherence between harvest policy and habitat objectives and focus the attention of the joint ventures on conservation actions designed explicitly to enhance continental carrying capacity (K). The principles of coherence and integration should be central components of the next Update of the Plan.
3 The federal wildlife services, together with interested partners, must commit to enhancing technical capacity for the Plan's Science Support Team (NSST) and the AHM Working Group as both will have key roles in the technical integration of waterfowl management.
4 The waterfowl management community must focus more scientific efforts on reducing the key ecological uncertainties surrounding current models of population dynamics (e.g., density dependence) and the relationships between waterfowl vital rates, carrying capacity (K), and landscape properties that habitat managers strive to manipulate.

5 The FWS and CWS directorates should convene a waterfowl management policy summit soon to address the decisions necessary to develop the policy framework for habitat and harvest management integration.
6 A new Human Dimensions Working Group should be convened to refine an assessment of waterfowl stakeholder values and the approach for incorporating this information into management decisions.

The JTG also acknowledged uncertainties in key areas. First, the nature of density dependence was unclear. We observe a negative correlation between population size and population growth at large scales, but exactly how it operates remains uncertain, as does the commonality of the process among species. There is also the possibility of system change over time. In other words, a change that prevents observed long-term relationships between, say, populations and habitat from persisting. The most obvious possibility might lie in long term change in wetness or agriculture in the Prairie Pothole Region with climate change, but other nonequilibrium changes are possible. There remain numerous uncertainties in the demographic impacts of the habitat work being carried out by the JVs. Finally, the authors acknowledged evolving attitudes of the public about waterfowl harvest and stakeholder support for what waterfowl managers do.

Deliberations of the JTG and reactions to the report paved the way for the pivotal 2008 Minneapolis Summit on the future of waterfowl management that I will take up next. In hindsight, we had reached the point of "double looping" in our adaptive management cycle (Pahl-Wostl 2009, Chapter 14) when after some progress with planning, delivering, evaluating, and revising our delivery efforts we had come to a broader question. Were our objectives right? Were our objectives broad enough to encompass the social-ecological system we had been trying to manage for 20 years? We didn't realize it yet, but we had just taken a step toward the first fundamental revision of the North American Waterfowl Management Plan in 2012.

References

Anderson, Michael G., F. Dale Caswell, John M. Eadie, Jeffrey T. Herbert, Min Huang, Dale D. Humburg, Fred A. Johnson, Mark D. Koneff, Seth E. Mott, Thomas D. Nudds, Eric T. Reed, James K. Ringelman, Michael C. Runge, and Barry C. Wilson. 2007a. *Report from the Joint Task Group for Clarifying North American Waterfowl Management Plan Population Objectives and Their Use in Harvest Management.* USFWS Unpublished Report. 57pp. (https://nawmp.org/timeline/joint-task-group-report).

Anderson, Michael G., John M. Eadie, Min Huang, Fred A. Johnson, Mark D. Koneff, James K. Ringelman, Michael C. Runge, and Barry C. Wilson. 2007b. "Harvest potential and habitat are inextricably linked." *Transactions of the North American Wildlife and Natural Resources Conference* 71: 275–89.

Johnson, Fred A. 2007. "Harvest management and the future of adaptive harvest management." *Transactions of the North American Wildlife and Natural Resources Conference* 71: 211–19.

Nichols, James D., Michael C. Runge, Fred A. Johnson, and Byron K. Williams. 2007. "Adaptive harvest management of North American waterfowl populations – recent successes and future prospects." *Journal of Ornithology* 148: 343–49.

Pahl-Wostl, Claudia. 2009. "A conceptual framework for analysing adaptive capacity and multi-level learning processes in resource governance regimes." *Global Environmental Change* 19, no. 3: 354–65.

Runge, Michael C., Fred A. Johnson, Michael G. Anderson, Mark D. Koneff, Eric T. Reed, and Seth E. Mott. 2006. "The need for coherence between waterfowl harvest and habitat management." *Wildlife Society Bulletin* 34: 1231–37.

17 The Future of Waterfowl Management Summit 2008

In its final report, the Joint Task Group wrote,

> We think the single most useful step that could be taken in 2007 would be for the Directors of the FWS and CWS, in collaboration with IAFWA, the Plan Committee, and the National Flyway Council, to convene a waterfowl management policy summit where implementation of the specific policy and technical recommendations in this JTG report might be debated, refined and acted upon. Creative solutions are needed, and the entire waterfowl community should be challenged to contribute novel ideas for solutions.

Helpfully, the notion of a Summit had been kicking around for some months before the JTG report was finished. The need for fundamental improvements also had been identified in the just-tabled NAWMP Continental Progress Assessment report (Assessment Steering Committee 2007) and by the IAFWA Adaptive Harvest Management Task Force. Flyway members in that Task Force had been pushing the FWS to consider the people-related assumptions implicit in harvest objectives (Andy Raedeke pers. comm. April 2025). That effort led to creating a first *ad hoc* Human Dimensions Working Group (HDWG) in 2005 with leadership from the Mississippi and Central Flyways and provided additional impetus for a Summit gathering.

A Summit planning team[1] was assembled including technical people and administrators, charged with developing the goals, expected outcomes, format, participants, location, and timing of the workshop. The Committee consulted widely across the waterfowl management community to ensure the idea had broad support. The services of D.J. Case and Associates were retained to help with planning, conducting a pre-workshop survey, facilitating the workshop, and coordinating the report (D.J. Case and Associates 2008).

The planning team hoped that the Summit would achieve a collective understanding of how a coherent framework for waterfowl management could produce benefits for administrators, managers, and hunters. We also sought to identify the key uncertainties, both technical and policy related, and how they might be addressed. We hoped to reach consensus on how to advance recommendations of both the JTG (Anderson et al. 2007) and the Continental Assessment reports (ASC 2007) and to identify what policy decisions should be considered by agencies with authorities over waterfowl management.

In Dave Case's action-oriented phrase – we hoped participants would leave the Summit understanding enough about a coherent framework *to know* how it could benefit administrators,

DOI: 10.4324/9781003732310-17

managers, and hunters; *to feel* that they have a stake in and responsibility for the outcome; and *be committed* to helping support actions needed within their own organizations.

In May 2008, D.J. Case administered a web-based survey of pre-registrants of the Workshop. The results helped refine the meeting objectives and the final agenda. Of 188 Workshop pre-registrants contacted, 162 (86%) completed the survey so interest was high!

Workshop Overview

The Workshop was held in Minneapolis, MN on August 26–28. In total, 192 individuals representing a cross-section of waterfowl management administrators, technicians, scientists, and other stakeholders participated (Appendix B of the workshop report). The National Flyway Council, the North American Waterfowl Management Plan Committee, and the Association of Fish and Wildlife Agencies sponsored the Workshop. Ducks Unlimited, Inc. served as the host. D.J. Case provided facilitation and recording for the Workshop, including facilitating full group discussions and overseeing breakout groups.

We began with four major presentations reviewing the work of the JTG and more:

1 From Concept to Implementation: Moving Towards Coherence in Waterfowl Management (Jim Ringelman, DUI)
2 Harvest Management in an Integrated Framework (Mike Runge, USGS)
3 Habitat Management in an Integrated Framework (John Eadie, UC-Davis)
4 Human Dimensions in an Integrated Framework (Andrew Raedeke, MO Department of Conservation)

The breakout sessions that followed were intended to identify areas of agreement, disagreement, or misunderstanding. Four groups of participants discussed the following questions in three breakout sessions:

Breakout session 1:

What are the goals of waterfowl management?

Breakout Session 2:

What are the technical challenges to integration (i.e., "coherence")?
What are the policy and structural issues with integration?

Breakout Session 3:

How do we move forward?
What needs to happen to achieve the benefits of coherence?
What recommendations do you have?

Reports of the breakout groups were next shared with all attendees and included in the workshop report. The final morning was devoted to full-group discussions of the issues and recommendations from the breakout reports. Ralph Morgenweck (FWS) and I each gave presentations that assembled commonalities from the breakouts into elements of consensus, insights about progress, and remaining challenges. Based on these and the ensuing discussions, a final series of live survey questions was presented to participants at the end of the Workshop. A significant majority of participants agreed that three recommendations born in Workshop discussions should be pursued as next steps:

- A group or venue be created to continue the work of the Human Dimensions Work Group (94% agreed or strongly agreed)
- The next NAWMP Update should be used to develop more coherent goals for waterfowl harvest and habitat management (88% agreed or strongly agreed)
- NAWMP goals should be revisited (76% agreed)

Though on a less optimistic note:

- Many (67%) either disagreed or were neutral on their personal understanding of the benefits of a more coherent waterfowl management system.
- And only 14% left willing to initiate work for coherence in their own agency.

Where to Next?

As the Workshop unfolded a small group of us met to consider steps toward the next Plan Update. Clearly, given the desire for greater consultation on NAWMP objectives and the issues at hand at the Workshop, 2010 or 2011 seemed like a more realistic completion target than 2009! My own files on what eventually became the 2012 Revision of NAWMP began naively with the label 2009 Update!

After the Summit, towards the end of 2008, Seth Mott and I suggested a path forward for Plan Committee consideration. We posited that the Summit helped elaborate the institutional needs and policy decisions necessary to build a more coherent framework for waterfowl habitat and harvest management. By "coherent", or "unified", we meant mainly a more structured approach for making certain waterfowl management decisions that would integrate current knowledge of waterfowl population dynamics, our ability to provide adequate habitats, and consideration of human elements such as hunter satisfaction. An improved framework would thus require agreed-upon objectives that ensured the connection of goals for habitat conservation, harvest management, and stakeholder support, and would help Plan partners use our human and financial resources as efficiently as possible.

Results from Minneapolis and the Joint Task Group's analyses and recommendations provided a basis for improving the framework for waterfowl management, but much work remained to move from a conceptual model to an operational system. Seth and I argued that the core of the next Update should be revised Plan objectives that corresponded to population levels sought under optimal harvest strategies, realistically achievable habitat conditions, and satisfactory levels of hunting. We concluded that *The complexity of determining the balance among these goals will necessitate a thorough dialogue within the entire waterfowl management community in Canada, the United States and Mexico.* Participants at the Minneapolis workshop had emphasized that. Unlike previous NAWMP updates, this one should feature direct input from technical experts in harvest management and human dimensions, as well as from the NSST, the JVs, the Flyway Councils, and NGOs. Picking up the unfinished recommendations of the

IAFWA hunter recruitment task force should also be considered. We urged that, *We should strive for clear, unambiguous objectives that provide coherence, transparency, and accountability and facilitate joint assessment of harvest and habitat programs* It was an ambitious and optimistic call to action.

An Update Steering Committee

To move forward, the Plan Committee formed a broad-based Steering Committee (SC) to serve as the focal point for gathering, vetting, and synthesizing ideas from the waterfowl management community and to advise the PC regarding the Update (Chapter 18). The SC would require the support of a writing team and technical advice from harvest, habitat and human dimensions scientists. The initial task would be a scoping process to accomplish five things:

1 To identify additional technical work and resources needed to advance waterfowl management modeling frameworks to the point that they could serve as the keystones for the Update. This would, of necessity, include a consensus-building process involving mainly the SC and representatives of the NSST and the Adaptive Harvest Management Working Group. The new Human Dimensions Working Group (HDWG) would also be consulted.
2 To review the Continental Progress Assessment report and identify major topics, in addition to those advanced at the Minneapolis workshop, that should be developed in the Update.
3 To propose a process for consulting with stakeholders on alternative waterfowl management objectives that would be developed during the Update process. This proposal must consider the country-specific requirements of the United States, Canada, and Mexico.
4 To propose an approach for reviewing existing waterfowl management processes and institutions considering the new unified framework to be developed during the Update.
5 To propose a work plan and schedule for completion of the Update, recommending additional appointments or subcommittee structures as necessary.

SC membership should reflect a wide range of expertise and institutional affiliations including individuals with technical expertise and experience in policy and administration.

Launching the Update

Steering Committee members were appointed by the Co-Chairs of the PC in January 2009, and the SC's initial scoping document was delivered by March to enable review at the 2009 North American Wildlife and Natural Resources Conference. It turned out, of course, that this was much more complicated than we first anticipated. The consultation challenge was massive, and the range of issues huge. Before we arrived at the deceptively simple theme of the 2012 Revision – People Conserving Waterfowl and Wetlands – we had a long journey ahead. The trail from Minneapolis to the 2012 Revision is the story ahead.

Note

1 Michael Anderson (Ducks Unlimited Canada), Ken Babcock and Dale Humburg (Ducks Unlimited, Inc.), Phil Bowman (LA), Bob Clark and Kathy Dickson (CWS). Bob Ellis (VA), Ralph Morgenweck and Seth Mott (USFWS).

References

Anderson, Michael G., F. Dale Caswell, John M. Eadie, Jeffrey T. Herbert, Min Huang, Dale D. Humburg, Fred A. Johnson, Mark D. Koneff, Seth E. Mott, Thomas D. Nudds, Eric T. Reed, James K. Ringelman, Michael C. Runge, and Barry C. Wilson. 2007. *Report from the Joint Task Group for Clarifying North American Waterfowl Management Plan Population Objectives and Their Use in Harvest Management.* USFWS Unpublished Report. 57pp. (https://nawmp.org/timeline/joint-task-group-report).

Assessment Steering Committee. 2007. *Continental Progress Assessment. North American Waterfowl Management Plan.* Final Report USFWS, CWS, SEMARNAT. 98pp.

D.J. Case and Associates. 2008. *The Future of Waterfowl Management Workshop. Framing Future Decisions for Linking Harvest, Habitat, and Human Dimensions.* Summary Report. 60pp. (https://nawmp.org/sites/default/files/2017-12/future_of_waterfowl_mgt_workshop_final_report.pdf).

18 A Fundamental Revision of NAWMP – People Conserving Waterfowl and Wetlands

Scoping the Update

The Plan Committee met in Sacramento in January 2009, now with Paul Schmidt, Basile van Havre, and Martin Vargas Prieto as the US, Canadian, and Mexican Cochairs, respectively. Seth Mott and I summarized some thoughts for the Cochairs after the meeting. We urged that the first part of the scoping process should consist of the Steering Committee and the Plan Committee, focusing on what coherence entails and what our long-term objectives respecting coherence ought to be. Then, in pursuit of that vision, decide what we could hope to accomplish within this next update. For instance, what decision frameworks could be erected? For how many species? In what timeframe? Clearly, the journey toward coherence would not be achieved in a single step. Plus, there were all the process questions about how we might accomplish the preparatory work. What would the consultation process entail? What would the decision process be? And then there were the 27 recommendations from the Continental Progress Assessment!

Complicating these questions, Andy Raedeke (MO) reflected (pers. comm. April 2025) was that at the same time for State and Flyway folks, three major issues were demanding attention – the JTG report and its challenge around coherence, the AHM Task Force recommendations about harvest management, and preparations for the next Supplemental Environmental Impact Statement on waterfowl hunting. The latter two were time-sensitive policy matters and essential for continuing to manage annual waterfowl harvests.

In March, the Plan Committee announced the formation of an Update Steering Committee (USC; later rebranded as the Revision Steering Committee, RSC)[1] to begin the work. In May, eight additional members from Mexico were designated for the Committee.[2] The Update would propose the development of coherent waterfowl management objectives and lay out a process for developing decision frameworks that could be used as the common basis for harvest and habitat management.

Ahead of the first USC meeting, Seth and I again urged that our first focus should be articulating a "clear description of current circumstances in harvest and habitat management that justify action and a clear and explicit statement of the desired outcome(s)". We urged that the USC state expressly what we wish to achieve in the long term and why, and what we ought to accomplish through this iteration of the Plan. We needed a clear statement of purpose for *this* Update, this first step.

The USC met face-to-face for the first time in Minneapolis in April. Interestingly, the new group quickly went to a question that the Joint Task Group debated years earlier, namely: "Where's the Fire"? So, what if there is a lack of coherence in waterfowl management? USC members soon recognized that many people in the waterfowl community remained unconvinced of the need for change. The "yield curve" concepts seemed opaque, or at least not compelling to

DOI: 10.4324/9781003732310-18

many nontechnical folks. Recall that 67% of attendees at the Minneapolis Summit polled on the final day indicated they felt neutral or weaker in their own understanding of the putative benefits of a linked, coherent management system. Many (43%) also felt neutral about their personal readiness to work toward coherence within their agency. So, the USC debated at length its own positioning and the communication and consulting work that would be needed to achieve unity about the path forward. Because several members of the USC were previously engaged in the IAFWA task force on hunting, the new group also dug into thoughts about hunter objectives and their relationship to habitat conditions and harvest regulations. The USC concluded that this update should be characterized as evolution rather than revolution to help bring people along. The former JTG members reminded us that no single body oversees all these systems. Thus, a "champion" or some confluence of working groups would have to become the node where all these ideas converge. Consultation with the NAWMP Science Support Team (NSST) and AHM Working Group seemed needed to ascertain what they could tackle during the next few years.

At the end of the meeting, the USC agreed that it was time to revise the Plan and that we needed to pursue a comprehensive consultation process in scoping the effort. It would likely incorporate aspects of both hunting and habitat conservation. They also supported putting all this in an adaptive management context to enable learning over time. One major technical advance seemed to present an opportunity. Structured decision-making frameworks (National Research Council 1996; see too https://www.structureddecisionmaking.org/) provided for the first time a robust way to consider multiple quantitative objectives and the trade-offs among them. Such tools allowed the assignment of various weights to various objectives which, among other things, could enable consideration of the effects of harvest or habitat actions on people more explicitly.

Box 18.1 Update or Revision?

In the many documents generated by Plan partners between 2008 and 2010, the work embarked on was variously described as a 2009–2010–2011–2012 Update or Revision. After reviewing the new material from the USC, the PC finally decided that, henceforth, this effort would be labeled the first Revision of the Plan (whether 2011 or 2012), because we were intending to reconsider the fundamental goals of the 1986 Plan, clearly a "double-looping" step in the parlance of adaptive management, perhaps even "triple-looping" if we opened the topic of governance as well (Pahl-Wostl 2009).

In the USC's first meeting, there emerged the first specific suggestion that an integration oversight group may be needed to help advance the ideas emerging during the Revision – the first allusion I'm aware of that an Interim Integration Committee (IIC) would be born as a product of the 2012 Revision. They also concluded that the Revision ultimately would need explicit objectives for populations, habitat (K, net landscape change), and stakeholder participation (hunter numbers) – presaging the objectives that emerged in the 2012 Revision. The USC also wanted to capture the cultural aspects of waterfowling, biodiversity concerns, provision of ecological goods and services, and economic aspects of NAWMP's work. With this expanded notion of stakeholder engagement, the USC decided that an optimistic target for a "final" draft would be March 2011, which indeed turned out to be optimistic. Action items and individual assignments followed, including developing a consultation plan. An outreach plan and the first of several communiques soon followed.

Another immediate need was scoping the required technical work and how it might be accomplished. To tackle this, a technical subgroup[3] met at Patuxent in June to frame the technical challenges for revising NAWMP. They concluded that the waterfowl community needed to reach a collective agreement on the desired future state of waterfowl management, comprising:

- Clearly, articulated objectives for populations, habitat conservation, harvest opportunity, and hunter participation with acknowledged trade-offs among them.
- Integrated modeling frameworks for harvest, habitat, and hunter management decisions.
- Integrated monitoring and assessment protocols.
- Sufficient progress in creating this framework to develop and sustain the momentum needed to conclude and fully implement it over time.
- An informed estimate of the timeline and resources needed to complete the coherent integration of waterfowl management.

They posited that the critical first step was a clear articulation of the fundamental goals of waterfowl management, i.e., at the highest level, what does the waterfowl conservation community want to achieve for waterfowl populations, habitats, and people? Consensus likely would require a new approach to stakeholder consultation. No venue nor institution existed to facilitate such deliberations, and there was limited expertise to lead such a consultation.

Faced with the need to examine harvest, habitat, and hunters in a more integrated manner, and believing that the yield curve (Chapter 16) could be used as a framework to support that work, the technical subgroup offered that the three highest order decisions seemed to be:

- Harvest policy: Where on the shoulder? Many inherent trade-offs.
- Habitat: What do we want Continental K to be numerically, and what would that mean on the land?
- Participation by hunters: What number of hunters is desirable?

Answers to those questions would affect many other things and breeding population goals could be derived from various combinations of them. For instance, we could choose, perhaps, a 90% shoulder point, K at 115% of present conditions, and stable hunter numbers as concrete goals. That would set up trade-off analyses where many permutations would be possible, along with key questions such as analyses for which species? My own scribbled notes at the end of the meeting concluded that we had a large iterative consultation process ahead to resolve this – perhaps well beyond this Plan Revision.

With this input from technical experts, the USC produced a status report for the Plan Committee. It recommended statements of purpose and principles, a process for moving forward, and an ambitious consultation plan for building consensus among Plan partners. The USC also included a draft first communique for stakeholders.

The Plan Committee reviewed the work of the USC in Calgary in August 2009. There were some awkward moments involving two new Cochairs unfamiliar with long-standing waterfowl institutions and processes, and what Paul Schmidt (the cochair with deep experience) described as the "long lead-up to where we are with the pending Plan Revision". But over two days, we worked through it. The USC's draft of desired outcomes led to a tense discussion about hunter participation objectives that were embraced by all except CWS (van Havre). With some minor wordsmithing, the rest of the USC recommendations for the first communique were endorsed.

Schmidt summarized that we were embarking on a significant revision; we would examine the fundamental goals of waterfowl conservation; we would incorporate a more explicit vision of waterfowl management; there would be movement forward and the PC would take on a greater leadership role. Acknowledging the volume of work ahead, he offered that completion in 2011 could no longer be assumed, more likely 2012, but the date seemed less important than what we could accomplish.

The PC also agreed that a single coordinator would be needed for the effort – a dedicated professional who could organize and coordinate the working groups, the consultation workshops, and perhaps the writing team, somewhat like Dave Paulin did during the Continental Assessment (Chapter 15). Fortunately, Jim Ringelman was willing to take on this daunting task and was designated for this role in January 2010. Lastly, the PC agreed to issue periodic communiques to the Plan partnership to keep everyone apprised of progress.

The Plan of Attack and a First Communique

The first public reveal from the USC and the PC (https://nawmp.org/sites/default/files/2017-12/Plan_Revision_Communique-01-01-10.pdf) was shared broadly with the Plan community in December. The anticipated questions, "Why Change?" and "Why Now?" were addressed by offering that harnessing advancements in decision analyses should improve the efficiency and effectiveness of conservation actions, improve the allocation of resources within the waterfowl management enterprise, and be responsive to the desires of stakeholders and funders.

The Communique urged that the 2004 Plan Purpose be made more inclusive. *The proposed expanded PURPOSE of the Plan is to sustain abundant waterfowl populations while preserving the traditions of wildfowling and achieving broad benefits to biodiversity, ecosystem processes, and the people of North America. Plan goals will be accomplished by partnerships that conserve habitats and sustain populations, guided by sound science.* The vision for integrated waterfowl management will entail: *Seeking a unified system of waterfowl conservation that features explicit and coherent objectives to guide habitat and harvest programs, and the means for coordinated actions to realize those objectives.* Such a management system would feature: *(1) a set of widely supported fundamental goals for waterfowl conservation that reflect stakeholder values; (2) a decision framework that allows managers to understand and balance trade-offs among these multiple goals; and (3) managers using that decision framework to efficiently allocate resources to achieve those goals.*

To accomplish all this, Plan partners should commit to the following desired outcomes for the Revision, to:

- Achieve a broad consensus on the fundamental goals of waterfowl conservation.
- Reach agreement on clearly articulated objectives and an integrated framework for making population, habitat, harvest, and hunter management decisions.
- Create momentum needed to finalize and implement the decision framework by developing an informed estimate of the timeline, resources, and institutional support needed for completion.
- Review progress made on the recommendations of the 2007 NAWMP Continental Assessment and define actions needed to advance recommendations that remain of high priority.

The communique described an ambitious and comprehensive agenda beginning with stakeholder workshops to first solidify fundamental goals at a strategic level and then address the means for taking actions flowing from those goals. All waterfowl managers, technicians, and administrators were encouraged to participate. They also announced that a webpage had been created by D.J. Case to house the reams of expected documents and data that would be available to stakeholders. That website (NAWMP.org) is still active today as a valuable, accessible repository of Plan documents.

Consultations Begin in Portland

The first stakeholder consultation workshop focused on NAWMP objectives was held in conjunction with the AHM Working Group in Portland, OR, in December. This first session was led jointly by Fred Johnson, Scott Boomer, Dave Case, and Jim Ringelman.

> **Box 18.2 Fundamental vs. Means Objectives**
>
> Workshop leaders characterized fundamental vs. means objectives like this:
> A *fundamental* objective has value "in and of itself".
>
> > Ask "why" you want to do something. If there is no reason other than "just because", then it's a fundamental objective.
>
> A *means* objective serves to accomplish something else.
>
> > Ask "why" you want to do something. If it is to accomplish something else, then it's a means objective.

The workshop began with a plenary discussion of the overall approach for the Plan Revision and the purpose of these initial workshops. A discussion of fundamental vs. means objectives was then followed by breakout groups that reviewed a pre-circulated table of 31 candidate objectives. People were asked to work through the handout individually, classifying each objective as fundamental or means, and adding any vital objectives they felt were missing. Then each breakout group discussed the list to reach agreement on whether each was a fundamental or means objective, to reword objectives for clarity if necessary, and suggest one or more measurable attributes for each fundamental objective. D.J. Case and Associates (2011) reported the objectives selected for each workshop.

After comparing results among all the workshop sub-groups, the top-scoring fundamental objectives were then diagrammed by each breakout group to show how these were supported by means objectives. The breakout groups also discussed what data might be useful in monitoring those linkages. These products were then shared in another plenary session and passed up to the USC team via workshop reports. As this whiteboard product (Figure 18.1) from Scott Boomer's group illustrates, these questions got complicated quickly! In defense of Scott, and as full disclosure because I was in his breakout group, ours was among the clearer depictions of objective linkages drafted that day!

After the Portland meeting, Jim Ringelman summarized the take-home points and began adapting the agenda for future workshops. Presciently, he identified several issues that continued through all the consultation workshops, some of which remain unresolved even as I write this in 2025. For example, Jim observed the need for a clearer articulation of a concise "problem statement" about contemporary waterfowl management. It became obvious during

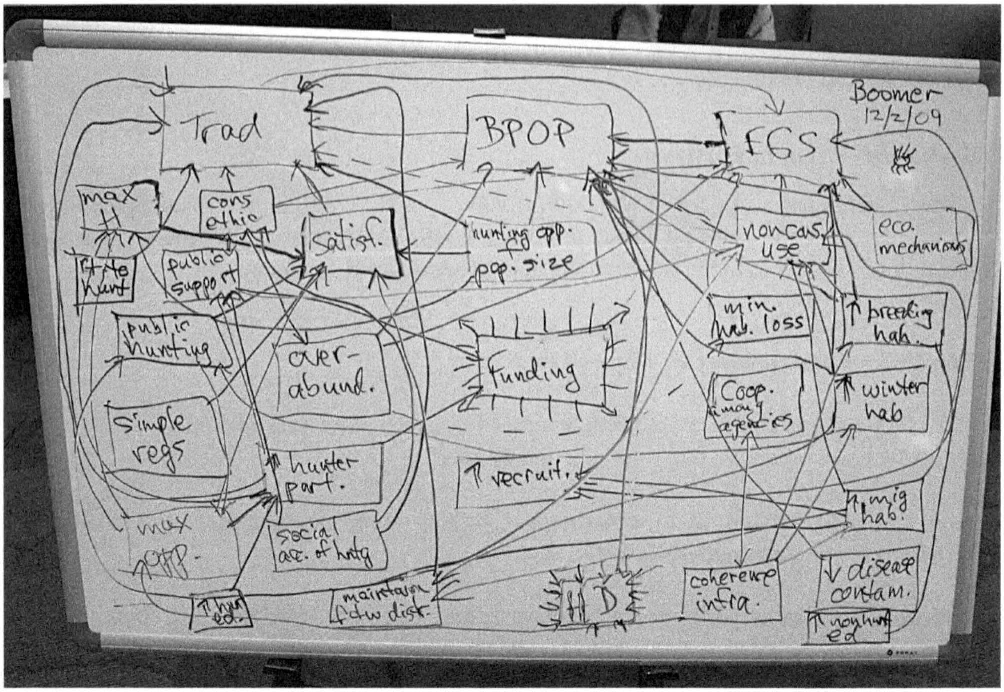

Figure 18.1 One conception of how means and fundamental objectives for waterfowl management might be linked. These were challenging new ideas and our depictions hopelessly complex at the outset.

the workshop that our pre-circulated Purpose Statement, while broadly endorsed, was not sufficient. Drawing from the 2008 Summit and the pre-summit survey, Jim listed examples of problems identified:

- Resources dedicated to conservation are not allocated optimally.
- Too much time is spent setting annual harvest regulations.
- Monitoring and evaluation need to be enhanced.
- Federal and state activities to conserve waterfowl and their habitats have declined.
- Too few resources are directed toward understanding waterfowl hunters.
- Universities and federal agencies are less attentive to waterfowl science and monitoring/evaluation [than previously].

Scott Boomer offered a draft problem statement, which Jim massaged a little and concluded should be discussed along with the problem list above at future workshops:

> The waterfowl management community is not in consensus on the fundamental objectives of waterfowl management, the means to achieve those objectives, nor the framework necessary for integrating multiple decisions in a way that efficiently allocates resources and coordinates actions.

Ringelman's second observation was there seemed to be a lack of agreement on the role of the NAWMP Revision in fostering coherence across waterfowl management. Most participants in Minneapolis agreed that the next NAWMP update should be used to develop more coherent goals for waterfowl harvest and habitat management, but many of the harvest managers in Portland speculated that few folks there realized what that statement meant, and that the process would necessarily be much more complex. The August 2009 open letter from the PC to the waterfowl management community stated plainly that *This revision of the Plan will seek to implement a unified system of waterfowl conservation that features explicit and coherent objectives to guide habitat and harvest programs, and the means for coordinated actions to realize those objectives.* Many harvest managers in Portland pushed back arguing that NAWMP should continue to focus on habitat issues and that it would be "overstepping its bounds" to create new processes or institutions that infringe on harvest management. So, what degree of involvement by the Plan community might make sense? Should NAWMP stick to just habitat issues or create an expanded forum to discuss harvest and human dimensions in addition to habitat? And in the end, should the NAWMP document be just about habitat; provide some architecture for coherence but stop short of recommending a detailed process; or become the primary vehicle to achieve coherence, seeking explicit objectives and new processes? The sense of the Portland meeting favored some modest middle-ground approach.

In a post-meeting memo, Dave Case astutely reminded us that development of a structured process is complicated by the broad scope of waterfowl management that operates across multiple spatial scales and has multiple jurisdictional elements (International, Federal, State, Regional) and legal mandates. Development of shared objectives and operational frameworks to integrate decision-making implied agreement broadly across the waterfowl management community. Case anticipated presciently that the 2012 Revision would be but one step in a sequence of decisions necessary to achieve eventually a coherent system of waterfowl management.

To Edmonton and Beyond

In January 2010, Jim Ringelman agreed to take on major responsibility for coordinating the remaining workshops. The next workshop was in Edmonton, AB, with broad attendance of government and nongovernment people, mostly engaged in the Canadian Joint Ventures. The tone of those discussions was dramatically different from those in Portland with the US harvest management technical community. Canada did not have an adaptive harvest management system, and more pluralistic views of human stakeholders. Thus, the focus was much less on hunter management. Each Round One workshop would, in turn, have its own flavor affected in part by the meetings the workshops were tagged to for travel efficiency.

Between late 2009 and early 2010, three more Round One workshops were held in the United States: Memphis, TN, in conjunction with an NSST meeting; Sacramento, CA; and in Milwaukee, WI, in conjunction with the North American Wildlife and Natural Resources Conference. Two were held in Canada in Edmonton, AB, and Ottawa, ON. Following the experience in Portland, discussing and refining the "problem statement", was added to each agenda. Otherwise, the process described above for Portland carried through all six of the Round One workshops. As in Portland, Dave Case and Associates provided facilitation and meeting support. Ginny Wallace of the Case team helpfully engaged in Edmonton and at every workshop that followed. Results of the workshops were ably summarized by Case and Associates (2011) and are available for viewing at NAWMP.org.

A total of 266 participants attended workshop Rounds 1 and/or 2 and others provided written input. In brief, Round One workshop participants settled on three fundamental objectives for waterfowl management from the list of 31 candidate objectives (up to 39 drafted in some workshops). Those fundamental objectives were to:

- Maintain landscapes capable of sustaining waterfowl populations in perpetuity.
- Perpetuate the tradition of waterfowling.
- Maintain healthy waterfowl populations as part of the North American fauna.

Also captured were ideas about which means objectives were important and, graphically, how people viewed them linked to the three fundamental objectives and other means objectives. Several interesting contrasts and commonalities emerged. Both the US and Canadian workshop participants thought that landscapes capable of sustaining waterfowl populations in perpetuity was fundamental (United States 83%, Canada 93%). Likewise, maintaining healthy waterfowl populations had universal support (United States 94%, Canada 91%). In contrast, on perpetuating traditions of waterfowl hunting, US workshops scored this as 84% fundamental and 14% means while Canadian results were 41% fundamental and 54% means. Interestingly, many respondents favored including distributional objectives (61%) for waterfowl species, perhaps motivated by concerns about harvest opportunity, a discussion point that harkened back to the US National Waterfowl Management Plan (Chapter 2) that predated NAWMP. Given the array of uncontrollable factors that affect distributions annually, however, this was not viewed by most as a practical objective.

Workshops in Mexico

Three additional workshops were held in Mexico using a Round 1/Round 2 hybrid approach. Mexican partners took a typically broader view of NAWMP objectives and recommended six high-level goals:

Habitat conservation: I. Ensure long-term conservation of wetlands and establish measures to carry out restoration projects in those exhibiting deterioration, as essential elements of the landscape that will maintain healthy populations of waterfowl for the benefit of society. II. Establish mechanisms to generate information needed to improve understanding of ecological mechanisms that cause changes in the abundance of waterfowl and their habitat.

Human dimensions: Enhance the culture and knowledge of waterfowl by promoting public education and hunter ethics by ensuring their participation in natural resource conservation and habitat.

Waterfowl populations: Maintain waterfowl populations in perpetuity favoring the recovery of those that are depleted to the point that their numbers represent danger to the heritage of North America.

Consumptive and nonconsumptive use: Ensure the use of waterfowl in a sustainable manner with appropriate promotion and monitoring.

Capacity: Develop technical support and capacities of the various actors involved in management, conservation, and use of waterfowl and their habitats in a manner that promotes understanding the specific circumstances of each country and their forms of administration.

Round Two Workshops

Round Two workshops were designed to share Round One results, to further clarify the chosen fundamental objectives and associated measurable attributes, to discuss how best to formulate objectives in the Plan Revision, to initiate discussions of institutions and processes that could facilitate integrated waterfowl management, and to provide any other feedback to the Plan Committee as they continue advancing the Revision.

In-person sessions were held again in Edmonton and Ottawa plus at five venues in the United States tagged to various other meetings. These were in Grand Rapids, MI, New Orleans, LA, Denver, CO (twice), and Portland, OR. As in Round One, after initial presentations, participants worked in facilitated breakout groups. The final exercise involved a discussion of institutions and processes of the waterfowl enterprise, including how well were we positioned to address our fundamental objectives and what adaptations might position us better.

Participants were also asked to consider a fourth fundamental objective that had been added between rounds by the Revision Steering Committee: namely, "Sustain opportunities for the public to view and enjoy waterfowl and waterfowl landscapes". Not surprisingly, this new objective engendered much discussion and the least resolution. Round Two participants also discussed the nature of potential objectives. For example, 97% of participants agreed that the Plan should include numeric objectives and 84% thought it made sense to have quantifiable objectives for each of the four fundamental objectives (D.J. Case and Associates 2011). But interestingly, only about half thought that the current NAWMP population objectives were inadequate to guide waterfowl conservation. Some 70% still thought that average population sizes over a period of years was the best kind of metric, while some two-thirds favored "no net loss" as an appropriate habitat objective. For hunter objectives, 39% favored numbers of waterfowl hunters and/or days afield as the metric, while 27% were disinclined to include any numeric waterfowl hunter objectives. One-third felt that participation in waterfowl viewing was an appropriate objective, yet nearly as many (30%) thought NAWMP should not include objectives for viewing.

When given four "votes" to indicate the fundamental objectives for which it was most important to have numeric objectives, workshop participants collectively identified populations as the highest priority (181 votes), followed by Landscape Conditions (165), Hunting (48), and Viewing (14). Scaling these choices to 1.0 equals weights of 0.44, 0.41, 0.12, and 0.03, respectively. Clearly, people were advocating most strongly for numeric population and habitat objectives as the Plan had featured from the beginning.

Supplementary to the facilitated workshops, the Flyway Technical Sections and Councils provided written feedback on the same consulting materials. Most were supportive although cautions were offered. The Atlantic Flyway Council said the Revision agenda was too broad, particularly attempting to do too much in harvest management. The Central Flyway opined much the same – that much more consultation would be needed if NAWMP were to get "back into harvest matters", cautioning that setting population and harvest objectives is a very complicated matter. The Pacific Flyway was also concerned with the lack of clarity about how NAWMP might help shape harvest management. The overall message was that NAWMP was straying "out of its sandbox", and that if we were headed toward more formal integration of harvest and population management, the Flyway institutions wanted to be heavily engaged. Feedback from the Mississippi Flyway was more nuanced. Their technical group remained interested in bringing people metrics into the equation. They also believed that if they were going to address declining hunter numbers, it would have to address more than just harvest regulations (Andy Raedeke, pers. comm. April 2025).

Box 18.3 Reflections on the 2012 NAWMP Revision

By Jim Ringelman, Ducks Unlimited Inc., Retired

It was 2008 and I was getting antsy. With about five years to go until semiretirement, I caught wind of intentions to update the NAWMP. My career had long involved me in various dimensions of the Plan: harvest management through the Central Flyway, working with the public as the state waterfowl biologist, or delivery of habitat programs – first in Colorado and later in the Great Plains Region with Ducks Unlimited. I was also deeply involved with the PPJV and other Joint Ventures, and committees including the Continental Progress Assessment team (Chapter 15) and the Joint Task Group (Chapter 16). Knowing that the NAWMP Revision would consider all these facets of waterfowl management, I figured that my experience positioned me to help. So, one fateful day, I approached Paul Schmidt (PC Cochair) and told him that I could probably carve out some time to assist with the Revision. He quickly accepted my offer, saying "his prayers had been answered", because he had been fretting over who would lead the update committee. Ah, Paul, I said I could carve out a little time to assist. That should have been my first clue about what I was getting into!

Discussions within the PC began to outline the scope of the proposed Update. Meanwhile, the "duck heads" were wrapping up the JTG report (Chapter 16) tackling how to bring harvest management and habitat management into alignment because each camp set objectives largely independently of the other. This desire for coherence emerged as a key element of the forthcoming Update, and people started to realize that this could be a big deal. It became apparent that the Update would be a complex undertaking – so big that one day Paul and others proclaimed that we were no longer talking about an Update to the Plan, but a REVISION. And thanks, Jim, for volunteering to lead it!

The surprises didn't end there. Because the Revision would address all facets of waterfowl management, stakeholders needed to have input, and the PC saw the need for an ambitious series of consultation workshops. I thought "Huh, somebody's going to have their hands full going to all of those workshops". Then one day, Dale Humburg says to me, "you know Jim, for continuity's sake, you should help lead all of the workshops". Turns out he was right. And so transpired 15 workshops, all of which I helped lead except the three in Mexico. Thank goodness for Dave Case and Associates.

There were many insightful discussions at the workshops, and even a few revelations. Two really stuck with me. The first concerned how participants with different affiliations viewed establishing species population objectives. Many people employed by state or provincial wildlife agencies considered the existing population objectives for some species unrealistic. They would often cite the population objective for northern pintail as an example, arguing that habitat in the core of the pintail breeding range had been inexorably altered. There was never any chance of it coming back, so penalizing duck hunters by imposing small bag limits was unfair, because "everyone" knew that a reduction in harvest could not materially increase the pintail population. Therefore, they argued, the population goal for pintail should be markedly reduced to a "sensible level" more reflective of the current carrying capacity of their breeding habitat. Not coincidentally, such a change might bring about more liberal harvest regulations. They saw the world mainly in terms of their constituent base, the duck hunter.

In contrast, participants affiliated with the Joint Ventures saw the world differently. Although unsaid, JVs participants in regions that were critical to species whose populations were far short of NAWMP goals enjoyed some advantage in securing funds. Rather than recalibrate population goals to bring them more in line with contemporary carrying capacity, they wanted to retain stretch population goals.

Despite the history of close collaboration within the waterfowl management community, it struck me that people would often look out for the interests of their organizations and stakeholders. I expected that going forward, reaching agreement on revised population goals would be a slow process.

The other revelation was a new topic that was on the back burner when the revision process began. Initially, the imperative to achieve coherence between harvest and habitat management objectives was front and center. This was perhaps in part because several people working on the Revision were the same people who were fresh off meetings of the JTG and the Minneapolis Summit. We were on a roll with yield curves, but the rest of the waterfowl community, not so much. That became evident in the first few workshops when, instead of focusing on coherence, participants were focused on hunters and public support for conservation. Hunter numbers were in decline, agency budgets were getting tight, and society was becoming more disconnected from nature. When participants were asked to consider fundamental objectives, triangles began to pop up on workshop whiteboards. They all were a bit different in detail, but had the same general themes: waterfowl populations, healthy habitat, and people. People?

Thus began a long series of discussions on the role of people in waterfowl management, a topic that was previously in the shadows but quickly moved into the light. "People", in various forms, would become an explicit and integral part of NAWMP. The personal turning point for me came with a letter from Karla Guyn, then Director of Conservation Planning at Ducks Unlimited Canada. By late in 2011, the first draft of the NAWMP Revision had been out for review, and a large stack of comments were piled on my desk. Most remarks were along the lines of "yep, looks good, needs a few tweaks" but Karla's letter to the Plan Committee stood out. I thought she had been rather hard on us on several issues (by then, as the leader of the writing team, I was taking things too personally). But toward the end of her letter, she crystallized the issue that had been nagging me. She wrote, "the greatest challenge to waterfowl conservation in the next decade will not be whether we have an integrated approach to waterfowl conservation, but rather if we have the needed public support for waterfowl management". Yes! In the next draft, the subtitle of the Revision changed from "Responding to Change: A Vision for Integrated Waterfowl Management" to "People Conserving Waterfowl and Wetlands". Karla's comment, and a plethora of triangle drawings, sent us back to the drawing board.

The 2012 NAWMP Revision covered a lot of topics, and I am proud of the contributions it made to waterfowl management. I believe that the explicit inclusion of people in the waterfowl management equation was the most noteworthy contribution of the Revision. Our drafting team took plenty of grief about that decision. In one memorable discussion, a senior and well-respected biologist who was involved in writing the original 1986 Plan told me that it was a boneheaded move to include "people" as an objective in a waterfowl plan. "The plan should focus on waterfowl populations and habitat; people are the beneficiaries" he told me, in no uncertain terms. I suspect that most biologists would rather focus on ducks than people. But as all the workshop "triangle figures" attested, "people" were very much on the minds of a broad range of stakeholders, and inclusion of human objectives was crucial for positioning NAWMP for the future.

Concepts Evolving

The harvest managers at the Portland workshop produced a detailed conceptual schematic of fundamental and means objectives with "maintaining a hunting tradition" at the apex. This hunter-centric concept was rather like the more mechanistic triangle model offered by the JTG in their final report (Fig. 6.3 in Anderson et al. 2007b). A very different model with conserved "landscapes" at the apex emerged at the meeting of mostly Canadian JV people in Edmonton. There the other points of the triangle were "waterfowl populations" and "support". The "maintaining a hunting tradition" objective from the Portland model morphed into simply "support" along the base of the Edmonton triangle – looking more like a means objective than a fundamental one. Other triangles sprang up on whiteboards throughout Round One consultations highlighting waterfowl populations, habitat, or hunters and harvest management in different ways. Eventually, as it appeared in the 2012 Revision, Figure 18.2 was a more nuanced representation of relationships including other habitat values and people beyond only hunters.

NAWMP and Harvest Management Teams Struggle to Find Common Ground

In December 2010, Jim Ringelman invested several days with the HMWG seeking agreement on a path toward integrated objectives. Bottom line? Little progress. There seemed to be little motivation within the working group to change the way US harvest was managed. Reporting

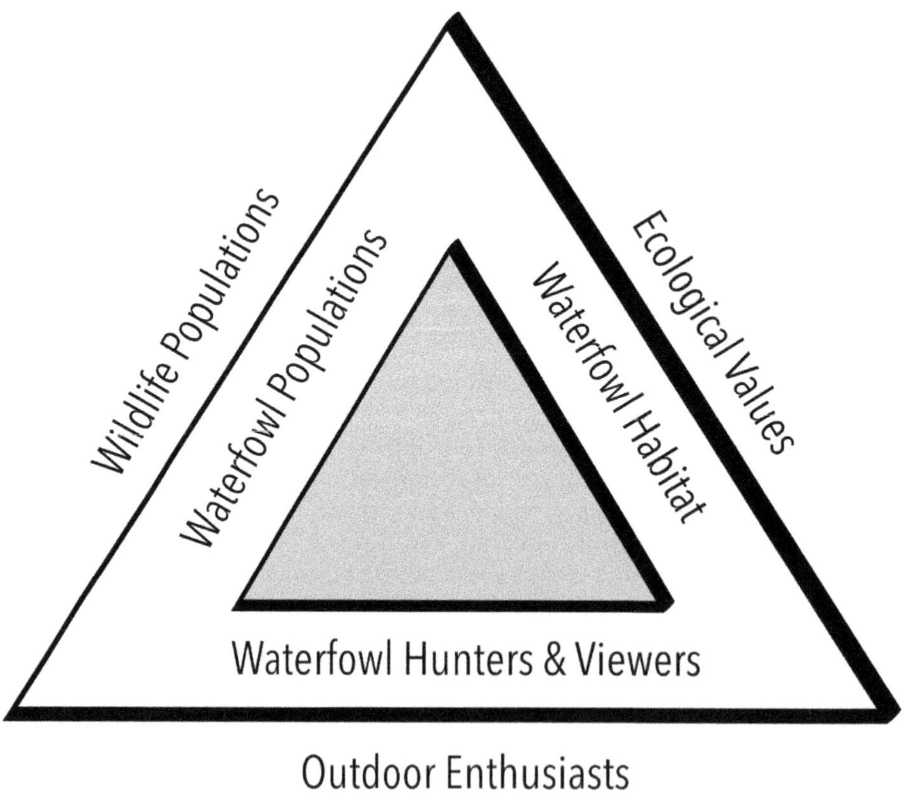

Figure 18.2 A broader depiction of human interests and values from the final draft of the 2012 NAWMP Revision (NAWMP Committee 2012a, page 18) with minor editing by the author.

back to the RSC and the Plan Committee, Jim noted that AHM had resulted in 15 years of stable liberal harvest regulations in the United States, during which time there was significant annual variation, but no consistent trend in midcontinent mallard populations. As for the "adaptive" aspect of AHM, a decade earlier, they had learned that weak density-dependent recruitment and additive mortality were the best supported models of mallard population dynamics and that had not changed with additional years of data. Harvest managers also learned that with partner agreement in advance on a limited set of regulatory packages, and an objective process for selecting a package based on annual data, they had avoided the annual bickering that characterized the pre-AHM decades. They were also learning that the degree of management control they had over harvest rates, population size, or habitat conditions was meager, but their regulatory system was "working".

The Joint Task Group had urged us to bring waterfowl management into a formal, coherent framework, but reflecting on his discussions with the HMWG Jim continued,

> …but after struggling more deeply with approaches to do so via Structured Decision Making, Bayesian Belief Networks and other formal approaches to decision analysis, many of us became convinced that our enterprise is too complex and operates at too many spatial and temporal scales for such an approach to be feasible now.

Jim went on,

> I think the key is setting our quantitative objectives under the fundamental goals in a manner that considers the tradeoffs and interrelationships among all our goals. For instance, maybe the JTG yield curve model can help us reach agreement on balancing our desires for harvest and habitat carrying capacity at a large scale…Then, given the need to maintain continental "K" at a certain level, our step-down habitat goals at the JV level might become something like "no net loss" of functionality compared to some historical time period, and progress might be assessed by estimating net landscape change going forward.

Similarly, we might test promising hypotheses about factors affecting hunter participation.

> As long as some oversight body, with adequate technical support, were in place to steward the whole package of actions, monitoring results, and adaptive re-jigging, then we would have an integrated approach to waterfowl management. Would it be elegant? No. Would it require change? Yes. Is it achievable? I think so. But we need to figure out what such a management system would require, while avoiding building a duck version of a Rube Goldberg machine.

Paul Schmidt and the RSC concurred with Jim's assessment and turned to investing more immediate effort in the Human Dimensions arena where it seemed gains might be made sooner.

Also in December, the HMWG (formerly the AHM Working Group) penned a widely circulated open letter to the Plan Committee (Harvest Management Working Group. December 17, 2010, Memo to the NAWMP Committee) expressing concerns with the scope and direction of the Revision. Specifically, they recommended that,

> the integration of harvest and habitat objectives should be the focus of the current Plan Revision, in keeping with recommendations of the Joint Task Group…However, … formal incorporation of human dimensions into an integrated waterfowl management

> decision framework is premature…because of the difficulty of articulating quantifiable, measurable attributes and the technical challenges of developing modeling frameworks to support the formal consideration of this [people] objective. The course that we recommend is to concentrate on formally integrating harvest and habitat management, while continuing to work on the technical aspects of human dimensions that need to be more fully developed.

Up Next: The Writing Team

With the workshops completed and >300 people having offered advice during these sessions or via written feedback (>400 pages), it was time for the Writing Team[4] to get to work, targeting July 2011 for a first public draft.

The Plan Committee met in Portland in January 2011 with several members of the RSC to digest all this feedback and provide guidance to the writing team. They reviewed the workshop summaries provided by Ringelman and the D.J. Case team with a focus on concerns expressed and constructive advice received. They reviewed a draft outline from the writing team, projected timelines, and a communications plan. Among many details, they agreed on simplified language for the fundamental objectives. The PC also concluded that the best approach would be to produce a strategic-level Revision document as soon as possible for signatures by the three countries, followed by a more detailed Action Plan – the same approach followed in 2004, but in reverse order.

The PC committed to continuing close coordination with the HMWG, the NSST, and the Flyways as well as formalizing a Human Dimension Working Group. The writing team also committed to staying connected with the species' JVs and requested that the PC ensure vigorous stakeholder communication throughout the process.

The RSC was forthright with the PC about the resistance encountered from the harvest management community, whose primary concerns were who would set the various objectives and how. Then, assuming they were set, who would monitor progress and adapt the system? Both technical and administrative bodies would be needed, and if resources needed to be reallocated, who could do that? No single agency or even one country had that authority. The RSC was told simply, "stick to habitat". Few saw a material problem with the status quo of harvest management.

Shortly after the meeting, Paul Schmidt, who had served ably as US Cochair of the Plan Committee for three challenging years, retired from the FWS to take a conservation leadership position at Ducks Unlimited Inc. Greg Siekaniec (FWS) replaced Paul as US Cochair for a short time (until May 2012) before being recruited by DU Canada. As the only DU person on the Plan Committee, I took some grief over this plucking of FWS talent! But I can assure you there was no DU "conspiracy" involved – just two independent searches for talented leaders! Jerome Ford (FWS) subsequently replaced Greg before the 2012 Revision was finished and still serves in the US Cochair role in 2026.

The writing team divided up the workload for fleshing out the outline with all members contributing – sharing drafts, editing, conference calls, and multiple meetings. I remember it as a creative, challenging, and eventually satisfying process. We were simultaneously working on the Action Plan too, with much creative attention from John Eadie, Bob Clark, Mark Koneff, and others. We also worked on terms of reference for the new Interim Integration Committee (IIC) and the HDWG. Ringelman somehow kept track of all the parts and kept the train on the rails and the Revision Steering Committee continued to help by providing constructive feedback on the pieces the writing team kept sending them.

In the fall of 2011, a complete draft of the Revision was sent widely to the Plan partners seeking critical feedback. We were hoping for consensus not only on the main elements but also constructive advice. Scores of responses were received, so the writing team continued to develop the drafts in concert with the RSC and the PC. It was a challenging but productive process.

The Revision Is Signed

The final draft Revision sent to the Plan Committee in December 2011 was titled: North American Waterfowl Management Plan 2012: People Conserving Waterfowl and Wetlands. We had debated the tag line at length but felt this brief subtitle conveyed the essence of what the Plan was about. The 2012 Revision (NAWMP Committee 2012a), signed by the three countries in May 2012, set forth three fundamental, overarching goals for waterfowl conservation:

Goal 1: Abundant and resilient waterfowl populations to support hunting and other uses without imperiling habitat.
Goal 2: Wetlands and related habitats sufficient to sustain waterfowl populations at desired levels, while providing places to recreate and ecological services that benefit society.
Goal 3: Growing numbers of waterfowl hunters, other conservationists, and citizens who enjoy and actively support waterfowl and wetlands conservation.

Goals 1 and 2, in various expressions, had always been foundational to the Plan. Goal 3 was new, explicitly highlighting the importance of people as a driver for what we do, as fundamental to the success of waterfowl conservation, and born out of concern for the ongoing loss of hunters. It also recognized the opportunities presented by people who pursue waterfowl with cameras and binoculars, and the recognition that the Plan can succeed only if waterfowl conservation is relevant to the public broadly. From the outset, the people goal was both fundamental and clearly recognized as a means for achieving Goals 1 and 2.

Additionally, the 2012 Revision recognized that waterfowl harvest management and waterfowl habitat conservation had evolved as separate institutions that lacked coherent, interrelated objectives, and that neither stream had formulated explicit objectives for people. It did note the desire to recruit new hunters, to increase the efficiency of existing programs, and to enhance broad public support for conservation. However, the explicit objective of achieving a coherent and linked decision system, so prominent in early discussions about the Revision in 2009, was noticeably muted in the final version. In large part, this was because of reticence of the US harvest management community to change, and growing recognition of the technical and institutional barriers to achieving a linked system. Yet, there were elements in the recommendations that spoke to a desire for a more integrated approach eventually.

The 2012 Revision provided seven recommendations:

1 Develop, revise, or reaffirm NAWMP objectives so that all facets of North American waterfowl management share a common benchmark.
2 Integrate waterfowl management to ensure programs are complementary, inform resource investments, and allow managers to understand and weigh trade-offs among potential actions.
3 Increase adaptive capacity so structured learning expands as part of the culture of waterfowl management and program effectiveness increases.
4 Build support for waterfowl conservation by reconnecting people with nature through waterfowl, and by highlighting the environmental benefits associated with waterfowl habitat conservation.

5 Establish a Human Dimensions Working Group to support the development of objectives for people and ensure those actions are informed by science.
6 Focus resources on important landscapes that have the greatest influence on waterfowl populations and those who hunt and view waterfowl.
7 Adapt harvest management strategies to support attainment of NAWMP objectives.

The Revision included a discussion of global change pressures, threats to bird populations and habitats, strengthening the pragmatic and emotional ties of people to waterfowl and wetlands, responding to social and environmental change, the value of learning while doing, better integration of the parts, growing support, and important new frontiers for learning. Appendices featured interim population objectives, an updated map, and descriptions of high-priority landscapes for waterfowl throughout the continent.

The Action Plan

As the Revision document was being readied for pro forma national statements, translations, and the frontispiece for ministerial signatures that add a few months to the process, the writing team turned its full attention to the Action Plan (NAWMP Committee 2012b). New elements added were Terms of Reference for the IIC and the HDWG – two legacy products of the Revision. The team also developed thoughts around what eventually became a Public Engagement Team with a marketing focus. The writing team for the Action Plan was supplemented by an informal group of technical experts[5] who provided valuable guidance and served as a sounding board for the contents. Consultations on some technical elements continued. Eadie, Ringelman, and I presented the Action Plan draft to the NSST in Charleston in January 2012 where we received astute and helpful feedback on ideas about the proposed working groups. The PC and the writing team had one last thorough kick at the document in person in Quebec City before it was finalized in December.

An important feature of the Action Plan was to elaborate "key actions" under each of the recommendations offered in brief in the Revision. These were detailed, targeted to specific committees or institutions, and included target completion dates between one and five years ahead. These are all available for review at NAWMP.org. Also featured were technical appendices with suggestions about adaptive management and triple-loop learning, weighting multiple objectives, alternative modeling approaches for integration, monitoring programs, decision tools, and approaches for multi-scale focusing of resources. Appendices also included draft terms of reference for the committees being created out of the Revision. Those follow-on initiatives are the focus of the next chapter as the partners sought to make progress in this broader, more complex conception of the NAWMP.

Notes

1 Paul Schmidt (USFWS), Basile Van Havre (CWS), Ken Elowe (ME), Tom Hauge (WI), Randy Kreil (ND), Ken Mayer (NV), Michael Anderson (DUC), Garry Donaldson (CWS), Pat McGregor (CWS), Ken Babcock (DUI), Steve Williams (WMI), Mark Koneff (USFWS), Dale Humburg (DUI), Seth Mott (USFWS), Mike Runge (USGS), Jorge Coppen (USFWS), Tim Jones (USFWS).
2 Roberto Aviña Carlín (SEMARNAT), Humberto Berlanga (CONABIO), Edward Peters (INE), Oscar Ramírez (CONANP), Javier Sosa (PROFEPA), Joel González (PROFEPA), Eduardo Carrera (DUMAC), Alberto Lafon (Univ. Chihuahua).
3 Mark Koneff (USFWS), Dale Humburg (DUI), Mike Runge (USGS), Seth Mott (USFWS), Min Huang (CN), Guy Zenner (IA), Andy Raedeke (MO), Jorge Coppen (USFWS), Tim Jones (USFWS),

Mark Vrtiska (NE), Pat Devers (USFWS), Rex Johnson (USFWS), Dale Caswell (CWS), Fred Johnson (USGS), and me.

4 Jim Ringelman (Chair), Michael Anderson (DUC), Bob Clark (CWS), John Eadie (UC Davis), Mark Koneff (FWS), Greg Soulliere (FWS), Andy Raedeke (MO).

5 Scott Boomer (FWS), Mike Brasher (DUI), Dave Case (D.J. Case), Kathy Dickson (CWS), David Howerter (DUC), Dale Humburg, (MO), Fred Johnson (FWS), Seth Mott (FWS), and Ken Richkus (FWS).

References

D.J. Case and Associates. 2011. *Stakeholder Consultation Process Results North American Waterfowl Management Plan Revision.* 40pp. (https://nawmp.org/document/stakeholder-input-workshops-results-rounds-1-and-2).

National Research Council. 1996. *Understanding Risk: Informing Decisions in a Democratic Society*. The National Academies Press. Washington, DC. (https://doi.org/10.17226/5138).

North American Waterfowl Management Plan Committee. 2012a. *North American Waterfowl Management Plan: People Conserving Waterfowl and Wetlands.* U.S. Department of the Interior, Environment Canada and Environment and Natural Resources Mexico. Department of the Interior. Washington, DC. 48pp. (nawmp.org/document/2012-nawmp-plan-people-conserving-waterfowl-and-wetlands).

North American Waterfowl Management Plan Committee. 2012b. *North American Waterfowl Management Action Plan: A Companion Document*. December 2012. U.S. Department of the Interior, Environment Canada and Environment and Natural Resources Mexico. Department of the Interior. Washington, DC. 61pp. (https://nawmp.org/sites/default/files/2018-01/NAWMP%20Action%20Plan%20Dec%202012-final).

Pahl-Wostl, Claudia. 2009. "A conceptual framework for analysing adaptive capacity and multi-level learning processes in resource governance regimes." *Global Environmental Change* 19, no. 3: 354–65.

19 Pursuing Multiple Objectives – Finding a Path through Complex Terrain

The time between the 2012 Revision and the second Future of Waterfowl Management Summit in 2017 was a period of diverse and heightened activities by NAWMP partners. As the goals of the Plan expanded in 2012, so did the structure and complexity of the committees and task groups assembled to tackle those recommendations. Progress was made, though not as quickly nor completely as many had hoped in 2012.

The years from 2013 to 2017 saw working groups formed, populated, interacting with one another and striving to help the Plan partnership progress on a much more complex "playing field" than anyone imagined in 1986. There is much information on the work of each team, some of it cataloged at NAWMP.org, but rather than elaborating all their twists and turns, I will offer a compact summary of their progress and the challenges encountered to set the stage for the next comprehensive look at the Plan in 2018.

Adding people to the set of NAWMP objectives was compelling but much more complicated than first imagined. A key question, first raised during the Revision workshops but still not entirely settled in 2025, is whether it is most appropriate to regard people objectives as fundamental or as means to achieving habitat and population objectives? Perhaps they are both? Partners' views of key stakeholders also vary from place to place – from Inuit people in the far north, to canola farmers in Saskatchewan, to ranchers in Oregon, to rice farmers in California, to duck clubs in Missouri, to birders on the Laguna Madre of Mexico. To what extent ought these people be engaged in the Plan? Which objectives related to people should be addressed at what geographic scale? Nationally? By Joint Ventures? By States?

A Human Dimensions Working Group

Emerging as an idea from the 2008 Future of Waterfowl Summit, and bolstered by consultations for the 2012 Revision, a Human Dimensions Working Group (HDWG) was anticipated to become a vital social science advisory body for NAWMP. It was conceived as parallel in many ways to the NSST for waterfowl and wetland biological matters. The initiative signaled a fundamental shift from treating people as passive beneficiaries of harvest and habitat management to engaging them as one of three essential goals.

Formalizing the HDWG meant expanding institutional capacity. Work on people-related matters had accelerated during 2002–2005 in the US harvest management community. IAFWA created an Adaptive Harvest Management Task Force that in 2005 proposed new surveys, focus groups, more communication, and most notably laid out a challenge to develop a national hunter recruitment and retention strategy. That later morphed into an "R3" strategy (adding reactivation) and remains active in 2025. A first *ad hoc* HDWG came together to oversee that work. Much of the impetus for this initiative came from state agencies, most of

DOI: 10.4324/9781003732310-19

which depended on hunting license revenue for their operations. The Joint Task Group report (Chapter 16) further reinforced the need to study and incorporate human dimensions science in waterfowl management. So, the stage was set well ahead of the consultations leading to the 2012 Revision and it's no surprise that plans for a standing HD Working Group consolidated during that time. More on the mission, structure, and composition of the HDWG follows later in this chapter.

The Interim Integration Committee

The 2012 Revision called again for progress in integrating the major elements of waterfowl conservation, including harvest management, habitat conservation, and goals related to people. An Interim Integration Committee (IIC) was envisioned to provide such leadership. It would report to the Plan Committee and be focused on technical solutions, process and institutional matters, and marketing related to integration. Technical work would be pursued with existing harvest and habitat working groups and the new HDWG. Because the IIC was envisioned as transitional to something more permanent, it was given an ambitious completion target of December 2014.

The IIC began work in early 2013 populated with members from the Plan Committee, the Flyway Councils, the NSST, the HDWG, and the Harvest Management Working Group.[1] Their work would require participation and support of the entire waterfowl community if aspirations for integrated systems were to be achieved. The habitat Joint Ventures had been engaged in the Plan Revision and were expected to continue working with IIC as well. Dale Humburg, a longtime Missouri waterfowl biologist and current DU senior biologist with broad waterfowl management experience, was recruited as the Chair.

The IIC began to develop plans and seek resources for integrating waterfowl management elements. But first, they, along with the NSST, were charged with specifying a fresh set of Plan objectives which had not been materially altered since 1986. The IIC and the Plan Committee struck a temporary NAWMP Objectives Task Group and sought input from Plan partners.

Sharing the Game Plan

In January 2013, Dale Humburg and I organized a series of presentations about implementing the 2012 Revision for the 6th North American Duck Symposium in Memphis, TN, and summarized the session in a paper in Wildfowl (Humburg and Anderson 2014). After reviewing the three-year Revision journey, we described seven major action items that should be pursued:

1 Develop, revise, or reaffirm NAWMP Objectives
2 Integrate waterfowl management
3 Increase adaptive capacity
4 Build support for waterfowl conservation
5 Launch an HDWG
6 Focus resources on important landscapes
7 Adapt harvest management strategies to support NAWMP goals

For the IIC, the most urgent of these was revisiting population objectives established in 1986 because so much else would depend on them. Regional population and habitat objectives would have to follow along with strengthening an understanding of the influence of regional habitat conditions on population vital rates.

If an integrated management system could be constructed, it would inform resource investment decisions by allowing managers to weigh trade-offs among potential actions. Such an approach would require linked models, decision tools that could help inform resource allocations at multiple spatial scales, coordination among management authorities and decision nodes, and monitoring and assessment to track progress and enable adaptation. Two immediate challenges were evident. First, how would multiple objectives for waterfowl management be established? Could that be accomplished by existing institutions, or would this require some new entity with overarching coordination functions? Whatever the process, it should be iterative and adaptive. Second, who and how would someone monitor progress toward achieving the objectives and adapt actions considering those results? In 2013, no entity possessed responsibility for all interrelated decision-making that might emerge – not the Flyway Councils, not the Service Regulations Committee, not the Plan Committee, not even any single country.

During development of the 2012 Revision, a technical team tried but abandoned efforts to develop a single formal structured decision-making framework for waterfowl management. They recognized many decision nodes, many decision-makers, and decision cycles operating at multiple spatial and temporal scales (Figure 19.1). The 2012 writing team therefore advocated instead "linked decision processes" and a continuing commitment to adaptive management. Just how to "link" the various nodes and scales remained unclear.

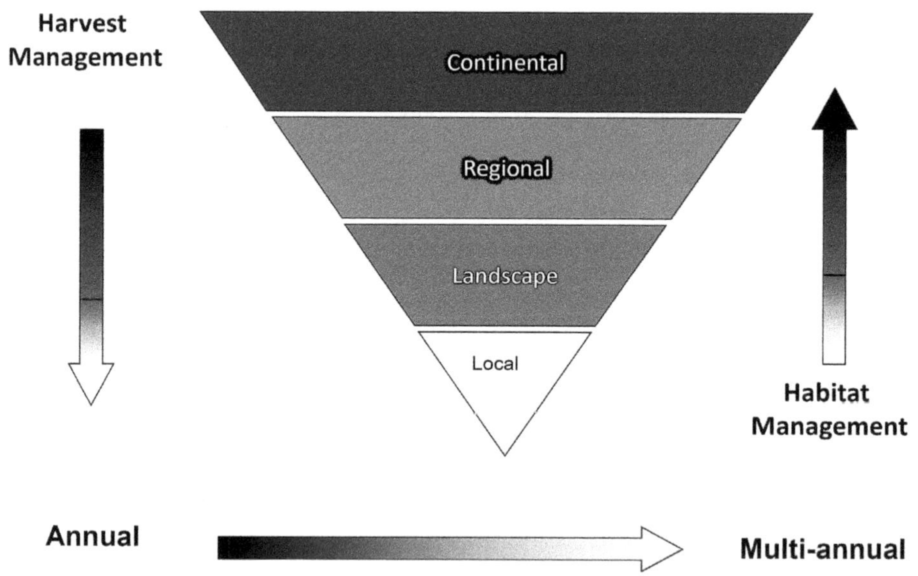

Figure 19.1 Schematic representation of waterfowl management decisions which are made by different decision-making bodies at multiple spatial and temporal scales. Linking these decisions to bring strict coherence to the overall management of waterfowl populations is fraught with challenges. (Illustration by John Eadie, University of California-Davis, used with permission).

The 2014 Addendum

By July 2013, Humburg's IIC had developed a work plan and shared it for partner reactions. Then the team focused on elaborating the three broad goals adopted in the 2012 Revision. The declaration of revised objectives was completed in 2014 (NAWMP Committee 2014). Labeled as an "Addendum to the 2012 Revision", the brief report laid out new objectives for populations, people, and habitats. Falling short of an "integrated" set of objectives, the Addendum nonetheless offered thoughts on how progress might be made in that direction.

2012 NAWMP Goal for Waterfowl Populations: Abundant and Resilient Waterfowl Populations to Support Hunting and Other Uses without Imperiling Habitat

Population objectives: Two notable changes were recommended to the duck objectives that had guided NAWMP since 1986. First, the average of the 1970s was set aside and replaced with longer term average duck population levels in both the mid-continent surveyed area (TSA) (1955–2014) and for the first time, the average populations estimated in the Eastern Survey Area (ESA) (1990–2014). Second, the team recommended a parallel objective for each species set at 80% of the historical range of population estimates. These were proposed as *dual* objectives, not alternative objectives, recognizing that because of the annually dynamic nature of waterfowl habitats, some years of exceptional reproduction (or winter carrying capacity) are needed to occasionally boost the population, recognizing that leaner years are inevitable. The notion applied readily to the prairie pothole region where natural wetness varies strongly from year to year; it proved less straightforward for some migration and wintering areas where providing super-abundant habitat might require greater annual expenditures. The clause above, "without imperiling habitat", recognized current concerns about the expanding population of white geese that were causing damage to arctic ecosystems requiring unprecedented actions by governments (Chapter 7).

Because long-term population trends had varied among species, these "stretch" objectives were aspirational for some (e.g., scaup, northern pintail, American wigeon) but not for others (e.g., gadwall, northern shoveler) whose populations had been trending upward. We will return to this dual objective concept in the 2024 Update (Chapter 23) as it proved helpful for a subset of JVs but vexing for others.

The addition of objectives for eastern populations proved useful for eastern JV conservation planning and for linking to adaptive harvest management models for American black ducks and Eastern mallards. Population objectives for less well-surveyed species (e.g., many sea ducks) and geese were updated in the 2012 Revision wherever better data existed. Thus, in the 2014 Addendum, only the ten most common species plus total ducks in the TSA, and six species plus total ducks in the ESA were included.

Caveats noted were that population objectives should reflect societal desires and values, but these had not been quantified and would likely vary from place to place, among stakeholders (e.g., farmers, hunters, birders) and among waterfowl species. Lacking such data, the planners simply assumed that if continental waterfowl populations were managed to meet the desires of waterfowl hunters, the desires of viewers and the public would also be met. Even for hunters, uncertainty remained about whether maximum harvest was considered more important than opportunities to "see lots of ducks", which could affect management strategies. Surveys proposed by the HDWG would soon probe these and related ideas (e.g., season lengths, bag limits, viewing opportunities). Debate also continued about whether harvest management should or could have a role in achieving waterfowl population objectives.

2012 NAWMP Goal for Waterfowl Supporters: Growing Numbers of Waterfowl Hunters, Other Conservationists, and Citizens Who Enjoy and Actively Support Waterfowl and Wetlands Conservation

Without more data on hunter satisfaction, desires of birders, or others who might value waterfowl, planners were limited in their options for human objectives and settled on essentially objectives of no-net-loss:

> Revised Objective: Increase waterfowl conservation support among various constituencies to at least the levels experienced during the last two decades.

Numbers of active hunters in the United States and Canada, between 1999 and 2013, were used to establish these numeric objectives. Other indicators selected were numbers of US waterfowl viewers who traveled a mile or more from home to view waterfowl (14.4 million annually during 1996 to 2011), numbers of US waterfowl viewers who traveled out of state to view waterfowl (4.6 million during 1996 to 2011), and numbers of self-declared birders in Canada (4.7 million). Also benchmarked were the sales of Migratory Bird Hunting and Conservation Stamps in the United States that averaged about 1.6 million with average revenue of $23.5 million; and the sale of Migratory Game Bird Hunting Permits in Canada at 178,000 with annual revenue of ca. $3.2 million. The Addendum also urged developing a metric for increased numbers of landowners participating in habitat conservation programs but implemented at the JV scale.

2012 NAWMP Goal for Waterfowl Habitat: Wetlands and Related Habitats Sufficient to Sustain Waterfowl Populations at Desired Levels, While Providing Places to Recreate and Ecological Services That Benefit Society

The Addendum concluded that,

> Unlike objectives for waterfowl populations and supporters, which can be expressed in large-scale terms, quantitative objectives for waterfowl habitat are largely unique to specific landscapes. Certainly, broad goals, such as "no net loss" capture the intent of habitat conservation, but the actions that account for waterfowl habitat delivery are most relevant at local and regional scales.

Thus, Joint Ventures require habitat objectives specific to their location and stakeholders, which was much the way NAWMP had operated since the 1980s.

> Revised Objective: Conserve a habitat system with the capacity to maintain long-term average waterfowl population levels, to periodically support abundant populations, and to consistently support resource users at objective levels

Conservation planners were thus urged to maintain the landscape features that produce periodic duck abundance (e.g., keep wetland basins intact even in dry periods). So, the Addendum reaffirmed that planning to support goal-level populations of birds, on any Joint Venture landscape, depends on what the birds need at that time and place and the "tools" and policies at a manager's disposal. These are essentially local planning activities so Flyway, National, or Continental habitat objectives only make sense as an accumulation of needs at local scales. Insightful habitat planning will never be a top-down continental process. Logically, this made sense, but in terms of direction to the JVs it was scarcely more helpful than advice from earlier NAWMP Updates.

The HDWG Launches

The HDWG had its first formal meeting in Bluffton, SC, in May 2013, chaired by Cal Dubrock, a state wildlife administrator with a passion for building social science knowledge to inform management decisions. We spent three days debating the initial work of the group and how it would function. Andy Raedeke (MO) served as the working group's Technical Chair, and Dave Case as the meeting facilitator.

The group finalized a work plan, developed internal operating procedures, and settled on its first initiatives. The dialog featured lots of "storming" from different points of view – harvest management, habitat management, States, Flyways, Canada, and the United States. The group expressed determination to link closely with the IIC, NSST, and PET, and we revised the group's draft Terms of Reverence for approval by the Flyway Councils and the Plan Committee. That dual reporting relationship to the PC and the Flyway Councils proved challenging but ultimately useful in generating support for the Group. Quoting from the Group's first report:

HDWG Mission

> To provide strategic guidance to the bird conservation community to integrate human dimensions into waterfowl and wetland conservation and to assist in developing adaptive frameworks to develop, implement, and evaluate programs and policies that contribute to maintaining or increasing participation in waterfowl hunting and viewing as well as increasing public support.

HDWG Roles and Responsibilities

1 Engage social and biological scientists to iteratively conduct baseline assessments, establish objectives, delineate models, identify and implement actions and monitoring programs that will contribute to maintaining or increasing participation in waterfowl hunting and viewing as well as increase support for wetland and waterfowl conservation.
2 Assist with the integration of human dimensions considerations into waterfowl population, harvest, and habitat decision-making.
3 Facilitate human dimensions research and monitoring to measure success and accelerate learning about how to best achieve NAWMP objectives.
4 Assist and support the Public Engagement Team (PET) to strategically target communication, marketing, and other actions to meet participation, support, and awareness of NAWMP objectives.
5 Provide support to improve coordination of human dimension activities among state, federal, and nongovernmental partners.

HDWG Organizational Structure

Designed to engage an array of expertise, the HDWG was populated with human dimensions technical specialists, representatives from each Flyway, people from several JVs, the USFWS, CWS, HMWG, NSST, NABCI, AFWA, the PET, and a national NGO. Scientists and others with people-management experience were sought to ensure both perspectives. A six-member Leadership Team connected to the Flyway Councils (2), the Plan Committee (2), the USFWS (1), and a Joint Venture (1) was struck to facilitate progress between meetings. All members were charged with helping to identify needed work and liaising with other NAWMP technical groups.

The HDWG did not have resources to fund projects directly. Funding for projects would be developed by partners on an individual basis. Nevertheless, the Group soon stimulated coordinated national surveys of hunters, viewers, and the public, and a novel choice experiment to inform deliberations at the next Future of Waterfowl Workshop scheduled for 2017.

The PET

Years earlier, reacting to the 2007 Continental Progress Assessment (Chapter 15), the Plan Committee created four internal working groups and assigned pertinent Assessment recommendations to each group: Science, Policy, New Funding, and Leadership & Communications. The groups worked independently from 2007 to 2010 but a single Leadership, Funding, and Communication Working Group (LFCWG) merged in 2011. One product of their work was a first NAWMP Value Proposition (https://nawmp.org/sites/default/files/2022-03/Value_Prop_March-sm.pdf) which made a broad case for conservation of wetlands and touted the breadth and success of the Plan partnership as a means for achieving even more. It concluded with a call to the agriculture, forestry, and energy sectors, and governments from the municipal to federal levels to join in.

Activities of all the PC working groups slowed as the attention of members turned to developing the 2012 Revision, but the need for long-term strategies for funding and communications remained. As the 2012 Work Plan coalesced, the PC decided to create two people-focused working groups: The HDWG just described would concentrate on social research. A PET would expand on the work of the internal PC working group. The hope was to entice a diversity of people – legislators, teachers, youth, urban residents, etc. – to build appreciation, and ultimately political and financial support, for waterfowl and wetlands conservation.

With the NAWMP goal of "growing numbers of waterfowl hunters, other conservationists, and citizens who enjoy and actively support waterfowl and wetland conservation", it was imperative to identify specific target audiences, whether they controlled lands vital for waterfowl, or were people who might be most receptive to an engagement campaign. Conserving and restoring waterfowl habitat provides numerous ecological benefits including biodiversity conservation, improved water quality, moderation of flooding, and carbon capture. These notions would be part of engagement efforts.

Box 19.1 Ecosystem Services

For our purpose, "ecosystem services" are the benefits people obtain from ecosystems. These include provisioning services such as food, water, timber, and fiber; regulating services that affect climate, floods, disease, waste, and water quality; cultural services that provide recreational, esthetic, and spiritual benefits; and supporting services such as soil formation, photosynthesis, and nutrient cycling. Often referred to simply as "multiple benefits", this conveys the multiplicity of values associated with habitat conservation.

The PET was intended to report to the PC and comprise a staff from state/provincial agencies, JVs, federal agencies, and NGOs, with at least one member from the HDWG. Reciprocally, at least one PET member would serve on the HDWG. Members should have expertise in communications, marketing, and education. A staff coordinator was also envisioned, but that didn't happen initially for budgetary reasons.

As these events unfolded, Gray Anderson (TN) began serving as Chair of the PC's LFCWG in 2013. PC member David Goad (AR) briefly chaired the new PET before passing away suddenly in 2014. Gray Anderson stepped up to Chair the PET as well. By 2016, its membership had expanded considerably.[2]

The PET's operating premise was that engagement begins with understanding the purpose and direction of Plan actions and how implementing the Plan could enhance collective conservation efforts. Support needed to be built at the highest levels in partner agencies, with elected officials, and in the private sector. The PET first met in person and jointly with the HDWG in St. Louis in 2014, and with facilitation by D.J. Case and associates, developed a prototype public engagement strategy that was refined through 2015, along with deciding how the PET and the HDWG would work together. A few years later, the two groups chose to merge.

Struggling to Build Adaptive Capacity

A predictable consequence of this proliferation of committees and the call to integrate the components of waterfowl management was a crunch in capacity to deliver – both in expertise and funding. In advance of a Plan Committee meeting at Oak Hammock Marsh in August 2013, and in concert with the chairs of the IIC, NSST, PET, HMWG, and HDWG, Seth Mott and I petitioned our fellow Plan Committee members to recognize that the ambitious goals of the 2012 Revision could only be achieved if we approached meeting the science, coordination, and communication work of the Plan differently than we had in the past.

We tried to visualize a process for identifying key decision contexts, linking them, developing adaptive management frameworks to allow learning, and specifying the monitoring and evaluation work needed to support the journey. We concluded that the job of "increasing adaptive capacity" had two main components: (1) developing the technical framework and plans to achieve this and (2) mustering the political and financial support to ensure that the work happened. We thought that existing technical working groups (NSST, HMWG, HDWG, IIC, PET) could eventually devise a workable adaptive decision framework. What remained would be for the Plan Committee and Plan partners to effectively advocate for implementing the framework. That would require international leadership to generate political support, financial support, and perhaps some adjustments in processes among institutions to ensure that needed adaptive loops would function. In short, how might we muster the capacity to set up an adaptive process that would move us effectively and efficiently toward achieving the 2012 goals?

While waterfowl conservation had been successful, it was still based on a mid-1900s model, working mostly in institutional silos. With the adoption of population, habitat, and human goal statements, the 2012 Plan carried with it the need for better integration of institutional support systems as well. The PC had been adequately structured for its initial tasks of overseeing creation of the Joint Ventures, coordinating with the Flyway Councils, and generally guiding the initial development of the 1986 NAWMP. With the creation of the NSST in 1999 and its unfunded science mandate, needs began to grow. Appointments of JV science coordinators in the United States and their part-time assignments to work on the NSST brought some much-needed capacity, but the NSST still struggled to deliver what the PC asked it to do. Coupled with the work of temporary task groups such as the Continental Assessment team, the NSST helped guide the Joint Ventures but fell short of its plans and full potential. Funding for research/planning activities that over-arched multiple JVs remained particularly difficult to find.

The broadened vision of the 2012 Plan had moved the Plan Committee into a new realm that required support for social as well as ecological science. We had conceived a system represented by a metaphorical "three-legged stool" of people, populations, and habitat but to support that new vision required thinking in transformative terms. We argued that greater adaptive capacity under NAWMP had become "mission critical". Though it was unclear how the PC might respond to these concerns, at a minimum, it seemed vital to develop the support of the most senior managers in our various agencies. We acknowledged going in that managing conservation programs in challenging fiscal circumstances involved difficult choices. The revised NAWMP needed to be seen as a high-value option for scarce resources. We thought that Plan partners had great stories to tell. Perhaps what the PET needed to do first was ensure that our existing partners better understand what we are doing and its continuing importance.

We asked the PC to consider what we could be doing NOW to help further implementation of the 2012 Revision. Results of the subsequent discussion were muted. Several PC members noted that few of our stakeholders perceived that there was any current crisis in waterfowl management, unlike during the 1980s. They reported more difficulty in getting the attention of their senior managers or elected officials despite the conceptual appeal of the 2012 Revision. In the 1980s, there were broadly recognized problems (drought, habitat loss, fewer ducks), a new and hopeful concept, crusading leaders, and not nearly the competition for government conservation funds perceived in 2013. Surprisingly, one senior PC member flatly rejected seeking additional federal funds above existing budgets for building NAWMP capacity!

More helpfully, Gray Anderson, chair of the PET, offered to poll PC members to identify key messaging to partners, to be followed by a fiscal plan to address the most urgent needs. Some members urged that we try to link up with the developing America Outdoors Initiative or maybe update our Value Proposition for use by the JVs in their annual crusades on Capitol Hill. In the end, however, little action emerged from the Plan Committee.

In hindsight, it is unfortunate that just as the Plan partnership chose to broaden its goals, the motivation of many agencies to support NAWMP seemed to have plateaued. Headwinds were perhaps the product of a few things. North America was still recovering from the 2008 global financial crisis. Unlike in the 1980s, there was not a widespread sense of crisis in the waterfowl world. Sure, habitat loss continued, in many places more than countering conservation gains, but that was mostly apparent to biologists in the field. For hunters, still the primary advocates for waterfowl, times were good. Liberal hunting frameworks had prevailed for many years. For agencies, their most urgent calls from the public were about other issues – nuisance urban Canada Geese or eagles and wind towers, to name just two.

By 2013, NAWMP was widely viewed as a success story – crisis solved – time to move money and attention to more urgent matters. The apparent success of the Plan, coupled with partially fortuitous high duck numbers (>45 million total ducks in the TSA from 2011 to 2017), begat complacency just as the partnership had set its sights on new work – new committees, new processes, all of which required incremental resources. It was a problem that would remain for more than a decade requiring entrepreneurial cobbling together of resources for any work the NSST, HDWG, IIC, and PET were to tackle…but cobble they did.

The Working Groups Forge Ahead

In an update to Plan partners in May 2017, the PC Cochairs reflected that, "Adding goals and objectives for supporters increased potential relevance of the Plan; however, it also presented a level of complexity that was not entirely anticipated". Indeed!

HDWG and PET Progress

Nonetheless, by 2017, the HDWG had networked with the Flyways and several social scientists to commission surveys of US and Canadian hunters, birders, and the broader public. They hoped that the results would further inform harvest management and public engagement efforts. A series of workshops for engaging agencies in recruitment, retention, and reactivation (R3) of waterfowl hunters was initiated during 2016. Next steps would involve several states implementing R3 efforts in a coordinated and hopefully adaptive manner. D.J Case and Associates were also about to take the pulse of waterfowl organizations and waterfowl technical experts ahead of a second summit on the Future of Waterfowl Management.

The PET in 2016 had produced a revised Value Proposition for NAWMP (https://nawmp.org/sites/default/files/2022-03/NAWMP%20Value%20Prop%20Abbreviated%20version%202016%20FINAL_sm.pdf) and made progress on some elements in its 2014 work plan. The recommendation to build support for waterfowl conservation by reconnecting people with nature through waterfowl, or by highlighting the environmental benefits associated with waterfowl habitat conservation, had seen little progress at large scales but some small-scale engagement experiments by a few JVs had progressed.

Task groups had been established under the PET for engaging waterfowl hunters, viewers, and landowners, though availability of staff and funding limited expansion. The PET had begun engagement with the birding community in 2013 including recruiting members to the group, some as subcommittee chairs, and birders organization were serving in reporting roles to various AFWA committees and other venues. This included the adoption of the PET Engagement Strategy by NABCI in their own HD objectives. The PET also succeeded in attracting several birding community leaders to the second Future of Waterfowl Summit in 2017.

Planners had concluded that engaging private landowners needed to occur at local or JV levels in the context of specific landscapes, land use, and conservation programs. PET efforts to engage JV leaders, USDA, and Provincial agricultural programs at larger scales had struggled. Encouragingly, however, at least 11 regional-scale pilot "experiments" to address drivers of hunter and viewer recruitment, demonstrate ecological values and services of wetlands, or incorporate social science into habitat planning and delivery had been initiated.

NSST Progress

In the years between the 2012 Action Plan and the second Future of Waterfowl Summit in 2017, the NSST pushed ahead on several fronts. Membership of the team grew after the Revision so by 2016 comprised a rather unwieldy 30 members,[3] but had the advantage of broad geographic and subject expertise. Mike Brasher (DU Inc.) served on the NSST from 2005 to 2025, initially as a JV technical person, and for five years on the executive leadership group. I asked him about this period that seemed especially productive. He thought that high productivity was stimulated mostly by the 2012 Action Plan and all the technical needs that flowed from it. He also reflected on the constructive synergies among a core group of members, many connected to JVs, that had been on the team for several years including Brian Sullivan, Greg Soulliere, Bill Uihlein, Diane Eggeman, Barry Wilson, and Mark Petrie matched up with nationally focused professionals like Kathy Fleming, Pat Devers, and more. There was a high degree of trust and mutual accountability among them that stimulated progress despite limited funding and competing job responsibilities.

Progress on the seven main recommendations of the 2012 Action Plan had been uneven, but constructive, nonetheless. After the PC approved revised objectives in 2014, follow-up work by the NSST provided guidance on "stepping down" continental population objectives to Joint Ventures for conservation planning during the nonbreeding period (Fleming et al. 2017). This was advanced further in 2019 after incorporating Canadian harvest/distribution data and winter population data for Mexico (Fleming et al. 2019).

Some JVs had strengthened the translation of regional population objectives into habitat plans. Concerns lingered about consistency in approaches across Joint Ventures and variability in the quality of data used. One NSST subgroup focused on developing daily ration models to help estimate carrying capacity of nonbreeding habitats. Another hosted a landscape change workshop seeking ways to work with the new (though short-lived) US federal Landscape Conservation Cooperatives structure.

Responding to FWS priorities, the National Wildlife Refuge System expanded land acquisition efforts and in 2013 sought guidance for targeting from the NSST and other bird conservation groups. Kathy Fleming (FWS) was involved in that effort, along with Keenan Adams (Refuge Inventory and Monitoring), Sean Fields (PPJV), and Ken Fowler (Refuge Realty). The project (later named TRACT) was built to replace existing NWR acquisition guidance that had been in place since 1987. It comprised three tools: one to prioritize acquisitions to benefit NAWMP using duck stamp funds, one to prioritize acquisitions for other birds of conservation concerns identified by Partners in Flight, the US Shorebird Conservation Plan, and the North American Waterbird Conservation Plan, and one for threatened and endangered species. Both of latter categories would use Land and Water Conservation Act funds. Kathy led a group of NSST members through a structured decision-making (SDM) exercise to assist. Pat Devers (BDJV), Mike Brasher (GCJV), Josh Vest (IWJV), Jorge Coppen (FWS), and Greg Yarris (CVJV) were involved and after the SDM exercise, continued meeting to develop the Refuge application form for NAWMP criteria. The TRACT tool (USFWS 2014) is still used to guide NWR land acquisitions. For NAWMP, TRACT incorporated the geographic and species prioritization rankings developed for the 2004 Update. New considerations related to the 2012 Revision, such as human dimensions or climate change projections, were not sufficiently developed to include at the time.

An earlier NSST committee had developed a map of geographic priorities, based largely on expert opinion, that was appended to the 2012 Revision (Appendix B, pp. 41–42). Subsequently, a new Priority Landscapes Committee (Chapter 21), working with USGS researchers, began to develop a scalable, data-driven decision support tool to assist in identifying landscape priorities.

The findings of NSST committees are often conveyed as reports (i.e., "white papers") not typically suitable for peer-reviewed journals due to their narrow purpose and target audiences. This often led to limited distribution of the reports and eventual obscurity, despite their value to Plan partners. The Team needed a better way to archive its technical products and raise the visibility of the NSST's work. Their goal became to create a citable repository of NSST technical reports and raise the visibility of these products. In 2013, the Team organized and upgraded the rigor of its technical reports (Brasher et al. 2013) ensuring that the work of the Team and its subcommittees would be preserved and made more accessible to others. The guidelines also described what papers would qualify for the archive, a process for internal peer review, and a process for content resolution and approval. The valuable NSST Library can be accessed at: https://nawmp.org/documents?field_document_type_target_id=52.

NSST to Tri-Initiative Science Team to Unified Science Team and Back

The notion of a multiple-initiative science team, comprising landbird, shorebird, and waterbird technical teams, was proposed about 2010. Many JV science coordinators had responsibilities for all birds and, thinking there might be synergies, or at least efficiencies to be gained, they began informally adding a session after NSST meetings to talk about science needs for other birds. In February 2014, the three groups agreed to formally pilot a Tri-Initiative Science Team (TriST) for two years.

Ideas kept evolving and at a NABCI workshop on South Padre Island in January 2015, where many NSST members participated, there emerged the idea of creating a more formal joint NSST/Partners in Flight/Waterbird/Shorebird technical body to pursue common information needs. By the following March, a proposal to create a Unified Science Team (UST) emerged with the objectives of sharing expertise and experiences and helping to leverage greater capacity for science support for all birds. Again, a two-year trial was undertaken to consolidate these teams with a first task of identifying and prioritizing projects to be pursued.

Although these might have become huge meetings, the common experience was seldom more than 25 participants comprising the usual NSST folks and a few from other teams, so the synergy initially sought only partially materialized. After three years or so, the NSST decided to return to its initial focus on NAWMP matters. While the UST trial resulted in some coalescing of thoughts among the taxonomic groups, there did not appear to be marked advantages. One idea that did emerge was exploring broader use of eBird data and a postdoctoral project at Cornell University, funded mainly by Ducks Unlimited, appears to have found potential for helping to parametrize migration location and chronology of waterfowl. Another useful paper from this time focused on landscape change assessments (Coppen et al. 2017).

Progress by the IIC

The NAWMP partnership had been committed in principle to using adaptive management since the late 1990s as evidenced in the 1998 and 2004 Updates, by establishment of the NSST in 1999, and throughout the 2005–2007 Continental Progress Assessment (Chapter 15). The 2012 Revision also urged that integrated management would be most effective if implemented within a rigorous, adaptive framework. This seemed particularly true for decisions involving codependent objectives like population targets. Such linked decisions still needed to be explored in greater technical depth, but even if technical solutions emerged, progress would remain conditional on creating an institutional environment and culture willing to act in a coordinated manner.

After specification of new objectives in the 2014 Addendum, a next key challenge was to explicitly link those objectives to help refine JV habitat objectives and conservation strategies. It was becoming clear by then that habitat objective linkages were unlikely to be practicable at continental or maybe even Flyway scales. Spatial variation in waterfowl populations, habitats and people, and their interrelationships across the annual cycle, seemed to necessitate more regionally based integration and planning approaches.

The notion of tight integration in waterfowl management where elements like hunters, harvest, population size, and habitat objectives are linked in decision support models so that each element is managed understanding the likely impact on all other elements had been a desirable goal since 2007. Making management decisions in such a coherent manner seemed logically desirable, but doing this was clearly much more complicated, technically, but especially politically, than we imagined in 2008. It was a panacea that still seemed far off in 2014.

In February 2015, the NSST and IIC met jointly at DU Headquarters in Memphis to take stock of lessons learned in integrating JV-scale objectives and hopefully build consensus about future approaches. The gathering was the first joint meeting of the IIC and the NSST and participants were a diverse mix of technical people.[4] The meeting report (Clark et al. 2015) documented early steps taken in the integration process, lessons learned, and suggestions that might be helpful in further work. An overarching challenge was to make more explicit how habitat programs/policies might be linked with objectives for waterfowl populations and people at the scale of JVs. People in this case were assumed to include hunters, other conservation supporters, and the broader public. Specifying explicitly the linked habitat-people objectives and associated management decisions was difficult. Yet, all discussion groups made some progress, and identified trade-offs associated with decisions that could most benefit birds, people, or both target groups. Participants, however, recognized high levels of uncertainty regarding assumptions and trade-offs involving human values, and that progress in linking objectives would be slow due to these knowledge gaps and limited JV capacity for more studies. Participants, nonetheless, found value in the dialog, in identifying challenges and possible solutions, in exposing their "beliefs" in an explicit manner, and specifying potential benefits of greater support for wetland and waterfowl conservation. The workshop helped to reveal the obvious and often not-so-obvious relationships among decisions within and across programs.

Recommendations for continuing work on integration included developing and sharing tangible examples of integrated objectives at useful scales, especially some that engaged multiple and diverse stakeholders in their creation. The general advice was to start simple and expect to learn and adapt along the way.

To explore scenarios faced by JVs, the workshop posed three general cases, developing ideas for integration in: (a) high-density waterfowl breeding area; (b) high-density waterfowl wintering area; and (c) low-density waterfowl wintering area. Participants were challenged to identify management decisions that most required consideration of multiple objectives and they were encouraged to consider a variety of approaches (quantitative models, maps, decision trees, influence diagrams, etc.) as methods for doing so. Lastly, the workshop report offered recommendations (pp. 7–9) for how a Joint Venture or a State/Province might productively set up and conduct similar workshops.

Presciently, the workshop concluded on a point that was reiterated again in the 2024 NAWMP Update – that is, it is okay for each JV to approach integration as it deems appropriate given its individual ecological/sociopolitical circumstances. While that may have disappointed some decision-theory purists, such flexibility could alleviate apprehension and encourage more trials. Integration may look different among JVs. The workshop report also cautioned that some early integration approaches may fail and that will be part of the process of learning to manage for multiple objectives well.

Moving on from Memphis, the IIC came to realize that improved coherence of waterfowl management decisions might still be accomplished through focus on a few linked decisions at certain spatial and temporal scales. For instance, harvest management frameworks are developed annually and mostly separately in Mexico, the United States, and Canada. While the Joint Task Group recommended a continental approach, initial progress might be made by linking harvest, population, and habitat goals at the national level, especially in the United States, where the bulk of the harvest occurs. It also seemed that user/supporter objectives might most usefully and easily be integrated with State/Provincial (or perhaps Joint Venture) habitat objectives. Developing a minimal set of multi-criteria decision problems that usefully and pragmatically incorporate the main linkages among populations, habitat, and people could be valuable even if incomplete. Still pending, of course, was resolving what decision authorities and processes would be required to oversee such linked decisions, and how responsibilities might be shared for funding adaptive capacity.

As of 2017, limited capacity and lack of compelling immediate need for closer integration were still hampering progress vs. that envisioned in the 2012 Plan. Also, supporter objectives established at national levels had received only initial discussions at smaller scales. Routine use of social science information in management decisions remained desirable but still had far to go in most jurisdictions. While the IIC provided near-term guidance on integrative efforts, continuation would require further development in the 2018 Update and ultimately, institutional arrangements to achieve integrated decisions and adaptive management well beyond 2018.

Toward 2018

In summary, the broadened vision of the 2012 Plan including objectives for populations, habitat, and human users/supporters expanded the partners' views of decision complexity and support requirements. Long-standing uncertainty about harvest management objectives and the theme of "coherence" introduced by the JTG remained with little progress. Plan objectives for mallards still seemed to offer the most hopeful starting point, although any link to hunter objectives remained elusive. The yield-curve approach, while theoretically compelling, still left unanswered how carrying capacity "K" from a yield curve could be applied to NAWMP habitat programs. In Mexico and Canada, the formal linkage between harvest and habitat management was of little interest.

Changes in the culture of waterfowl management institutions had been progressing slowly. Among these developments were recognition that: (a) small rather than large scales offer the best opportunities for innovation and learning; (b) useful integration of waterfowl management might be accomplished initially through a few linked decisions at selected spatial and temporal scales; and (c) there remained a need to clarify decision authorities and processes to oversee any linked decisions, and responsibilities for ensuring related adaptive capacity.

Clearly, integration of population, habitat and people objectives, and actions had proved more challenging than envisioned in 2012. The specificity needed around decision contexts, species, scales, and processes were daunting and the paucity of information about human desires limited the range of models informed by data. Adaptive capacity remained severely limited but there were a few encouraging examples of linking populations and habitat such as the PHJV Assessment, and nonbreeding JV energetic models.

Three-way integration of populations, habitat, and people turned out to be technically very difficult. Such mathematical optimization is fraught with challenges. Some skilled analysts like Scott Boomer and Dave Howerter worked hard on this. It seemed most tractable to anchor on one variable (habitat? people?) and then model other goals contingent on the anchor goal. Then look for spatial synergies and trade-offs. Decide on weights (equal?). But then, how to choose to spend money where it is predicted to be most effective is a policy question beyond modeling!

The recommendations in the 2012 NAWMP Action Plan still seemed relevant and provided a foundation for a second Future of Waterfowl Management Summit (FOW2). Collectively, however, we were thrashing in a tangle of committees, with no new funding except what the teams could cobble together project by project. Reflecting on trying to corral the HMWG, NSST, IIC, HDWG, PET, the Flyway Councils, and the Plan Committee, it's a wonder the system didn't collapse from the coordination challenges! We clearly needed FOW2 to help frame the next Update. It also seemed clear that the integrative functions of the IIC would need to be vested in a more permanent body after the 2018 Update.

Today, partners are still trying to reconcile some of the complexities introduced in 2012. We are still working through a decade-long "mess in the middle" as Dave Case is fond of describing such periods in planning cycles. Many of us felt this complexity and the glacial pace of progress at the time, but I hadn't stepped back then to consider what we inflicted upon ourselves, with all good intentions, about people objectives, coherence, elegant decision processes, and more! The second Future of Waterfowl Management Summit and the 2018 Update would hopefully provide a clearer path through the complexity we had embraced in 2012.

Box 19.2 If It Weren't for People, This Job Would be Easy!

By Dale Humburg, Missouri Department of Conservation and Ducks Unlimited, Retired

Sufficient habitat, a bit of luck with the weather, and ducks will produce more ducks. Hunters will be happy. Obviously, this is oversimplified; however, it captures the general perspective of waterfowl management during much of the last century. Harvest management accounted for much of biologists' management efforts.

The last 40+ years have seen the purposeful expansion of waterfowl and wetland conservation planning and implementation. The biological foundation was strengthened during the first decade of the new millennium as a Continental Assessment focused on the strengths of waterfowl and wetland management and the work remaining (Chapter 15). Assumptions about hunter satisfaction were tested too as the management community formally added a third element to the Plan goals – people (Chapter 18). And people meant not just waterfowl hunters. Trying to satisfy all hunters is difficult, just as providing for different species in various habitats and across seasons complicates landscape conservation. Setting seasons to "keep everyone happy" is an impossible task, so, adding preferences of nonhunters, and provision of ecological goods and services to diverse constituencies doesn't seem very smart on the surface!

Yet, for the Plan to remain relevant, it was important to evolve. Management at a continental scale requires knowledge of habitat conditions, bird numbers, and factors that affect production and survival. Habitat management at different latitudes is directed at seasonal waterfowl biology and associated requirements. Then a few ambitious biologists (OK, me among them) decided to add "people" objectives to the mix. Different attitudes, perceptions, and objectives further complicated an already challenging conservation and public use picture. "I just want to see ducks" vs. "I want to shoot ducks" presents managers with a balancing act, and these competing objectives play out mostly at a more local scale. Pursuing people objectives also required new professionals – those with human dimensions training – who up to that point had been participating on the margins. Public engagement became a central component of waterfowl management.

Management of waterfowl in North America over the past 60 years is largely a success story, but that will continue only with ongoing vigilance and conservation investments. This "system" is not just administrative, but includes the interrelated elements of waterfowl populations, their habitats, and the relationship with people. A basic premise from the 2012 Plan was that effective integration among conservation programs and partners should occur at all scales at which decisions are made and management actions implemented. We now believe that this scale is more often local than previously thought.

Notes

1 Dale Humburg, Chair (Ducks Unlimited), Lyle Saigeon (SK), Ron Bjorge (AB), Natalie Sexton and Ken Richkus (USFWS), David Goad (AR), Michael Anderson (PC), David Duncan (CWS), Mike Brasher (NSST), Mike Carter (PLJV), Diane Eggeman (FL), Andrew Raedeke (MO), Jim Gammonley (CO), Dan Yparraguirre (CA), and Dean Smith (AFWA).

2 Gray Anderson, Chair (TN), Tasha Sargent (CWS), Dean Smith (AFWA), Beth Huning (SFBJV), Jennie Duberstein (SJV), Roxane Bogart (USFWS), Dave Smith (USFWS), D. Reynolds (ACJV), Joe

Benedict (TN), Rachel Levin (USFWS), Ruth Ostroff (CVJV), Ashley Dayer (VT University), and Andy Raedeke (MO).

3 Jorge Coppen (Coordinator, USFWS), Joe Fuller (NC), Mike Brasher (DUI), Tim Bowman (USFWS), Andre Breault (CWS), Carol Beardmore (USFWS), Luke Naylor (AR), Pat Devers (USFWS), Josh Vest (USFWS), Kathy Fleming (USFWS), A. Ryba (USFWS), Bob Clark (CWS), Gregg Yarris (USFWS), Tim Jones (USFWS), Kevin Kraai (ND), Anne Bartuszevige (PLJV), Dave Howerter (DUC), Stuart Slattery (DUC), Mark Petrie (DUI), Greg Soulliere (UMRGLJV), Blair Stringham (UT), Dana Varner (RWBJV),Dave Olson (USFWS), P. Blank (ABC), John Eadie (UCD), Anne Mini (ABC), C. Joos (ABC), Sean Fields (USFWS), Josh Dooley (USFWS), and Michael Anderson (ex officio as PC science subcommittee lead).

4 Bob Clark (CWS), Diane Eggeman (FL), Todd Arnold (Univ. MN), Barry Wilson (GCJV), Anne Mini (LMVJV), Kathy Fleming (USFWS), Mike Brasher (LMVJV), Dana Varner (RWBJV), Dave Gordon (USFWS), Gray Anderson (TN), Jim Gammonley (CO), Sean Fields (PPJV) Greg Yarris (CVJV), Joe Fuller (NC), Greg Soulliere (UMRGLJV), Tim Jones (ACJV), John Eadie (UC Davis), Erik Osnas (USFWS), Anne Bartuszevige (PLJV), Dave Howerter (DUC), Jorge Coppen (USFWS), Dale Humburg (DUI). Dave Duncan (CWS), Ken Richkus (USFWS), Rob Holbrook (EGCPJV), Michael Anderson (PC), Luke Naylor (AR), and Josh Vest (IWJV).

References

Brasher, Michael G., Anne Bartuszevige, Carol Beardmore, Jorge L. Coppen, Mark J. Petrie, Christina Sloop, and Gregory J. Soulliere. 2013. *Guidelines for the North American Waterfowl Management Plan Science Support Team Technical Report*. NSST Technical Report 2013-01. 4pp.

Clark, Robert G., Michael G. Anderson, Anne Bartuszevige, Michael G. Brasher, Dale D. Humburg, and Josh Vest. 2015. *Towards Integrating Multiple Joint Venture-Scale Objectives to Achieve NAWMP Goals: Lessons Learned*. Memphis, TN. 46pp. (https://nawmp.org/sites/default/files/2018-02/FINAL%20Integration%20Workshop%20Report%20-%20Lessons%20Learned%20about%20Integration%20Steps%20for%20People%20and%20Habitat.pdf).

Coppen, Jorge, Cara Joos, Anne Mini, and William Vermillion. 2017. *Information Needs to Inform Net Landscape Change Assessment and Cost-Effective Habitat Allocation Decision-Making*. NSST Technical Report 2017-02. 14pp. (https://nawmp.org/sites/default/files/2024-07/2017.02_net-landscape-change-assessment-information-needs.pdf).

Fleming, Kathy K., Michael G. Brasher, Dale D. Humburg, Mark J. Petrie and Gregory J. Soulliere. 2017. *Derivation of Regional, Non-Breeding Duck Population Abundance Objectives to Inform Conservation Planning*. NSST Technical Report 2017-01. 32 pp.

Fleming, Kathy K., Michael K. Mitchell, Michael G. Brasher, John M. Coluccy, J. Dale James, Mark J. Petrie, Dale D. Humburg, and Gregory J. Soulliere. 2019. *Derivation of Regional, Non-Breeding Duck Population Abundance Objectives to Inform Conservation Planning—2019 Revision*. NSST Technical Report 2019-01. 44pp.

Humburg, Dale D. and Michael G. Anderson. 2014. "Implementing the 2012 North American Waterfowl Management Plan: People conserving waterfowl and wetlands." *Wildfowl, Special Issue* 4: 329–42.

North American Waterfowl Management Plan Committee. 2014. *Revised Objectives: An Addendum to the 2012 North American Waterfowl Management Plan September 2014*. 8pp. (https://nawmp.org/sites/default/files/2017-12/NAWMP_Revised_Objectives_North_American_Waterfowl_Management_Plan_Final_9-22-14.pdf).

U.S. Fish and Wildlife Service. 2014. *Draft Land Acquisition Ranking Criteria for the National Wildlife Refuge System*. Technical Report. Version 1.1. U.S. Department of Interior, Fish and Wildlife Service, National Wildlife Refuge System. Falls Church, VA. 111pp.

20 Another Future of Waterfowl Summit and the 2018 Update

NAWMP in 2017

The 2012 Revision of the North American Waterfowl Management Plan signaled a fundamental shift to managing toward multiple goals for bird populations, habitats, and people. The partners also aspired to do this in a way that would enable them to learn and improve over time.

The Revision launched an unprecedented, if sparingly funded, flurry of actions by a new Interim Integration Committee (IIC), the long-standing NAWMP Science Support Team (NSST), a new Human Dimensions Working Group (HDWG), and a closely aligned Public Engagement Team (PET). Along with associated scientists, contractors, and more, they explored more holistic and integrated approaches to waterfowl management. The five years following the Revision were a busy and challenging time for the partnership. We learned much not only about our stakeholders but also that North Americans' wildlife values were changing (Manfredo 2008; Kellert 2012; Kellert et al. 2017) from a mostly consumptive, utilitarian viewpoint to more affiliative attitudes of wildlife as companions. With every social survey completed, new questions kept emerging. Some of the results challenged our preconceptions about public attitudes.

Meanwhile, on the breeding grounds, duck populations were generally abundant, although there were still concerns about species persistently below objective levels (northern pintails, lesser scaup, spectacled eiders, etc.) or so poorly monitored (e.g., many species of sea ducks) that we knew little about their status. Since 2004, species' priorities had been based on their importance in the harvest and population trends from 1970 to 2003. What to do about species where we did not know their status was unclear. How much of NAWMP's resources should be devoted to addressing deficiencies in those criteria? Complex challenges posed by overabundant species continued too, most notably the population explosion of Snow and Ross's geese and temperate-nesting Canada geese, causing concerns about degradation of breeding habitats along with human conflicts in urban areas and with crop depredations.

Net habitat loss was continuing in many regions, despite accomplishments by Plan partners. The combined influences of a growing global human population, increased demands for agricultural commodities, changing agricultural practices, and perhaps climate change continued to deplete the upland and wetland resources on which North American waterfowl depend (Alexander et al. 2015). Annual tillage was again on the increase in prairie regions at the expense of grasslands (Morefield et al. 2016; Statistics Canada 2017). Wetland loss through drainage and filling had continued affecting both breeding and nonbreeding regions (Dahl 2011; Watmough et al. 2017). Water management decisions for endangered fish threatened to limit water availability for rice agriculture in California, and changes in rice agriculture in California and Texas were expected to reduce the availability of this critical winter food habitat for birds (Miller et al. 2010; Fleskes et al. 2012). Sea-level rise was affecting coastal habitats while the frequency of

DOI: 10.4324/9781003732310-20

droughts in the Western United States seemed to be on the rise. Managers realized that favorable moisture conditions in the prairie pothole region were unlikely to persist. Yet, in aggregate, there seemed to be little reasons to depart from the Plan goals and objectives elaborated in 2014 despite the difficulties in achieving them in the short term.

Complexities of integrating decision systems for people, birds, and habitat turned out to be more daunting than first imagined, both for decision support modeling and institutional processes. Decisions linkages across spatial scales from continental to Flyway to Joint Ventures to states/provinces to counties remained elusive too. One useful though still vague notion had been emerging: integrating more of our work at smaller scales. That is, focusing on tractable circumstances for integration and not chasing a chimera of perfect coherence at large scales.

By 2016, almost all the habitat Joint Ventures were in place and maturing at various rates. Most had formed effective regional conservation partnerships. Monitoring and evaluation were still deficient in many, but others were adapting splendidly to new data and delivery experience. Pilot ecological goods and services projects in several JVs were encouraging, but by 2017, there were still only a few success stories in hand. The focus around people was still mostly on recreational users (hunters, birders, etc.) but partners were beginning to sense that may not be enough. In some places engaging new partners might depend less on birds than on municipal water supplies or flood abatement.

Another subtle change was playing out as long-serving waterfowl-focused staff were retiring or moving to new positions. As this happened, the diversity of skills represented in new JV hires began to broaden and we began to see increasing diversity of JV operations. While healthy in many respects, the effectiveness of national guidance lost some traction.

Prioritization of landscapes remained difficult too. The notion of prioritization is welcome if you are in a high-priority landscape, but not so much if you are not! These ideas remained contentious…difficult to achieve consensus… and who can make the decisions to refocus work to other areas? Among activities, yes, several JVs had done that, but among jurisdictions was another matter. Matching money for projects may not be portable. And priority allocations should be based on what? All ducks? Certain ducks? Other birds? Species at risk? Wetland trends? Recreation opportunities? Some combination of these? This matter of spatial prioritization comprised a vexing challenge in 2018 and remains difficult today for every partnership and even for the few organizations that invest in many places (e.g., USFWS, Ducks Unlimited) and should theoretically have an easier time redistributing resources.

In the five years since 2012, new issues were also emerging. One was "How big a partnership 'tent' do we desire"? Where might be the "sweet spot" of scale for gaining synergies while retaining the ability to make progress toward waterfowl goals effectively and efficiently? Said another way, who needs to be at the "table" and at what geographic/political levels, for achieving effective outcomes?

Concerns about the future supply of waterfowl and wetland professionals were emerging too because of changes in academic institutions that formerly trained many of the people now in mid-career. This led some NAWMP thought leaders to pose the question, "Who will mind the marsh?" so that the education pipeline of well-trained professionals persisted (Chapter 21).

The Future of Waterfowl Management – Round Two

With so much activity bubbling within the NAWMP committees, and numerous studies probing the knowledge and desires of Plan stakeholders, the community agreed that it was time again to "take stock" of the enterprise before another update. The first summit in Minneapolis in 2008 was widely viewed as helpful for launching the 2012 Revision, so a "Future of Waterfowl 2" team[1] was struck to plan for a gathering at the hospitable National Conservation Training Center in Shepherdstown, WV, in the fall of 2017. The purpose, in simple terms, would be to assess progress toward achieving the goals and recommendations of the 2012 Revision, explore solutions to problems encountered, and help focus the next update.

While celebrating the accomplishments of NAWMP, participants would be challenged to consider more than just the next five-year window. They would be asked to look ahead a decade or more at contemporary social and ecological trends and provide guidance for the Plan. Most importantly, the summit would seek to identify actions that must be taken to achieve the goals and objectives of NAWMP. We hoped that participants would leave the summit feeling confident about the ability of the waterfowl community to influence those conservation outcomes.

Pre-summit Surveys

Prior to the workshop, the HDWG and D.J. Case and Associates conducted new surveys to elucidate current stakeholders' knowledge and preferences about waterfowl and wetland management. Various surveys targeted waterfowl management agencies, waterfowl and wetland professionals, waterfowl hunters, birdwatchers, and the US public.

The survey of 29 waterfowl management institutions (Flyway Councils, JVs, governments, NGOs) asked about actions they had undertaken to implement the seven recommendations from the 2012 Action Plan. All organizations had taken some actions, mostly small steps, to address some of those recommendations. Paraphrasing D.J. Case's summary of organization responses,

> We've not changed our business model in response to the 2012 Revision, but there has been a convergence of strategies, because we recognize the need to incorporate waterfowl population objectives, expand our support base, reconnect North Americans with nature, articulate socioeconomic benefits of conservation, and better understand what motivates the public.

The survey of individual professionals was designed to capture candid reflections on implementation of the 2012 Revision. Results comprised feedback from 367 respondents (61% reporting rate). Most had worked in some capacity to implement the Plan objectives. Most perceived waterfowl management to be performing well; however, they suggested changes were needed in stakeholder collaboration and university training of biologists. Relatively little success was noted with incorporating hunter satisfaction metrics into waterfowl management options. Similarly, little success was perceived in rallying conservation support among esthetic-oriented users, birders, or the general public. About half of the respondents agreed with the need to reallocate resources among waterfowl landscapes and increase attention to monitoring and evaluation. Respondents also felt that an inordinate amount of time was still being spent on hunting regulations, and that at federal and state levels, program emphasis on waterfowl and wetlands had declined. Respondents thought that more progress seemed apparent on habitat and waterfowl populations than on raising constituent support, awareness of the need for wetland conservation, or interest in waterfowl hunting. Although only moderate progress was perceived on the 2012 recommendations, each was still viewed as important to include in the 2018 Update. Finally, "don't know" was a common response in parts of the survey, indicating a need for greater communication among professionals about NAWMP matters.

The Flock Gathers in the Forest of Shepherdstown

Workshop participants journeyed to NCTC from the United States (141), Canada (23), and Mexico (4) representing federal, state, and provincial government agencies, nongovernment organizations, academic institutions, and industries with land management interests. The gathering included waterfowl scientists, managers, other bird conservation professionals, and social scientists.

The approach for the event was highly collaborative. The first day featured nine plenary presentations, followed by a first facilitated breakout session focused on *relevancy and public engagement* that was tackled through a SWOT analysis of NAWMP, and then, given each group's SWOT results, an effort to identify strategies for addressing each category of weaknesses, opportunities, and threats. The second day began with a short review of Day 1, followed by a second breakout focused this time on *institutions and integration*. Finally, a plenary panel summarized the breakout groups' conclusions. The workshop reports, available online, summarized the ideas generated in the breakout groups, the main insights from the pre-workshop surveys, as well as comments received by attendees following the summit. All the plenary talks are archived at www.NAWMP.org as well.

Synthesis Report: https://nawmp.org/sites/default/files/2018-02/FOW2%20Synthesis%2029%20January%202018.pdf.

Full Report: https://nawmp.org/sites/default/files/2018-03/Full%20report%20on%20The%20Future%20of%20Waterfowl%20Management%20II%20Workshop.pdf.

Primary strengths identified were the NAWMP partnerships, wide use of science in decision-making, land conservation experience, and supportive stakeholders (mainly hunters). Limited resources and capacity (especially human dimensions' expertise[2]), inertia in our institutions, and insufficient communications were seen as weaknesses. These breakout group observations generally mirrored those expressed by respondents in the pre-workshop surveys.

Every breakout group expressed that ecosystem services (ESs) attendant to waterfowl habitat conservation presented opportunities to show how such work benefits everyone. Ideas were offered for making the public more aware of these benefits. One was to tie wetland conservation to human health and safety (e.g., clean water, flood protection) or to sustainable agriculture (e.g., flood irrigation, rice straw management). All the breakout groups sensed opportunities for expanding partnerships; however, some participants also expressed concern that too many partners might lead to a reduced focus on waterfowl.

The *institutions and integration* discussion proved to be more difficult. Integration meant different things to different people, and that was reflected in the diversity of recommendations from these sessions. Ideas included using human dimensions information to help guide what types of habitats to restore and where, to benefit more people as well as birds. The importance of strengthening education of diverse young professionals – future "minders of the marsh" – was fully recognized; however, there were few specific suggestions about how to enhance this.

Elaborating and implementing strategies to achieve the goals of the 2012 Revision had challenged waterfowl partners for the past five years. While FOW2 did not result in breakthrough ideas, it did provide a venue to continue the dialog among diverse conservation partners. Workshop participants together generated a few final messages. An unwavering commitment to conserving habitats across North America remains a powerful, unifying element of the Plan, as does the passion of waterfowl people. But partners need to better communicate their successes and lessons learned. Offering messages from trusted sources about the diverse values provided by the habitats conserved by NAWMP should resonate with nearly all facets of society. Proof of these broader ESs was widely seen as a rallying opportunity. The waterfowl partners' deep experience and strong capability in habitat conservation delivery should be leveraged as we align waterfowl conservation with the needs of other birds and ES values. In some cases, it may be sufficient to remain focused on waterfowl population values when designing habitat projects while engaging others by quantifying the ES values of those habitats, but if not, we may need to consider possible trade-offs of deliberately broader actions designed for waterfowl and ES benefits jointly.

The full suite of goals from the 2012 Revision was still judged as important to pursue. The value of JV partnerships was reaffirmed, with JVs representing a tractable scale for advancing multiple, integrated objectives. JV efforts to broaden partnerships were strongly encouraged. John Eadie's mantra of "think continentally, integrate locally" had broad appeal at the Summit.

There was evident tension throughout the workshop between a desire to increase the diversity of partners and a concern about losing focus on waterfowl. This seemed to emerge both in dialog about all-bird initiatives and ES marketing, signaling more work was needed by the partners to reconcile such expansions.

Plan partners still lacked adequate human dimensions capacity and considered steps like hiring new staff or providing existing staff with appropriate training to address this shortcoming. While existing waterfowl management institutions were seen as functional, further review and possible restructuring seemed appropriate to consider. The silos of harvest management, habitat management, and social science need to be more porous. Areas for improvement included increased integration among policy-level committees, stronger linkage between technical working groups and the Plan Committee, and better coordination of adaptive management across institutions.

Emphases recommended for the 2018 Update included habitat protection and management, monitoring waterfowl habitat trends and conservation success, monitoring waterfowl population abundance and demography, policy efforts to conserve habitats, and engaging greater support from the public at large.

2018 and Another Update: Reinforcing the 2012 Revision

Over its 32-year history, the Plan had purposefully evolved through regular updates that reflected the partnership's commitments to learning and adapting its actions accordingly. In 2016, ahead of FOW2, the PC began to organize for the next update that would review progress in implementing the 2012 Revision's recommendations and identify priority actions for the next five to ten years. Again, an Update Steering Committee[3] (USC) was formed to begin scoping that task and in mid-2016 had a first joint meeting with the IIC to take stock of progress and consider the job ahead.

As usual with Plan updates, the USC and writing team[4] would be centrally involved, drawing on the work and advice of various temporary NAWMP committees[4], assembling a draft, and seeking reviews by Plan partners to improve the product leading eventually to approval by the Plan Committee and the governments of the three nations. The agreed-upon purpose of the 2018 Update was to assess achievements since 2012, digest lessons learned, provide fresh focus on what remained to be accomplished, and help rekindle the passions needed to advance NAWMP in the years ahead.

Given momentum from FOW2 and survey responses from Plan partners, Update leaders quickly decided to reiterate the goals and numerical objectives for waterfowl populations, habitats, and people adopted in the 2014 Addendum. One obvious advance in the previous six years had been using social science to better understand stakeholder desires with the long-term goal of increasing waterfowl hunters and other conservation supporters. Due largely to the guidance of the IIC, the HDWG, and the PET, that effort began while Plan partners continued to conserve waterfowl populations and their habitats.

While adding people-centered objectives had added complexity to decision-making, Plan leaders hoped that gaining a better understanding of the human dimensions of the enterprise would soon provide new ideas and approaches to furthering progress on all objectives. Clearly, more capacity would be needed for adaptive multi-objective management, but tools to assist this effort still were mostly in early development.

Case studies assembled in the 2018 Update described some new innovative partnership approaches by Joint Ventures that reflected the spirit and direction of the 2012 Revision. To illustrate:

- Based on hydrological studies of playa wetlands, the city of Clovis, NM, and the Playa Lakes Joint Venture (PLJV) formalized a partnership to help ensure aquifer recharge and water conservation for the city, marrying the PLJV's waterfowl habitat objectives with the City's water assurance plan.
- In Southern Oregon and Northeastern California, mutual interest in flood-irrigated wet meadow habitats for livestock and waterbirds stimulated new partnerships on working ranches.
- In California, the San Francisco Bay Joint Venture invested in an outreach strategy that helped influence Bay Area voters to support an annual parcel tax specifically for wetland restoration. The successful ballot measure generates about $25 million annually for wetland restoration and public access.
- Black Duck Joint Venture partners used waterfowl harvest, band encounters, and birdwatcher data from eBird to identify where people hunt or observe waterfowl in the Atlantic Flyway. Then, using New York and Georgia as case studies, they described characteristics of outdoor recreation sites available to waterfowl hunters and birdwatchers. The analysis suggested which characteristics had the greatest influence on where people chose to hunt or view birds. The method showed promise for incorporating human dimension knowledge into habitat delivery and understanding potential trade-offs vs. biological objectives (Devers et al. 2017).
- The 2012 Action Plan inspired the Upper Mississippi River and Great Lakes Joint Venture to develop a decision support tool for integrating social and biological objectives at a regional scale. They developed six model-based maps of factors such as habitat areas of greatest importance to breeding and nonbreeding ducks, retention and recruitment of waterfowl hunters and birdwatchers, and reducing watershed impairments contributing to Gulf hypoxia and lost function of Great Lakes coastal wetlands.
- Wetlands in the Canadian Prairie Pothole Region continue to be lost through agricultural drainage. Finding new ways to engage the public and communicate the value of habitat conservation had become a priority for the Prairie Habitat Joint Venture (PHJV). The PHJV invested in research to quantify the environmental costs of wetland drainage as well as the economic and societal benefits attainable from habitat conservation or restoration. Their data had begun to help inform wetland protection policies in Manitoba and Alberta (Pattison-Williams et al. 2018).

These initiatives collectively were encouraging. Yet, while the NAWMP partners had many years of experience in addressing waterfowl population and habitat goals, they faced a steep learning curve in addressing goals for engaging people in conservation. Further analyses of the growing body of HDWG surveys and choice experiments were still needed.

North Americans are increasingly urban, have competing demands on their time, and for many, their dependence on nature has become less obvious. Still, social studies tell us that many people retain an interest in spending time in nature. In *The Nature of Americans*, authors revealed that people value nature in diverse ways, regardless of age, race, ethnicity, residential location, educational attainment, income level, or gender (Kellert 2012; Kellert et al. 2017). Value systems affected include affinity and attraction, intellectual development, spirituality, and symbolism. Thus, the areas we see as waterfowl habitat might help unite many citizens, regardless of background, in working toward conservation.

The 2012 Revision acknowledged that people develop emotional ties to waterfowl and their habitats through a wide range of activities. The common thread though is personal experience. Thus, the waterfowl management partnership faces a two-step challenge to help strengthen people's connection with nature and transform it into support for conservation. The tasks seem to be to deepen human connections to nature while nurturing a greater understanding of the multiple benefits waterfowl habitats provide – hopefully leading to participation in conservation activities. Wetlands and associated habitats help recharge groundwater, sequester and store carbon, improve water quality by cycling nutrients, provide biodiversity services like pollination, attenuate runoff and reduce flooding, and provide places for recreational and spiritual enjoyment (Olewiler 2004), and we need people to internalize this. Plan partners also should strive to connect habitat conservation to issues of broad concern such as healthcare, safety, education, transportation, food systems, and energy. Decisions made in these areas will likely have large-scale impacts on society, land, and water.

Finally, the USC agreed to sustain in the 2018 Update previous NAWMP commitments to adaptive management and the use of model-based strategic planning and evaluation. Key steps in the process after assessing current conditions should include developing models that depict current understanding of the systems; establishing change objectives; predicting outcomes of alternative management actions; monitoring outcomes; and adjusting based what is learned. As a product of initial investments in social surveys, by 2018, partners were already learning things that challenged some widespread assumptions of waterfowl managers. For instance, condensed from the 2018 Update:

1 Waterfowl harvest management had long assumed that hunters desire maximum opportunity to take birds. Results from hunter surveys ca. 2017 suggested that many hunters may be more concerned about the "quality" of hunting experiences, and less about larger bag limits or more days afield. Both hunters and birdwatchers viewed increased travel distance as a disincentive to participation.
2 Waterfowl hunters reported being less engaged in conservation organizations and activities than many in the waterfowl profession had thought. There remain opportunities for greater penetration in this core group.
3 It is commonly assumed that birdwatchers are a larger and more stable population than waterfowl hunters. But the latest North American Birdwatching Survey indicated that the average age of birdwatchers is 59, compared to 47 for waterfowl hunters. While recruiting the next generation of birdwatchers still seems important, recruiting other younger nature enthusiasts may be equally important.
4 Many assumed that educating the public about the benefits wetlands provide may spur people to action. The NAWMP General Public Survey, however, indicated that while almost 90% of respondents felt wetlands play an important role in providing clean water and air, few reported engaging in activities to support wetland conservation. To develop public engagement strategies that compel action, we need to better understand what drives people to support conservation both monetarily and through their behavior.

The contrasts between these four commonly held assumptions and recent survey results highlight the value of social science investments for designing public engagement strategies.

While efforts to formulate and apply social objectives were just developing in 2018, more tangible progress had been made in linking regional population objectives to continental population objectives (e.g., Petrie et al. 2011; Fleming et al. 2019). Some JVs had also been advancing knowledge of the linkages of habitat actions to population impacts on waterfowl, a key step in enabling comparisons of returns on conservation investments.

Since its inception, the Plan had emphasized the need to target conservation resources to regions and habitats most important to waterfowl. The 1986 Plan included a map of "Waterfowl Habitat Areas of Major Concern". Each Update that followed included an iteration of a map identifying important waterfowl habitat areas (1994 and 1998) or areas of continental significance to waterfowl populations (2004, 2012 Revision). Having said that, the map attached to the 2012 Revision was, in large measure, still the product of professional opinion. By 2018, work underway by the NSST Priority Landscape group had begun to provide better data-based tools for assessing this.

Integration: More Challenging than Expected

Waterfowl populations, habitats, and the people who support and derive benefits from them are inextricably linked. Theoretically, then, efficient management actions should obtain from understanding how management decisions affect progress toward all three objectives. Specifically, *integration* in this context means considering the impact of management decisions on all objectives and coupling that with learning about the effects of those actions on the attainment of those objectives. Results then should support the adaptive modification of management decisions to improve returns on investments over time.

As evidenced during FOW 2, that work had proven much more difficult than expected. The Joint Task Group Report (Chapter 16) offered one possible approach to harmonizing NAWMP population and habitat management. However, the technical mechanics of also considering people objectives remained unresolved and the scale at which these interconnections most matter was uncertain (Chapter 19). Many implementation decisions are regional or local so integration may be more tractable at state, provincial, or JV scales, but in 2018 that assessment was still preliminary. Among the more promising opportunities was the work of the NSST and partner researchers in developing decision frameworks that would enable explicit incorporation of waterfowl and people objectives into spatial habitat targeting (Chapter 21). One of the more significant advancements was the demonstration of how spatial guidance for conservation decisions is sometimes highly sensitive to values (i.e., weights) that stakeholders place on the individual objectives being considered.

With reports from FOW2 and the associated surveys in hand, the writing team assembled a draft update for PC review followed by the usual reviews and feedback from the NAWMP partners before arriving at a final report adopted by the Plan Committee (NAWMP Committee 2018). It included eight major recommendations condensed below.

2018 Update Recommendations

1 The conservation of wetlands and associated upland habitats, in concert with managing waterfowl populations through monitoring and harvest regulation, is foundational to the NAWMP enterprise. Plan partners must remain committed to habitat conservation while also acquiring the knowledge and capacity to integrate social science in decision-making.
2 Help people understand the opportunities for outdoor recreation resulting from NAWMP activities and how society benefits from waterfowl habitat conservation.

3 Compel people to take action to conserve waterfowl habitat. Help transform care into participation in conservation.
4 Identify key geographic areas where the best opportunities exist to meet the needs of waterfowl and people.
5 Establish a process to review and update Plan objectives every ten years and provide guidance on implementation.
6 Share knowledge from all work to integrate the Plan objectives. Learn from our collective experience and adjust our guidance using an adaptive management framework.
7 Bolster training programs for future waterfowl management professionals through universities and colleges. Strengthening academic leadership is particularly critical in Canada and Mexico.
8 Better define the roles and responsibilities of the Plan Committee, how it is structured, and functions to help integrate the various technical work groups. Replace the IIC with a new system of connections between the Plan Committee and the working groups.

Unlike previous updates, the 2018 edition introduced only a few new themes. Rather, it largely reiterated the ambitious recommendations from 2012. While it embraced that broader suite of objectives, this was to be done without diminishing the focus on waterfowl and waterfowl habitats. It seemed reaffirming that the 2012 Revision had moved us closer to full awareness of the concepts needed to deliver NAWMP in an enduring manner.

Box 20.1 Reflections on the NSST

Michael G. Brasher, Ducks Unlimited Inc.

NAWMP leaders long insisted that regional planning and conservation programs be guided by the best available science, with gaps in understanding refined as new information became available. Most JVs relied on regional science committees to advise on planning models, habitat evaluations, and identification of science priorities. Technical guidance at continental scales had been offered first by the Continental Evaluation Team and then its successor, the NSST. The latter was populated in 2000, bringing with it greater representation of regional expertise (Chapter 13). Participation was sought from science staff of every NAWMP-sanctioned JV. Added capacity was provided through the participation of technical experts from across the NAWMP partnership. Thus, the NSST served as a forum where JV science staff shared their expertise, collaborated to address technical challenges, and interacted with other waterfowl experts to enhance the biological foundations of the Plan.

My participation in the NSST began in 2005 when I started as the Biological Team Leader for the Gulf Coast Joint Venture (GCJV). After leaving the GCJV in 2019, I continued to serve on the Team as a senior waterfowl scientist for Ducks Unlimited, Inc. Those 20 years comprise a broad basis for commentary, but rather than tell stories or catalog accomplishments, I offer here some lessons learned and a few ideas that might assist the NSST going forward.

Participation in the NSST is voluntary, and its leaders lack authority to compel members to undertake specific work. Progress then requires not only effective leadership but also work that is relevant beyond the committee. Personally, I found it easiest to serve on or lead a task if I knew it would benefit one or more of our GCJV (or DU) science priorities. One of the best examples was our collaboration with Kathy Fleming (USFWS) to produce regional population abundance objectives for JV planning (Fleming et al. 2017, 2019). This work was foundational to regional waterfowl planning during the nonbreeding period, and it continues to be used by the JVs while inspiring further advancements.

JV science staff routinely face technical challenges with regional planning or habitat evaluations and leveraging the NSST's capacity to address them brought added value. It may be useful for future NSST work plans to embrace a subset of shared regional challenges identified by NSST members in addition to priorities delivered by the Plan Committee or emerging from NAWMP Updates. This added relevance might help increase engagement of JV scientists.

The NSST should more frequently serve as a forum for information exchange and peer review of new planning efforts by individual JVs or other Plan partners. We occasionally fell into thinking that discussions at NSST meetings should relate only to our work plan. In fact, hearing about innovative ideas or new analyses being conducted by various JVs provided some of the most inspirational and useful exchanges. Those providing such peer review often benefited as much than those requesting it, often leading to new collaborations. Extending these observations, I believe the NSST could play a larger role in leading scientific discourse of issues affecting waterfowl conservation broadly. I've heard professional colleagues lament the recent decline of organized forums where bright minds assemble to tackle such challenges. The NSST could help fill that void.

A long-standing frustration of NSST members has been the lack of funding to hire external assistance with analytical work. Absent such resources, progress depended on the dedication of NSST members and others who volunteered their time. Accomplishments over the past two decades testify to the shared commitment of this group, yet recent experience revealed the rapid progress that can occur when added resources are brought to bear. Two examples are the work of Anastasia Krainyk and Jim Lyons at USGS, and Nick Masto and Orin Robinson at the Cornell Lab of Ornithology. In a multi-year project, Krainyk led the NSST Priority Landscapes Committee in developing spatial planning tools to identify areas of greater continental and regional significance for achieving multiple Plan goals. Because of its potentially far-reaching influence, this project attracted a large committee, involved numerous virtual and in-person meetings, and required specialized expertise in geospatial planning and structured decision-making. Substantive progress simply would not have occurred without full-time dedicated capacity that Krainyk brought to the team. Nick Masto's work made quick analytical progress in comparing waterfowl migration patterns derived from eBird data to those built from traditional waterfowl surveys. Masto's work provided a first look at the potential application of a rapidly growing new dataset to inform regional conservation planning.

In both cases, the NSST served an advisory capacity, creating a committee to engage with the collaborating researchers on technical details while ensuring the deliverables addressed conservation priorities. External funding for these projects was provided by NAWMP partners – the USFWS and USGS for Dr. Krainyk' s work, and DUI, the UMGLJV, and the Cornell Lab for Dr. Masto. These examples demonstrate the mutual benefits of this approach. Added technical capacity ensured timely and thorough completion of the work, while engagement with regional practitioners maximized the project's influence on conservation. The NSST should strive to secure more such incremental funds.

Additional technical capacity and leadership during my time on the NSST came from university or agency scientists that shared an interest in our work. These individuals were typically mid- or late-career professionals, bringing decades of scientific experience, and mentorship to our discussions. Notable contributors were Dale Humburg, John Eadie, Bob Clark, Mike Anderson, Jim Dubovsky, and Dave Howerter, several of these transitioning from regular NSST membership to *ad hoc* participation as careers advanced. Such leaders were valuable for helping to conceptualize challenges and forging useful paths. They also encouraged younger professionals to assume leadership roles on the NSST, were generous with their time, and embodied a contagious passion for scientific thinking. Many of these individuals have recently retired but Kevin Ringelman (U.C. Davis) deserves recognition for picking up this mantle as an academic leader for NAWMP. I hope for a future in which NSST priorities inform more graduate research questions, promoting tighter linkages of university scientists with Plan partners and the questions we face. A new cohort of professors specializing in waterfowl ecology is emerging across North America, with many holding endowed chairs. I encourage these and other academic and agency scientists to become involved in the NSST, elevating it as a forum through which their intellectual talents can make lasting contributions to waterfowl conservation (see John Eadie's commentary in Chapter 21).

I had the fortune of working with both Jorge Coppen and Tony Roberts in their positions as US NAWMP Coordinator. Although leadership styles differed, each found ways to motivate NSST members to volunteer their time. The structure of the NSST also provided for "leadership from within", as an Executive Committee worked closely with the US Coordinator to identify work priorities and politely nudge committees to make progress. Both sources of leadership were essential as the Executive Committee helped represent the collective voice of the NSST and the broader NAWMP science community.

It is rewarding to reflect on the opportunities the NSST provided for exchanging ideas, promoting consistency in regional planning, and helping shape the future of continental waterfowl conservation. Service on the NSST yielded lasting memories and some of the most impactful contributions of my career. The journey was more enjoyable because of friendships made along the way. These friendships become more than superficial, as the trust and mutual accountability they generated became the backbone of a community that not only enjoyed each other's company but also challenged each other to excel.

More often than not, we rose to the challenge, and the intellectual rewards were heightened because of the diverse skill sets and backgrounds that populated the NSST. I take great pride in knowing that, as a community of waterfowl scientists, we approached our collaborations with humility and always with the goals of the NAWMP as our north star. My hope is that the NSST continues to grow its scientific legacy by asserting itself as the lead forum through which today's pressing scientific challenges are tackled.

Notes

1 Gray Anderson, Michael Anderson, Eduardo Carrera, Mike Carter, Dave Case, Robert Clark (Cochair), Anne Glick, David Gordon, Howie Harshaw, David Howerter, Dale Humburg, Holly Miller, Silke Neve, Paul Padding (Cochair), Tasha Sargent, Paul Schmidt, Jay Slack, and Tim Sopuck.

2 The term human dimensions refers to how and why humans value natural resources, how humans want resources managed, and how humans affect or are affected by natural resources management decisions (Decker et al. 2001).

3 2018 USC: Bob Clark, Jim Devries, John Eadie, Diane Eggeman, Jim Gammonley, Eduardo Carrera, David Gordon, Karla Guyn, Sarah Mott, Paul Padding, Ed Penny, Lyle Saigeon, Barry Smith, Dave Smith, Dean Smith, Alberto Lafón Terrazas, and Dan Yparraguirre.

4 *2018 Update Writing Team*: Michael Brasher, Stephen Carlyle, Jim Devries, Kathy Fleming, David Gordon, Andy Raedeke, Christian Roy, Kyle Spragens, and Dan Yparraguirre.

Survey Development Working Group: Dave Case, David Fulton, Jim Gammonley, David Gordon, Andy Raedeke, and Dean Smith.

Organization Assessment Working Group: John Eadie, Ed Penny, Barry Smith, Dean Smith, and Dan Yparraguirre.

Objectives & Priority Landscapes Committee: Michael Brasher, Jennie Duberstein, James Dubovsky, David Gordon, David Howerter, Barry Wilson, and Mark Vrtiska.

Plan Committee Review Team: John Eadie, Diane Eggeman, Keith McKnight, Dave Morrison, Seth Mott, and Tasha Sargent.

References

Alexander, Peter, Mark D. A. Rounsevell, Claudia Dislich, Jennifer R. Dodson, Kerstin Engstrom, and Dominic Moran. 2015. "Drivers for global agricultural land use change: The nexus of diet, population, yield and bioenergy." *Global Environmental Change* 35: 138–47.

Dahl, Thomas E. 2011. *Status and Trends of Wetlands in the Conterminous United States 2004 to 2009.* U.S. Department of Interior, Fish and Wildlife Service. Washington, DC. 108pp.

Decker, Daniel J., Tommy L. Brown, and William F. Siemer. 2001. *Human Dimensions of Wildlife Management in North America*. The Wildlife Society. Bethesda, MD. 447pp.

Devers, Patrick K., Anthony J. Roberts, Scott Knoche, Paul I. Padding, and Robert Raftovich. 2017. "Incorporating human dimensions objectives into waterfowl habitat planning and delivery." *Wildlife Society Bulletin* 41: 405–15.

Fleming, Kathy K., Michael G. Brasher, Dale D. Humburg, Mark J. Petrie, and Gregory J. Soulliere. 2017. *Derivation of Regional, Non-Breeding Duck Population Abundance Objectives to Inform Conservation Planning*. NSST Technical Report 2017-01. 32pp.

Fleming, Kathy K., Michael K. Mitchell, Michael G. Brasher, John M. Coluccy, J. Dale James, Mark J. Petrie, Dale D. Humburg, and Gregory J. Soulliere. 2019. *Derivation of Regional, Non-Breeding Duck Population Abundance Objectives to Inform Conservation Planning—2019 Revision.* NSST Technical Report 2019-01. 44pp.

Fleskes, Joseph P., Brian J. Halstead, Michael Casazza, Peter S. Coates, Jeffrey A. Kohl, and Daniel A. Skalos. 2012. "Waste rice seed in conventional and stripper-head harvested fields in California: Implications for wintering waterfowl." *Journal of Fish and Wildlife Management* 3, no. 2: 266–75.

Kellert, Stephen R. 2012. *Birthright: People and Nature in the Modern World.* Yale University Press. New Haven and London. 264pp.

Kellert, Stephen R., David J. Case, Daniel Escher, Daniel J. Witter, Jessica Mikels-Carrasco, and Phil T. Seng. 2017. *The Nature of Americans: Disconnection and Recommendations for Reconnection.* DJ Case & Associates. Mishawaka, IN. 364pp.

Manfredo, Michael J. 2008. *Who Cares About Wildlife?* Springer. New York. 228pp.

Miller, Michael R., Jay D. Garr, and Peter S. Coates. 2010. "Changes in the status of harvested rice fields in the Sacramento Valley, California: Implications for wintering waterfowl." *Wetlands* 30, no. 5: 939–47.

Morefield, Phillip E., Stephen D. LeDuc, Christopher M. Clark, and Richard Iovanna. 2016. "Grasslands, wetlands, and agriculture: The fate of land expiring from the Conservation Reserve Program in the Midwestern United States." *Environmental Research Letters* 11, no. 9: 094005.

North American Waterfowl Management Plan Committee. 2018. *2018 North American Waterfowl Management Plan Update: Connecting People Waterfowl and Wetlands*. Environment Canada. Ottawa. 46pp. (https://nawmp.org/sites/default/files/2018-12/6056%202018%20NAWMP%20Update_EN16.pdf).

Olewiler, Nancy. 2004. *The Value of Natural Capital in Settled Areas of Canada*. Ducks Unlimited Canada and the Nature Conservancy of Canada. Stonewall. 36pp.

Pattison-Williams, John K., John W. Pomeroy, Pascal Badiou, and Shane Gabor. 2018. "Wetlands, flood control and ecosystem services in the Smith Creek Drainage Basin: A case study in Saskatchewan, Canada." *Ecological Economics* 147: 36–47.

Petrie, Mark J., Michael G. Brasher, Gregory J. Soulliere, John M. Tirpak, Duane B. Pool, and Ryan R. Reker. 2011. *Guidelines for Establishing Joint Venture Waterfowl Population Abundance Objectives*. North American Waterfowl Management Plan Science Support Team Technical Report No. 2011-01. 36pp.

Statistics Canada. 2017. *Canadian Censure of Agriculture 2016*. (https://www150.statcan.gc.ca/n1/daily-quotidien/170510/dq170510a-eng.htm?indid=10441-3&indgeo=0). Statistics Canada. Ottawa, ON.

Watmough, Michael D., Zhen Li, and Elizabeth M. Beck. 2017. *Canadian Prairie Wetland and Upland Status and Trends 2002–2011*. Prairie Habitat Joint Venture, Environment and Climate Change Canada. Edmonton. 100pp.

21 Enhancing Institutions, Training, and Targeting

Like all Updates to the Plan, the 2018 version highlighted several initiatives that the framers considered vital to advance in the years just ahead. In this chapter, I focus on three of these: (a) a review of NAWMP institutions, especially the Plan Committee itself; (b) concerns about academic training and availability of future waterfowl professionals; and (c) landscape targeting approaches for achieving multiple objectives simultaneously.

Institutional Reviews

As partners grappled with the expansion of Plan goals after 2012, the Plan Committee decided that it was time to review the partnership's technical and governance structures as well. Consequently, during development of the 2018 Update, a temporary Institutions' Committee[1] began to reassess the roles of the Plan Committee and its satellite groups. In November/December of 2017, they interviewed 24 current or former PC members and others closely engaged with the Committee. The team also reviewed records of PC meetings, recommendations and follow-up actions, and evidence of connections to associated working groups. The Plan Committee first embraced a "job description" in the 2004 Implementation Framework. By late 2017, those details had been lost by all but a few long-serving PC members and there was considerable uncertainty about the PC's mission, other than the general sense of leading Plan partners and overseeing periodic updates. The Institutions' Committee report, tabled in January 2018 (NAWMP Institutions Committee 2018), was endorsed by the PC in April, but work remained to consolidate the feedback and resolve initial actions. That work was finished in November 2019, followed in 2022 by a list of performance metrics to be monitored annually by the PC.

Recommendations

The Institutions' Committee offered both strategic and tactical recommendations for the PC to consider, based on the interviews, review of historical documents, and discussions among committee members. The report considered it imperative that the PC identify a small number of primary, specific responsibilities for itself and second, refocus energy and efforts on those areas. The PC alone has policy-oversight responsibility for monitoring and managing adaptively toward the achievement of the multiple objectives embraced in the 2012 Revision and the 2014 Addendum. The following were chosen as the most important "work elements" for the PC:

1 *Evaluation and adaptation* – Establish a process to regularly monitor progress with achieving Plan objectives, and recommend adaptive modifications as needed.

DOI: 10.4324/9781003732310-21

2 *Integration* – Continue integration of NAWMP goals, keeping front and center the goal of providing sufficient high-quality habitat to sustain waterfowl populations at desired levels, while providing places to recreate and ecological services that benefit society. Increasing and broadening conservation support remained a key focus for human objectives.
3 *Linkage of policy and technical efforts* – Success in the two areas just listed will require linkages across both policy and technical entities and between policy and technical bodies.
4 *Connections with JVs* – Enhance connections and communication with the JVs and also promote enhanced JV to JV, and JV to Flyway collaboration and planning. Do so while recognizing the diversity of JVs and evolutionary changes in these entities.
5 *Enhanced communication* with working groups, NAWMP partners, and the public to reaffirm the PC's role as the international voice and champion for waterfowl and wetland conservation.
6 *Funding* – Make greater efforts with federal and state agencies and other partners to develop additional funding support for NAWMP and the work of the PC.

Having urged these general things, the report further recommended that the PC identify specific objectives, tasks, and responsibilities related to the three fundamental goals of the 2012 Revision. Then, determine what, if any, changes in the institutional structures or processes of NAWMP might be helpful for achieving those. Some interesting findings from the Institution Committee interviews remained to be taken up including how to reenergize the PC given recent (2012–2017) increases in waterfowl populations and thus less agency angst about waterfowl numbers; how to recruit talented PC members willing to commit material time to the work of the Plan; how to make more time for strategic discussions vs. simply receiving progress reports; whether to broaden the membership or skill sets among PC members; and how to effectively manage the evolving relationships between the PC and the Joint Ventures. All of these were simply relegated to food for future thought. Energy, commitment, and active leadership remained as important attributes for PC members, and the Institutions Committee wondered if these might be enhanced by a clearer sense of responsibility for tangible outcomes.

An Integration Steering Committee

The 2012 Revision sought to "Integrate waterfowl management to ensure programs are complementary, inform resource investments, and allow managers to understand and weigh trade-offs among potential actions". The Interim Integration Committee (IIC), established in 2013, provided a forum for building consensus around linking objectives; however, by 2017 a more permanent focal point seemed necessary for continuity of this effort. The challenge was to retain a focus on integration but do so in a way that would be incorporated into ongoing operations of the Plan Committee and affiliated technical groups. Simplifying structural arrangements seemed desirable too. It also seemed evident that better communication and coordination were needed among the working groups and policy groups, so that partners might stay more aware of how the pieces fit together. The solution chosen was to assign working group liaison responsibility to individual PC members and to appoint ex officio representatives to the PC from those same four working groups (NSST, HMWG, HDWG, and PET).

Additionally, assigning primary responsibility for coordinating "horizontal" integration among the working groups and the Joint Ventures could be enhanced by a permanent position that would replace the former part-time IIC coordinator. Fortunately, the PC had a great candidate at hand for that role. Diane Eggeman, formerly of the Florida Fish and Wildlife Conservation Commission, had previously served on both the PC and the IIC and had deep Flyway experience. Diane was willing to step into the new role as NAWMP Integration Coordinator, and the new Integration Steering Committee (ISC) was structured to achieve the linkages desired. The initial appointees[2] achieved the desired vertical and horizontal connections and the individuals in these roles evolved as committee memberships changed. All PC and ISC members had explicit responsibilities beyond simply communicating, including attending PC meetings, advocating for the priorities of their working groups, ensuring connections across Joint Ventures and with the Flyway system, and connecting with each other during monthly virtual meetings.

In a report to the Plan Committee in early 2025, the ISC listed work underway on ten items including coordination with a revitalized NAWMP communications committee, assisting HDPET (the HDWG and PET had merged) and JVs with setting up people objectives, encouraging multi-objective decision support planning by JVs, helping to connect better the species and habitat Joint Ventures, and assisting harvest managers with developing social objectives. Despite frequent turnover, the cross-membership model between the PC, the ISC, and with the other working groups under the Plan remains promising.

The Plan Committee

Meanwhile, the Plan Committee and a renewed Institutions Committee tackled prioritizing the PC's work after further consultations with each of the NAWMP technical groups. The second report (NAWMP Committee 2019) provided the PC's response. In describing their roles and responsibilities, the PC retained the two categories of *Plan Leadership* and *Plan Management* established in 2004. Having described those, they identified five high priorities for attention, with anticipated timelines:

1 Improve the Plan Committee's leadership functionality.
2 Implement performance evaluations and reporting (for habitat, populations, people).
3 Adopt a comprehensive role for integrating actions across the NAWMP community.
4 Lead in communications and reporting NAWMP successes.
5 Align internal resources and secure new resources.

Performance Metrics

In December 2022, after lengthy debate and advice from several working groups, the PC, led by Cameron Mack and Jim Dubovsky, settled on performance metrics to routinely monitor and respond (NAWMP PC 2022). These were:

Population Metrics

- Total numbers of breeding ducks (TSA and ESA Annual Surveys)
- Proportions of Time Population objective are met.

Habitat Metrics

- Proportion of stepped-down NAWMP population goals that are currently supported by JV landscapes.

People Metrics

- Estimated numbers of active waterfowl hunters in the United States and Canada – baseline average from 1999 to 2013.

- Reported revenue from sales of Migratory Bird Hunting and Conservation Stamps – baseline average from 1999 to 2013.

Integration Metrics

- Number underway, and progress on human dimension efforts that are explicitly harvest management-related and are being tracked by the HMWG and HDWG-PET.
- Number of JV's using decision support tools to weigh impacts across NAWMP multiple goals/objectives.

Several of these performance metrics should provide annual "signals" of progress to the PC. Others, like for habitat, will take longer to collect and discern trends. The integration metrics are admittedly weak and at best lead indicators of progress to come. We should expect these metrics to evolve over time. Imperfections aside, the PC and its working groups had finally sharpened their foci and means of tracking progress toward Plan goals, hopefully leading to greater confidence and efficiency in the years ahead. As of December 2025, I am unaware of any comprehensive report from the PC on how these ambitious plans have played out. Timely reports to the Plan partnership on the status of these priorities, timelines for completion, or adjustments made may be useful.

The Need for Skills and Training

Waterfowl and habitat conservation had been supported historically by a close relationship between managers and applied researchers. Some of this occurred within agencies or NGOs, but much of it depended on cooperation with universities. Those academic institutions also graduated generations of hopeful waterfowl and wetland ecologists who brought their talents and passions to waterfowl conservation (Williams and Castelli 2012; Roberts et al. 2018). Alarmingly, the number of universities offering specialized training in waterfowl and wetland management has been in long-term decline (Kaminski 2002, 2013). There has also been a long-standing concern about the availability of waterfowl scientists and managers in Mexico (Chapter 15).

To address this concern, it seemed vital for academic institutions to be part of the solution. The 2018 Update called for a North American Waterfowl Professional Education Plan (NAWPEP) to be developed by yet another special committee[3] by 2020. Chaired by Diane Eggeman, the team developed a proposal for a new education initiative, terms of reference for a steering committee, and identified several early steps to assess current supply and demand of wetland/waterfowl university graduates. John Eadie and the NAWPEP committee provided a very useful summary of these ideas in a webinar archived at NAWMP.org, entitled "Who will mind the marsh?" (https://nawmp.org/content/nawmp-webinar-series).

The ambitious goal of NAWPEP, endorsed by the PC, was to

> engage and assist universities, colleges, and all NAWMP partners with establishing, sustaining, and enhancing academic and experiential programs in waterfowl science and management, in order that sufficient numbers of professionals representing human diversity from across North America are supported, available, and employed to sustain professional capacity and excellence of future waterfowl science and management.

The terms of reference for the NAWPEP Steering Committee were adopted in 2022 (Brasher et al. 2022), recommending a diverse composition of members from academia, the waterfowl science endowed chairs, each NAWMP signatory nation, government agencies, NAWMP implementers, and NGOs, while ensuring ethnic and gender diversity. The work of the group was to be supported by their individual organizations, and they would report progress at least annually to the Plan Committee. Annual reports of NAWPEP progress can be found at https://nawmp.org/.

One healthy debate by the group involved whether to include specifically wetland science and management broadly in the NAWPEP charge (Mike Brasher, pers. comm. June 2025). Ultimately, they decided that the scope could become too broad. While wetland conservation and management expertise are vital for NAWMP success, the original Plan was not conceived for wetlands writ large. Arguably, a broader focus on wetlands could expand to related elements like grassland management, sustainable agriculture, and more, so, the decision to focus on waterfowl ecology held.

Initial steps to help frame the work were to (a) obtain information on academic and practical credentials perceived necessary for waterfowl professionals to succeed in the profession; (b) estimate numbers of graduates with baccalaureate and postgraduate degrees required to fill anticipated employment needs related to waterfowl science and management in 2025; (c) survey universities and colleges to assess their current and future training capacities for waterfowl and wetlands specialists; and (d) compare needs to capacities for training and employment and communicate results to universities, agencies, other NAWMP partners, and professional organizations. Results were reported in Brasher et al. (2022).

The team also began gathering ideas from faculty on things that might be done to encourage more young people to consider a career in waterfowl conservation. They also networked with other groups with similar interests in The Wildlife Society, Pheasants Forever, the AFWA Waterfowl Working Group, and at the triennial North American Duck Symposium. As I write this in 2025, the NAWPEP has generated a lot of positive "buzz" in the profession and there seems to be widespread acknowledgment of the importance of paying attention to these needs for both education and science support from universities. The needs in Canada and Mexico are particularly great given the current paucity of training opportunities there. Of course, resources will be needed to sustain these programs which are a related matter on which NAWPEP is not presently focused.

Landscape Targeting for Multiple Objectives

Waterfowl biologists have long recognized that geographic regions and habitats within them vary markedly in their importance to waterfowl (Linduska 1964). Beginning with the first Plan in 1986, partners were urged to target most of their conservation investments toward regions that were believed to most affect waterfowl populations. This became increasingly difficult as Joint Ventures proliferated following the successful launch of the original seven, and considerations for diverse avian species emerged. Then, with the broadening of Plan goals in 2012, planners were charged with thinking about landscape priorities in a more complex and inclusive way.

In 2015, the NSST in collaboration with the IIC formed a Priority Landscape Committee[4] and appointed members from various organizations. This was a natural development built on previous work by the NSST Map Committee that in 2012 recommended a process for integrating multiple objectives. Specifically, they said, *The committee advocates a succeeding effort to develop products for guiding conservation at appropriate scales and addressing the three fundamental goals of the 2012 NAWMP Revision* (Soulliere et al. 2012).

This work accelerated when Seth Mott (USFWS) committed funds to the group and Jim Lyons (USGS) found resources to hire Ana Krainyk as a postdoctoral associate to lead the project. Multiple criteria were developed for ranking landscapes: demographic impacts on waterfowl while there, risks of habitat loss, human recreational opportunities, etc. By 2019, Krainyk and company had produced a spatial value model to support the identification of areas important to waterfowl during the breeding and nonbreeding periods and to resource users, mainly hunters (Krainyk et al. 2019a, 2019b). ArcGIS Desktop was used as the platform for managing, analyzing, combining, and displaying the spatial data. Thirty-eight spatial layers were developed, and several composite maps of North America were produced based on committee recommendations. Separate maps were produced showing the relative importance of geographies for breeding and nonbreeding waterfowl combined (20 variables), for waterfowl breeding periods only (10 variables), for nonbreeding periods only (10 variables), for conservation supporters only (5 variables), and for conservation supporters and waterfowl using breeding and nonbreeding periods combined (8 variables).

Not surprisingly, the composite maps were quite like the 2012 NAWMP maps developed by expert opinion that sought to depict areas of greatest significance to waterfowl. There were some differences; however, especially in Arctic regions and Mexico. The authors thought this was likely because of including social objectives, lack of population surveys in some regions, availability of expanded survey data in other regions, and use of model-based waterfowl population estimates for unsurveyed areas. The PLC intends to continue improving these models/maps over the long term but meanwhile they can be used by planners to aid conservation targeting decisions.

A significant advancement in Krainyk's modeling framework was the ability to assign different weights, reflecting stakeholders' "values", to individual objectives and then visualize how geographic priorities may differ as a result. This graphic depiction of trade-offs between, say, social goals and demographic impacts on waterfowl was eye-opening. It also affected the PLC's decision not to produce a single "updated NAWMP map". They realized it was impossible for a single map to reflect stakeholder values across the continent. It did demonstrate that the scale of prioritization should reflect the nature of the decision being made (Mike Brasher, pers. comm. June 2025). So, the idea emerged of a "family of maps" that could identify areas of greatest biological significance to the birds while also informing discussions about specific multiple decision problems. It seems that much promising work remains on this important topic.

Follow-on analyses dug deeper into regional applications for considering waterfowl impacts and human objectives combined for conservation targeting, including Palumbo et al. (2021) for the state of Wisconsin and Soulliere et al. (2017) for the Upper Mississippi and Great Lakes region. Using a different approach, but with similar objectives, Devers et al. (2017) examined waterfowl abundance, harvest, and eBird data to identify high-priority habitats in the state of New York. More recently, Krainyk and seven others from the PLC (Krainyk et al. 2022) offered a structured decision-making analysis to deal with the long-challenging problem of ranking NAWCA proposals, thinking that "it could be helpful to consider a transparent, repeatable, and data-driven framework for prioritization within regions". Ducks Unlimited incorporated several of these same factors in its latest continental conservation plan as well (Dave Howerter, pers. comm. April 2025). Collectively, these developments have strengthened data-assisted targeting of habitat actions for multiple objectives and should enhance returns on conservation investments.

The Krainyk et al. (2022) NAWCA models were tested using linear additive models and proposals submitted to the Upper Mississippi/Great Lakes JV in 2016 and to the Gulf Coast JV in 2017. The seven criteria used were those developed by the NAWCA Council with the JVs in 2016 (adapted from ACJV 2020, https://acjv.org/evaluation/; D. Gordon and K. Kriese, US Fish and Wildlife Service, unpublished data). Model-based rankings were then compared to those previously assigned by the JV management boards. Ranks assigned by the models differed from ranks assigned by the boards for the Upper Mississippi/Great Lakes JV, but not for the Gulf Coast JV. The weighted-value model seemed beneficial for ranking NAWCA proposals because it allowed JVs to tailor the ranking to their specific regional concerns. The challenge of ranking proposals among JVs remains, of course, but this seems like encouraging progress.

The main message from this growing body of work is that there is much potential for more objective, transparent, flexible, and repeatable methods to assist JVs in ranking habitat actions for multiple objectives – and some of these are already available. While this remains short of the fully integrated multi-objective decision process that many of us naively envisioned in 2008, it represents enormous progress, particularly at the JV or regional scales where most such decisions are made. I agree with Mike Brasher (pers. comm. June 2025) that most of us failed to fully envision the challenges inherent in expressing broad social goals as actionable objectives at large scales. The prognosis for such multi-objective strategic approaches seems much greater at smaller scales where human land use issues are more homogeneous. The next chapter will highlight additional areas of progress paving the way for the 2024 Update.

Box 21.1 NAWMP Integration: Progress and Challenges

By Diane Eggeman, NAWMP Integration Coordinator, Retired

The IIC served an important role following the 2012 Revision. It challenged the waterfowl community to wrestle with how its various parts should operate together. That was a timely and needed step to move the enterprise in the direction of integrated management called for by the Revision.

My own professional focus until that time was population and harvest management through work in a state agency and flyway. I didn't fully appreciate, much less champion, the connections and potential synergies with habitat conservation at the continental scale. I advocated that habitat was the most important piece of waterfowl conservation, but my preferred focus was on populations. By participating as a member of the IIC, and due to Dale Humburg's vision and leadership, I began to feel more responsible for the whole machine. Each piece was necessary and dependent on the others for effectiveness. As stewards of waterfowl on behalf of the public, we are charged collectively to take responsibility for it at that higher level. That meant employing decision-making in a way that accounted for impacts across all three NAWMP goals.

The IIC had assembled leaders from the Plan Committee, the Flyways, the NSST, JVs, the HDWG and the HMWG to tackle Plan objectives as a team. We explored how objectives may be linked, and ways to assess trade-offs. The IIC's work helped us envision how that might be done in a more explicit way. Despite only modest technical progress on models for integrated decisions, the effort improved the waterfowl enterprise by expanding the way we think about our work – less siloed and more connected.

The 2018 Update recommended that the integration effort continue, led by a successor group, the ISC, and I was recruited to chair it. Since 2019, the ISC has built on IIC advances. From the ISC's Terms of Reference (2019),

> The purpose of the steering committee is to improve vertical and horizontal communications and establish linkages across the technical and policy levels of the waterfowl management community. To address this need, the steering committee includes

ex-officio representation from the working groups to the PC…and liaison representation by PC members assigned to each working group.…Integration is recognized as key to advancing goals of the 2012 NAWMP Revision and as a formal and explicit element of NAWMP implementation.

Thanks to leadership by its members in their respective realms, the ISC has improved awareness and linkages across groups. Some individuals and groups who may have viewed their work as separate from NAWMP now are realizing the connections. Progress on formal integration, however, has been slow. On the positive side, various integration "building blocks" have been developed or improved, including decision support tools, integration metrics, monitoring, and improved databases related to people. However, in terms of developing explicit (measurable) and linked objectives, predictive models, and assessing trade-offs, the integration effort is still in early stages. There are isolated efforts underway, but ongoing, operational implementation is rare.

Momentum also seems missing. Some groups (i.e., NSST, HMWG, individual JVs) have been in place for 20 or more years and have made excellent and valuable contributions to the waterfowl enterprise, working toward their individual stated purposes. So, it's not surprising that asking these groups to contribute to the broader, NAWMP-wide goals, which now explicitly include people, is seen as unfamiliar and difficult. Adjusting decades of successful, if siloed, work patterns is a difficult, adaptive task. The Plan partnership is large, complex, and well established. Resources and leadership to accomplish the desired integration have not increased materially. It may be necessary to approach formal integration in a stepwise manner, more explicit and quantitative in some elements and more general in others. It seems likely, for instance, that linked decision-making will continue to develop at different rates in different Joint Ventures, if we continue to encourage this approach.

Yet, despite the hurdles, we are now better at imagining how traditional waterfowl conservation and management, focused on waterfowl populations and habitat, can benefit from also including a more explicit focus on people. This is evidenced in conversations, investments, planning efforts, and research across the NAWMP partnership. The ISC has helped advocate and grow this concept. Commitment to this shift remains strong under the 2024 NAWMP Update.

Box 21.2 Who Will Mind the Marsh? The Role of Academics in NAWMP

By John M. Eadie, University of California, Davis, Retired

Many of us began our journey into the marsh, clasping the hand of a mentor – a parent, grandparent, or friend – as we journeyed into the cattails for the first time. For those of us who made a career as a waterfowl professional, we also began our journey with a mentor – a wildlife manager or an academic. The list of outstanding mentors who helped shape the current generation of waterfowl professionals is long. They left a legacy and a challenge to those of us who followed. Who will continue to carry out this mission? Who will mind the marsh?

As a university professor for 36 years, I've had the privilege of being involved in training the next generation. But do we have the capacity and the commitment to maintain those training programs in waterfowl and wetland ecology? Teaching positions in waterfowl biology have been declining over the past two decades. In a 2023 assessment, Eggeman et al. (submitted) reported that only 44 of the 55 faculty positions dedicated to teaching waterfowl ecology and management remained from a survey in 2013 (Kaminski 2013). Most worrying is that for these positions, the departments reported only a low (40%) to medium (26%) likelihood of refilling the position with a waterfowl ecologist. Coupled with an aging faculty population, Eggeman et al. estimated that as few as 15 positions might remain soon. Who, indeed, will train the next generation of marsh minders?

With this comes another gap to be filled – sustaining and renewing a commitment to NAWMP. As conservation planning becomes more technical with complex tools and analyses, the engagement of the academic community may be more important than ever. I have been personally involved with NAWMP for a couple of decades and this engagement provided me with a unique opportunity to participate in the ever-changing, ever-growing scope of the Plan. The Continental Progress Assessment (Chapter 15) found me on a team of 15 waterfowl professionals who between 2005 and 2007 concluded the first international assessment of the accomplishments and knowledge base of the Plan including all the habitat and species Joint Ventures. It was a groundbreaking effort to dig deep beneath the surface, beyond acres reported, and look closely at the successes, failures, and challenges facing Plan partners, 20 years along. As an early career academic, this seemed a daunting task, eye-opening, and brave, and I became a believer in the integrity of those who built and sustained the Plan. Interestingly, only two of the 15 committee members were university academics, although eight had PhDs and the others were trained in wildlife programs with many years of management experience.

Around the same time, the Plan community sought to more explicitly consider integrating systems of sustaining waterfowl populations and habitat with harvest management (Chapter 16). These two arms of the waterfowl management enterprise ("habitat" and "harvest") often have different directives and serve different management structures. The Joint Task Group sought to develop and advocate for an integrated decision framework. This too was a formidable task for the Group, at times technical and not infrequently controversial. Nonetheless, the NAWMP community once again was willing to probe deeply, address thorny issues, and propose a coherent, unifying framework. Again, of the 14 members on the Joint Task Group, only two were university academics. Why were there not more university academics involved?

My personal involvement with the Plan continued with other committees engaged with the Future of Waterfowl Workshops (2008, 2017), Plan update committees, and one of the most fulfilling endeavors I have been part of as an academic – the 2012 Plan Revision (Chapter 18). That revision was a watershed – for the first time explicitly connecting all three "legs of the conservation stool": populations, habitat, and people. It was a bold proposition and though 13 years later we still struggle how to include people in our management equations it was an audacious effort to expand NAWMP's conservation mission. Human dimensions were explicitly recognized as an important component in the quest, and the idea of integrated decision-making continues to guide us. The 2012 Revision was a massive multi-year effort, led superbly by Jim Ringelman, Dale Humburg, Mike Anderson, and many others, engaging managers and professionals across three nations.

As a participant on the 2012 Revision committee, I have never felt so intellectually challenged. Echoing my refrain about the importance of mentors, I learned more, as a mid-career professional, working with Jim, Mike, and Dale than I ever did in any of my formal undergraduate or graduate training. Of the 20 or so individuals who served in multiple roles on the writing team, steering committee, or technical committee, only two were academics at universities. So, once again, I wondered, what should be the role of academics in NAWMP? What opportunities might we be missing for the Plan and for the academic community?

In a paper, led by Tony Roberts at the 2016 Duck symposium (Roberts et al. 2018), we asked why it was increasingly difficult to conduct waterfowl research with direct management application in universities? Some of the challenges are purely financial: reduced funding for applied research especially at the federal level, high overhead rates at academic institutions, and increased competition for limited management dollars in agencies. Resource limitation is indeed one concern, but I believe that there is another issue at play: applied research is often perceived as being of lower scientific value relative to "basic" theoretical research. Within many academic institutions (my own included), merits and promotions are closely tied to the perceived prestige of publications and grants received, and applied products are viewed as less prestigious. Accordingly, some university researchers are less inclined to pursue applied research, and along with that comes reduced training in practical research and management for students. It is not surprising that there is a growing dissonance between academics and agencies (Rupp 2012; Roberts et al. 2018).

So, how will we support NAWMP in the future, especially as conservation issues become more complex and interdisciplinary? We need habitat biologists, population ecologists, harvest analysts, human dimensions specialists, and university academicians engaged to improve ever-changing models and methods of monitoring, assessing, implementing, and updating the scientific underpinnings of the Plan. Perhaps as importantly, because we are free of the political constraints that come with a job in an agency or NGO, academics can speak their mind. We can ask hard questions that others may not be able to ask – are our assumptions correct? Do we have the right framework? Have we become complacent in our beliefs and processes after almost 40 years of success?

I think there are ways to strengthen the links between academics and agencies, such as (1) exchanges where university faculty and agency partners spend time in each other's departments and contribute to each other's programs; (2) cooperative agreements with fixed, low overhead rates to provide secure, reliable funding for applied research; and (3) encouraging academic institutions to reconsider metrics of performance and the value of applied research in faculty appraisals.

None of these ideas is new. Some could be difficult to implement (e.g., convincing universities to reduce overhead rates). We also need to establish more opportunities for academics in NAWMP technical and leadership roles. Recent updates of NAWMP have recognized this need. I have been privileged to serve on the NAWPEP for the past several years. Led ably by Diane Eggeman, the NAWPEP goal is to *Engage and assist universities, colleges, and all NAWMP partners with establishing, sustaining, and enhancing academic and experiential programs in waterfowl science and management….* This is a start, focusing on the training role of academic institutions. But how will we implement and sustain this continentally?

Perhaps it is time to revisit the motivation of the land-grant model established by Justin Smith Morrill in the Hatch Act of 1862. That act helped to establish new agricultural institutions in every state, to provide training and research so "the land could yield more and renew itself", and in doing so, new kinds of universities came into existence with a "public purpose" (Widmer 2015). Many of the wildlife departments in American universities were established in those land-grant colleges. Might we emulate Morrill's vision by minimally establishing in each administrative flyway, centers to sustain teaching, training, and research in waterfowl ecology and management? These could also serve as focal nodes in a national network. To some extent, that vision is already aborning with the creation of now 13 endowed chairs in waterfowl and wetland ecology in the United States and Canada.

But perhaps we need more synergy than just endowed chairs. Could we conceive of those nodes as endowed or partnered *programs*...more than a single position, with linkages to the research structures in existing JVs? My personal vision is a program in every flyway that comprises multiple scientists with different expertise (minimally, habitat management, population ecology, and human dimensions), collaborating on research and training related to key issues in that region, but connected continentally. These programs must have active partnerships with Joint Ventures and their member organizations. We cannot afford to have academic expertise barricaded behind university gates; we need more mud on the walls of the ivory tower and critical independent thinking in NAWMP forums.

Such a model would require establishing mutual goals and joint agreements among universities, agencies, and NGOs. There are already several universities and departments that function much like these envisioned programs, but they are scattered throughout the country and constantly at risk of turnover. My call is to strategically, formally, and financially develop a vision and a plan to ensure that NAWMP has the people, expertise, and resources to confidently evolve given increasingly uncertain social and environmental circumstances. We need to connect all partners in this effort and erode the barriers that have grown between applied and basic research.

Notes

1 Diane Eggeman (Cochair), John Eadie (Cochair), Michael Anderson, Brad Bortner, Dale Humburg, Chris Smith, and Jeff Ver Steeg. Facilitated by D.J. Case and Associates with interview assistance from Keith McKnight, Seth Mott, Dave Morrison, and Tasha Sargent.

2 HDWG/PET (PC Liaison Gray Anderson, now Stephen Carlyle, Ex officio David Cobb, now Barb Avers)
NSST (PC Liaison Karla Guyn, then Mark Gloutney, vacant 2025, Ex officio NSST Chair, Dana Varner, then Kyle Spragens, now Joe Lancaster)
HMWG (PC Liaison Jim Dubovsky, now Todd Sanders, Ex officio Kevin Kraai)
Communications (PC Liaison Joe Benedict, Ex officio Dean Smith)
Integration Coordinator Diane Eggeman, now Tony Roberts
USFWS coordination Jorge Coppen, then Dave Gordon, now Tony Roberts
Joint Ventures (ex officio) Sean Fields, now Catherine Wightman and Justyn Foth
Canadian province (ex officio) Glen McMaster, now Corie White

3 Diane Eggeman (Chair, now Tony Roberts), Michael Brasher, Eduardo Carrera, John Eadie, David Howerter, Richard Kaminski, J. Drew Lanham, Tom Moorman, Kevin Ringelman, and Frank Rohwer.

4 Anastasia Krainyk and Jim Lyons USGS Patuxent Wildlife Research Center, Michael Brasher, Dale Humburg, Greg Soulliere, David Howerter, John Coluccy, Mark Petrie, Stuart Slattery, Mindy Rice, Joe Fuller, Josh Vest, Sean Fields, Kevin Kraai, Kyle Spragens, Blair Stringham, Patrick Devers, Kathy Fleming, and Luke Naylor.

References

Brasher, Michael G., Joel Brice, Eduardo Carrera, John Eadie, Diane Eggeman, Justyn Foth, Adam Janke, Richard M. Kaminski, Shaun Oldenburger, Kevin M. Ringelman, Christian Roy, Lisa Webb, and Lauren Hernandez-Rubio Senn. 2022. *Estimated Capacity and Future Need for Trained and Educated Professionals in Waterfowl Management and Conservation.* North American Waterfowl Professional Education Plan Committee. 7pp. (https://nawmp.org/nawmp-udpate/north-american-waterfowl-professional-education-plan).

Devers, Patrick K., Anthony J. Roberts, Scott Knoche, Paul I. Padding, and Robert Raftovich. 2017. "Incorporating human dimensions objectives into waterfowl habitat planning and delivery." *Wildlife Society Bulletin* 41: 405–15.

Eggeman, Diane, Richard M. Kaminski, and Luke Laborde. (Submitted). "University Waterfowl programs and future professionals – still at risk?" *The Wildlife Professional.*

Kaminski, Richard M. 2002. "Status of waterfowl science and management programs in United States and Canadian universities." *Wildlife Society Bulletin* 30: 616–22.

Kaminski, Richard M. 2013. "An endangered academic niche? University-based waterfowl programs in U.S. and Canada." *Wildlife Professional* 7, no. 4: 58–61.

Krainyk, Anastasia, James E. Lyons, Michael G. Brasher, Dale D. Humburg, Gregory J. Soulliere, John M. Coluccy, Mark J. Petrie, David W. Howerter, Stuart M. Slattery, Mindy B. Rice, and Joe C. Fuller. 2019a. *Spatial Integration of Biological and Social Objectives to Identify Priority Landscapes for Waterfowl Habitat Conservation.* U.S. Geological Survey Open-File Report 2019-1029. 41pp. (https://doi.org/10.3133/ofr20191029).

Krainyk, Anastasia, Bart M. Ballard, Michael G. Brasher, Barry C. Wilson, Mark W. Parr, and Cynthia K. Edwards. 2019b. "Decision support tool: Mottled duck habitat management and conservation in the Western Gulf Coast." *Journal of Environmental Management* 230: 43–52.

Krainyk Anastasia, James E. Lyons, Gregory J. Soulliere, John M. Coluccy, Barry C. Wilson, Michael G. Brasher, Mohammed A Al-Saffar, and Dale D. Humburg. 2022. "Structured decision-making to rank North American Wetlands Conservation Act proposals within joint venture regions." *Journal of Fish and Wildlife Management* 13(2): 375–95. (https://doi.org/10.3996/JFWM-21-089)

Linduska, Joseph P. (Editor). 1964. *Waterfowl Tomorrow.* U.S. Department of Interior. Washington, DC. 770pp.

NAWMP Institutions Committee. 2018. *Report on Review of the Plan Committee – Achieving NAWMP Objectives.* Submitted to the NAWMP Committee January 15, 2018. 15pp. (https://nawmp.org/sites/default/files/2018-04/Plan%20Committee%20Review%20and%20Institutions%20final-April%20 2018.pdf).

NAWMP Plan Committee. 2019. *NAWMP Plan Committee Strategic Priorities and Actions Over the 2018 Update Period.* 15pp. (https://nawmp.org/sites/default/files/2019-11/Plan%20Committee%20Strategic%20Priorities%20Final%2011-15-19-EN.pdf).

NAWMP Plan Committee. 2022. *NAWMP Metrics Approved for Implementation by NAWMP Plan Committee.* 3pp. (https://nawmp.org/sites/default/files/2023-01/NAWMP%20Metrics-%20PC%20 Approved%20-final.pdf).

Palumbo, Matthew D., Jacob N. Straub, Mohammed A. Al-Saffar, Gregory J. Soulliere, Jason L. Fleener, Mitchell T. Bergeson, John M. Coluccy, Andrew Cruz, Taylor Finger, Drew N. Fowler, Brian J. Glenzinski, Rusty K. Griffin, Scott E. Hygnstrom, Greg Kidd, Nicholas A. Miller, Kent Van Horn, and Kurt Waterstradt. 2021. "Multi-scale waterfowl habitat conservation planning in Wisconsin, USA." *Landscape Ecology* 36, no. 10: 3207–30.

Roberts, Anthony, John M. Eadie, David W. Howerter, Fred A. Johnson, James D. Nichols, Michael C. Runge, Mark P. Vrtiska, and Byron K. Williams. 2018. "Strengthening links between waterfowl research and management." *The Journal of Wildlife Management* 82, no. 2: 260–65.

Rupp, Susan P. 2012. "Practical dissonance: Bridging the gap between academia and agencies." *Wildlife Society Bulletin* 36, no. 4: 796–801.

Soulliere, Gregory J., Benjamin M. Kahler, Timothy A. Bowman, Michael G. Brasher, Michael A. Johnson, Robert S. Holbrook, Mark. J. Petrie, Stuart M. Slattery, and Josh L. Vest. 2012. "Process for developing the 2012 NAWMP map – Geographies of greatest continental significance to North American waterfowl." *North American Waterfowl Management Plan Science Support Team Technical Report* No. 2012-01. 27pp.

Soulliere, Gregory J. and Mohammed A. Al-Saffar. 2017. "Targeting conservation for waterfowl and people in the Upper Mississippi River and Great Lakes Joint Venture Region." *Upper Mississippi Great Lakes Region Joint Venture, Technical Report* No. 2017-01. 16 pp.

Widmer, Ted. 2015. "Justin Morrill, the man behind America's higher education." *Boston Globe*, June 18, 2015. https://www.bostonglobe.com/ideas/2015/06/18/justin-morrill/OVsD2QnwPQJrcXCb7YqZZK/story.html#:~:text=Morrill%20was%20troubled%20by%20the,farms%20were%20already%20in%20decline.

Williams, Christopher K. and Paul M. Castelli. 2012. "A historic perspective of the connectivity between waterfowl research and management." Pages 155–78 in Joseph P. Sands, Stephen J. DeMaso, Matthew J. Schnupp and Leonard A. Brennan (Editors). *Wildlife Science: Connecting Research with Management.* CRC Press, Taylor & Francis Group. Boca Raton, FL. 333pp.

22 Communications, People, and Planning 2018–2023

As for most of the world, NAWMP had a period of reduced activity through the COVID-19 pandemic. Thanks to online communication tools NAWMP working groups were able to make some progress even without face-to-face meetings. By 2023, the partners were regaining some of that lost momentum. In the previous chapter, I described the NSST's advances in landscape prioritization. Also, the new NAWPEP was formed and moved its initial investigations forward. In this chapter, I will review some other impactful developments briefly during these years on the path toward the 2024 NAWMP update.

A New Communications Team

An explicit recommendation from the 2018 Update was to create a new communications team[1] charged with ensuring that more people, both inside and beyond the waterfowl community understand NAWMP and its partnership opportunities. Following on the premise that "developing a communication strategy that celebrates the accomplishments, emphasizes the needs and showcases the wide-ranging benefits of NAWMP is an important step in securing its future", the team adopted a new communications strategy in 2021 (https://nawmp.org/sites/default/files/2023-09/NAWMP%20comms%20strategy_8-24-2021.pdf).

Key challenges flagged were declining financial support for NAWMP partners as hunter numbers decline, loss of continuity when elected officials and senior bureaucrats change jobs, and a sense that many people do not understand NAWMP and its many positive outcomes (e.g., McGraw 2018). The lack of a clear and present "crisis" in waterfowl management had reduced NAWMP's visibility. Plan partners also felt they had received little recognition for the positive impacts of waterfowl habitat conservation ranging from water management to biodiversity to benefits for working lands. The thesis of the new strategy therefore became, "while waterfowl may be the heart of NAWMP's success story, they are only the beginning".

In 2022, D.J. Case and Associates (DJ Case) was contracted by the Association of Fish and Wildlife Agencies to identify marketing assets that could provide a foundation for future communication efforts. The project had two deliverables – an electronic inventory of existing marketing assets used in past campaigns, and recommendations for new assets. The inventory focused on elements created since 2017. The Case team applied four criteria when assessing message relevancy of those elements. These were (1) audiences think NAWMP is a valuable investment; (2) audiences think healthy wetlands and waterfowl rely on NAWMP funding; (3) audiences feel that through NAWMP, they are part of the solution to conservation issues; and (4) audiences engage and actively support the Plan enterprise. The inventory provided 227 links to assets, with 73% considered relevant to current NAWMP messaging, although only about 5% were specifically focused on the NAWMP brand. Assets were classified by country

DOI: 10.4324/9781003732310-22

of focus, target audiences, type (image, video, document, etc.), and perceived utility. Communication about NAWMP itself had been minimal. From the earliest beginnings of the JVs, most publicity had been focused strategically on delivery partners rather than NAWMP broadly.

Establishing best practices for recruiting financial and political support or new partners is still under development. Much like with improving habitat programs for ducks, the NAWMP partnership might gain from rigorous testing of marketing strategies. There should be opportunities to learn from trials by various JVs or by individual partners. The PC Communications Committee is engaged in these matters and seems keen to do more.

A New Marketing Plan

With these assets identified, the next step was preparing a new marketing plan, which was provisionally adopted in November 2023 (D.J. Case and Associates 2023). After more interviews of Plan partners, the marketing team achieved consensus among PC members that in the 2024 Update, the profile of the Plan should be that NAWMP is the United States, Canada, and Mexico working together to protect waterfowl, wetlands, and wildlife. NAWMP is also the biggest conservation success story in history. Partners look to NAWMP for funding, support, technical assistance, and credibility. NAWMP's most important audiences are decision-makers and conservation leaders of federal, state, provincial, territorial, and tribal agencies and conservation NGOs. A secondary audience of "waterfowl end users" was also flagged – these being hunters, birders, and other outdoor-oriented people.

NAWMP is a partnership and not an organization to which people "belong". The individual partner organizations have a greater identity with the public, e.g., USFWS, DUMAC, Audubon, and California Waterfowl Association. It hasn't been clear how to promote NAWMP *per se* but generate support for the individual partners, or *vice versa*. What we most need, it seems, is recognition of the NAWMP partnership, appreciation for its scope and initiatives, but support for the work of the partners. For both the primary and secondary target groups, the marketing plan developed key messages and strategies for engagement. It also explored the NAWMP brand and what changes might be helpful from logos to a rallying cry and criteria for their use.

Box 22.1 Updating the Logo, Tag Line, and Rallying Cry 2024

By Dean Smith, Association of Fish and Wildlife Agencies and
Joe Benedict, Tennessee Wildlife Resources Agency.

The NAWMP logo has been used inconsistently on websites and other communication platforms for many years. Plan Updates use the trilingual version, while most other platforms use the stylized pintail graphic or spell out only NAWMP in their documents. The Plan Committee concluded that a more standardized approach was desirable and approved an NAWMP Brand Standards and NAWMP Marketing Plan in 2023.

NAWMP has a peculiar branding challenge in that it exists as both a plan, "The North American Waterfowl Management Plan", and an enterprise – the collective work of the governments, agencies, and other organizations. The Plan Committee solved this issue by adopting a dual brand. Any document related only to the Plan should henceforth use the trilingual names in combination with the stylized pintail graphic (Figure 22.1 left or right). The NAWMP enterprise (federal, state, NGOs, and others) should be acknowledged by the new logo, which uses the word *Partnership* and the stylized pintail (Figure 22.1 center). Most of the communications and marketing messages will utilize this new branding. The phrase *Conserving Habitat Together*, placed below the word Partnership, was adopted as the NAWMP rallying cry.

NAWMP
PNAGS
PMAAN

NAWMP
PARTNERSHIP

North American Waterfowl Management Plan
Plan nord-américain de gestion de la sauvagine
Plan de Manejo de Aves Acuáticas de Norteamérica

Figure 22.1 The new NAWMP logos. See text for explanation of their varied uses. Used with permission of the Plan Committee and D.J. Case and Associates.

Human Dimensions Progress from 2018 to 2023

As the final draft of the 2018 Update was in production, surveys of Canadian birders and hunters emerged with national reports and individual reports by Flyways (Harshaw 2018a, 2018b). Then in 2021, more detailed and analyzed data from the US national surveys of birders and hunters were also published (Patton 2021a, 2021b). These fresh reports provided more nuanced insights for use by national and flyway agencies focused on hunter and supporter recruitment and retention. In broad terms, Canadians and Americans were similar in their perceptions of wetland environments. The waterfowl hunter and birdwatcher surveys highlighted similarities between these groups too, in terms of identity as conservationists and appreciating the value of being immersed in nature.

The US public survey indicated that most people were familiar with wetlands, lived in reasonable proximity to them, and the majority had visited them in the past year; however, this familiarity did not translate often into pro-environmental behaviors. Waterfowl hunters remain a small portion of the population, and their numbers are still declining. These findings reinforce the need to continue to build and diversify the spectrum of people who engage in NAWMP's conservation work.

The Human Dimensions Working Group and the Public Engagement Team (PET) formed after the 2012 Revision quickly recognized the value of working together and after a joint workshop in January 2018 operated in unison. In 2022, with the PC's blessing, they merged. The combined HDPET remains a forum for sharing new publications and social science data and encouraging public engagement efforts. They continue to develop engagement plans for three target audiences: (1) hunter recruitment, retention, and reactivation, (2) birders, and (3) landowners. Recent accomplishments of the group were reported in the 2024 Technical Report (Howerter et al. 2024).

NSST and Integrated Bird Planning

Following the 2018 Update, the NSST reengaged in probing elements of integration, issuing reports on coordination across bird taxa as well as integration with people objectives. NAWMP connections with other bird taxa were first recommended in 1998 and addressed to some degree in each update since then. Despite this mandate, the JVs had been left largely on their own to develop all-bird planning approaches suitable to their circumstances. It seemed timely to collect and compare the approaches employed by the JVs in doing this and thereby stimulate thinking everywhere about solutions. The Team surveyed all the current 22 habitat JVs to collect these experiences (Soulliere et al. 2022a). Two sets of survey questions followed by online interviews were held between the NSST committee and the JV teams (usually the coordinator, the science coordinator, and various others). Most (59%) JVs reported that their conservation projects included consideration of more than one bird group using common habitats within their planning units, but only three (14%) indicated their planning documents explicitly combined habitat objectives for multiple taxa. Some JVs had developed species-habitat models to assist landscape prioritization, often employing designated focal species or umbrella species as habitat representatives. The group most often linked with waterfowl was wetland birds (32%). As expected, waterfowl had the strongest perceived science foundation in most (82%) JVs. Nearly half (45%) of JVs used BCRs or subdivisions of BCRs as their primary planning units. Multiple bird-group initiatives, however, occurred on spatial scales ranging from individual projects up to entire JVs. USDA Farm Bill programs and NAWCA funding were most often accessed to support multi-taxa work.

As for limitations, data availability and varying approaches in adjacent JVs limited the opportunity for cross-regional planning, and no JV had a program in place to evaluate multi-taxa conservation actions other than site-specific monitoring. Some offered that larger initiatives like perhaps the JV8 Central Grasslands initiative might eventually enhance conservation efficiencies for grassland birds by multiple JVs working across the Great Plains.

Integrating People

Factoring social science knowledge into conservation planning has challenged Plan partners since committing to do this, in principle, in 2012. Partners in Flight (PIF), the North American landbird umbrella, is committed in principle to a similar approach. Ten years later, the migratory bird Joint Ventures were at widely differing stages of integrating peoples' desires and values into habitat delivery plans (Soulliere et al. 2022b). This assessment, parallel in many ways to the multi-taxa study just described, sought to benchmark progress of the JVs in people planning and uncover barriers that might persist. The study involved initial surveys of 22 habitat JVs (US and Canadian Pacific Coast groups were treated separately), followed by virtual interviews.

They found JV partners "keenly aware" of the social and environmental changes occurring in North America and the importance of using social science to help understand people within JV regions. Nevertheless, many JVs were struggling with how to implement more data-driven approaches to incorporating social science insights in their planning. Perceived limitations included JV cultural inertia, partnership composition, landscape characteristics, and especially JV staff capacity. Reported levels of social science engagement ranged from no work yet (11 of 22 JVs) to use of available data and literature to better target conservation, develop models, or learn about landowner decision-making (a few). Eight JVs had contracted human dimensions consultants but incorporation of what they learned in conservation plans was mostly still pending.

Regarding the question of humans as means or fundamental objectives, only one JV had established an explicit HD objective for waterfowl hunter abundance. Most JVs viewed social science objectives as means to attain biological goals. Adding social science expertise was a near-term desire for many JVs. Seven had staff members with at least some formal training in social science. Another JV had recruited a social scientist to their Technical Committee.

The interviews also explored the kinds of information that JVs thought might be most useful, although one JV opined that,

> we are so far from achieving our biological objectives for birds, that if we were to incorporate any other objectives that involve tradeoffs (such as additional objectives to meet the needs of people), it would put us even farther from the finish line.

More constructive ideas shared included:

- Each JV is unique, and issues vary among regions. Social science considerations and applications may differ among landscapes but regardless, sharing experiences across JVs might reduce false starts and unproductive efforts elsewhere. Purposeful sharing of successes and failures could be valuable.
- While habitat plans are largely based on biology, conservation delivery remains largely driven by social considerations.
- Understanding the motivations of various private-land communities is where HD evaluations are most needed.

Questions about the appropriate nature and scale of HD objectives and approaches to integrating them in waterfowl conservation planning were destined to carry over to the 2024 Update and beyond.

Species Prioritization Revisited and Expanded

When thinking about Plan population objectives, many people default to considering ten or so mid-continent duck populations plus Black Ducks in the east. But population objectives in the Plan include some 70 species or populations spread across the continent. Current population estimates vary widely in their precision, and some species are not well-surveyed at all. Plan partners face tough choices about how to prioritize actions for all these birds. Should we focus on developing the means to better assess populations? Should we favor those whose populations are below goal or showing sustained declines? Should we favor those species most important in the hunter's bag?

As the Plan launched in 1986, these were questions planners wisely chose to defer, but in the 2004 Update, the technical community took a first swing at prioritization criteria (Chapter 14). They settled then on two factors, equally weighted, comprising the importance of each species in the harvest – more taken, more favored – and population status relative to NAWMP objectives. Thus, a preferred species below goal (e.g., northern pintail) was listed highest in priority. Conversely, an abundant or increasing species of low hunter preference (e.g., northern shoveler) was ranked lowest. Priorities were established for the continent overall and for individual Waterfowl Conservation Regions, essentially congruent with BCRs.

Categories of population trends were defined as *Increasing*, *Stable*, *Unknown,* or *Decreasing* – with Unknown and Decreasing trends weighted the same – for the period 1970–2002. Harvest data were taken from the USFWS Waterfowl Parts Survey for 1980–1999. The Canadian Wildlife Service provided comparable harvest data for Canada. Mexican sport harvest was assumed to be inconsequential continentally. So, each species was fitted in a matrix (see 2004 Update, pp. 53–83).

Unlike for ducks, many goose and swan populations have management targets established by the federal agencies through the Flyway System. Because of their tendency to breed, migrate, and winter in well-defined traditional places, this is easier to do than with more vagile duck populations. This more intensive management of goose populations tends to ensure that increasing populations of geese that are above objective levels are harvested more liberally than declining or stable populations that are at or below objective numbers. For geese and swans, continental prioritization was based on a matrix of more recent (1993–2002) population trends and deviation from Plan population objective (2000–2002).

In practice, many species are managed as a guild (e.g., mid-continent dabbling ducks) so that what Plan partners did for mallards may have also benefited gadwall and northern shovelers, species of lower priority. Some species with distinctive habitat preferences (e.g., northern pintails in short-grass prairie, wood ducks in flooded woodlands) could be favored by where delivery agencies target habitat programs.

One consequence of developing these criteria was the incorporation of species priorities in the scoring criteria for NAWCA standard grants. In 2024, of the 100-point scoring system, up to 15 points depended on targeting higher priority species, 15 points on occurrence in NAWMP priority geographic areas, ten points for wetland status and trends, etc., which seems to have provided added incentive to plan for certain species.

This method served the Plan reasonably well, but by 2023, with 20 years of additional data on population trends, and new population estimates for additional species, it seemed timely to revisit the framework created in 2004. Moreover, some groups like sea ducks continue to be poorly censused in aggregate yet are important in the harvest of First Nations people, a factor not considered 20 years earlier. Led by Tony Roberts (FWS), a team of scientists[2] from the NSST and the Flyways undertook a revision (Roberts et al. 2023).

This new iteration was limited to a continental assessment of priorities, although the team intends to develop these at regional scales later (Tony Roberts, pers. comm. 2024). For the first time, this analysis relied heavily on the Avian Conservation Assessment Database (ACAD) created by Partners in Flight (2021). A primary source for many ACAD assessments included expert opinion of waterfowl managers, including the NSST members and associated JVs. For ducks, the waterfowl population objectives of the Plan were compared with trend information from ACAD, and the habitat objective was scored using ACAD threats to habitats. To address the human dimensions objectives, two criteria were used for ducks, total harvest from federal harvest surveys and observations by bird watchers using eBird.

For goose and swan populations, the objectives were based on the most recent ten-year trend estimates if those were available (e.g., Trumpeter Swans and some geese are not surveyed sufficiently to use the same criteria). The ACAD habitat threats scores were used to classify populations for the habitat objectives. For the human dimension prioritization, population objectives defined in flyway management plans were assumed to represent societal goals for management, and scores were based on the qualitative difference between management plan objectives and current abundance. All criteria were equally weighted.

The ranges of final scores among populations were subdivided approximately equally to obtain three levels of prioritization (high, medium, or low) for ducks and geese/swans. High priority was assigned for 12 of 38 duck species (40 populations), medium to 17, and low to 11. For geese and swans, 11 of 35 populations were classified as High priority. Because the rankings in this system rely on expert opinion and not data gathered annually like BPOP estimates of harvested birds, it will be vital for the NSST, presumably with ACAD people, to review and update these assessments periodically.

More Surveys Preparing for 2024

As was done in advance of the 2018 Update, Plan partners agreed to sponsor two new surveys ahead of work on the next update – one of waterfowl professionals and another of waterfowl conservation organizations – both led by D.J. Case and Associates (Robert Holsman and Cynthia Longmire for professionals; Jessica Mikels-Carrasco and Longmire for organizations).

Professionals

The online survey of waterfowl professionals in April 2023 replicated several questions previously asked in 2017 (Holsman and Longmire. 2023). Unfortunately, the response rate was low (n = 162) which limited the analyses. Most of the respondents (82%) were from the United States with 16% from Canada and only 1% from Mexico. Most were biologists. A strong majority endorsed all three of the 2012 goals as important to retain, although most judged progress to date as modest. All but one of the recommendations from the 2018 NAWMP Update were also endorsed as important to retain in the 2024 update. The exception was an outdated process question. Numerous comments, however, implied some frustration with the vague nature of certain 2018 recommendations and consequently uncertainty about actionable steps. The long-standing NAWMP efforts in habitat management and technical support were rated highest, both for greatest progress and, as highest priorities going forward.

Respondents identified collaborations with stakeholders as the biggest area in need of improvement to advance waterfowl management, but the conceptual understanding of how social science might help implementation remained weak. When asked to rank which priorities needed more emphasis in 2024, various stakeholder strategies ranked near the bottom. So, respondents thought they needed this but weren't sure for what. There remained widespread belief, however, that behavioral change would happen if people were presented with more information about the values of wetlands.

In their reflections, the DJ Case team surmised that the relatively high levels of support for Plan components were shallow and reflected only a nominal understanding of the mechanics of NAMWP. The COVID-19 pandemic obviously diminished opportunities for many professionals to be deeply engaged in the waterfowl community, so the Case team sensed a need to help reengage people in NAWMP committees and boards, especially early-career professionals. There also emerged a widespread sense that federal and state agencies were placing less emphasis on waterfowl management than in the past. NAWMP may be able to play a material role again in elevating wetland and waterfowl conservation among agency directors.

More attention to clarifying 2018 goal language was urged by many. The three major goals were assumed to be of equal importance, but survey respondents expressed a clear preference for strategies that prioritize habitat and waterfowl monitoring. Of course, this reflects the historic paradigm of the profession and their partnership with hunters, which may slow the partnership's ability to shift its culture to a broader suite of supporters who might assist more if sufficiently motivated. The DJ Case authors asked,

> Is the role of social science expected to serve the accomplishment of growing waterfowl, increasing habitats, and touting their recreational benefits (the old paradigm), or is the community open to adopting a new paradigm based on what future social science may tell us about people's values and desired benefits?

Indeed, this seems like an important question to ponder.

Organizations

The inquiry of conservation organizations was more open-ended and qualitative. Requests for input were sent to 102 organizations but only 20 responded in the time allotted. Responding organizations included state and federal agencies, Joint Ventures, and NGOs. As in the 2017 survey, there remained respect for NAWMP, but a sense that recent updates have not greatly affected what organizations are doing, perhaps because of competing priorities, or because of the relatively frequent five-year repeat cycle vs. the typical planning cycles of agencies.

Some organizations expressed that the typical "high-level" guidance from NAWMP had included little advice about implementation. While there remained support for the Plan, many organizations have broader mandates and are pivoting to prioritize other taxa, habitats, and user groups. This doesn't seem to mean abandoning hunters and wetlands but rather considering them as part of a larger suite of stakeholders and habitats. This trend may increasingly challenge NAWMP partners in the years ahead.

Box 22.2 Reflections on Communications and Lessons for the Future

By Dave Case, D.J. Case and Associates

In 2019, I was asked to share my perspective on NAWMP human dimensions and communications as a case study for the Trilateral Committee meeting in Victoria, BC. I opened my talk saying: "The NAWMP is a public engagement campaign cleverly disguised as a landscape scale waterfowl and wetlands conservation plan. I am, of course, exaggerating, but not by much". Since then, and especially after participating in the 2024 Update process, I've come to see this "exaggeration" as not an exaggeration at all. The NAWMP partnership is indeed a public engagement/communications initiative driven by a conservation plan. So, what would I say to future NAWMP partners about the past 40 years of NAWMP communications that might help them in the future? Here are four lessons that float to the top:

Be Effective Communicators

The NAWMP partnership will continue its success to the degree that waterfowl conservationists are effective communicators. The conservation professional ranks are too small, our time is too short, and the challenges to the landscape too great to think communications can be wholly delegated to others. We need every one of us to be the best communicators possible. We all should be both waterfowl scientists and advocates for waterfowl conservation.

Of course, that takes both discipline and a broader skill set. How do you do that? Well, that's the subject for another book, but over many years of observing conservationists in action, I offer three key things: listen, know your audience, and show you care (Christiano and Neiman 2018). Each time I recite that list, I think of Larry Jahn. He became President of the Wildlife Management Institute in 1987 and was a major influencer in the early years of NAWMP, as well as conservation provisions of the US Farm Bill. I attended many meetings with Dr. Jahn and other movers and shakers on the US conservation scene as they discussed and set direction for NAWMP. Discussions could go on for hours. Dr. Jahn would listen intently but usually not say much compared to the other participants. Eventually, though, Jahn would take the floor, summarize what he had heard from the folks around the table, and then offer his thoughts on how to proceed. His recommendation or some close version of it would usually be adopted. He listened carefully, he knew his audience, and he showed he cared.

Use Strategic Communications

Strategic communication means starting with clear-headed answers to questions such as (a) Who are the specific audiences or groups of people that need to be persuaded? (b) What do we want each audience to know, feel, and do because of our communications? Of course, this implies we know our audiences' attitudes currently. (c) What communications channels will reach our target audiences most effectively? Then, and only then, do we move on to deciding what messages will be most effective and what communication tools or actions are needed to deliver those messages.

Waterfowl biologists and managers as effective communicators are necessary but not sufficient. Two other groups of professionals have been integral to NAWMP success and will be more so in the future. First, NAWMP partners have embraced social science professionals to an increasing degree. Second, we also need professional communicators. Understanding human behavior and having knowledge about people's attitudes toward conservation issues are important. However, designing, delivering, and evaluating communications actions to *influence* how people think, feel, or act requires other expertise. This practice of communications includes a host of disciplines from public relations and marketing to citizen participation and environmental education.

Over the arc of the NAWMP enterprise, the sophistication of human dimensions and communications expertise and practice has increased dramatically. As evidence, today both the chair (Dr. Barb Avers, Michigan DNR) and staff coordinator (Dr. Cindy Longmire, DJ Case) for the HDPET have formal natural resource training along with advanced degrees in social science. And the current HDPET membership includes a breadth and depth of expertise not dreamt of in 1986.

Embrace Conflict

As I reviewed my notes from the 1980s and 1990s, I was struck by the number of conflicts, arguments, and disagreements that cropped up. Management and resolution of conflicts is an integral part of what communications people need to do. The job of communicators is not just "rah, rah" but listening and helping people work through issues in a systematic way. Effectiveness should not be measured by a lack of conflict but by how the conflict is managed. Harnessing the passion and energy of conflict in productive ways turns conflict into progress.

Joint Ventures Driving Communications

You cannot read this book without coming away with admiration for how the NAWMP Joint Ventures were conceived and implemented. They comprised a monumental accomplishment. That especially applies to communications. Joint Ventures are a driving communications force because:

- The people attracted to the Joint Venture coordinator and staff positions are invariably excellent communicators both internally and with the external publics the Joint Ventures are trying to persuade.
- The Joint Ventures deliver conservation actions at the state/provincial and project scale. They are "close to the ground" and reflect and respond to local needs. As a result,

their communications efforts are more targeted, more relevant, and more persuasive; and finally,
- The Joint Venture boards are not just advisory. They oversee the JV and bring their respective organizational communications expertise and capacity to the partnership.

Joint Ventures should continue to build their communications expertise and capacity and expand their role as a driving communications force.

"Work Worth Doing"

Teddy Roosevelt wrote, "Far and away the best prize that life offers is the chance to work hard at work worth doing". I was 28 years old in 1987 when I got my first NAWMP assignment. Looking back to that time early in my career to now at the end of my career, I know NAWMP is "work worth doing".

North Americans care about the conservation of nature. Their connections with people and nature are important, complex and should be at the heart of our persuasion efforts. The importance of human dimensions and communications as a vital part of the NAWMP enterprise by necessity will continue to grow (e.g., Dayer et al. 2019).

Whether we are engaging with students or early-career professionals; with farmers or ranchers; with small town or big city managers; with waterfowl hunters; with birders; volunteers; or elected officials…we must remember they all want that "best prize" in their endeavors. They want to be part of something important. Our fundamental communications challenge, and opportunity, is to understand their perspectives and communicate how NAWMP conservation efforts align with their "best prize".

Notes

1 2025–2026 Communications Committee – Dean Smith, AFWA (Chair), Joe Benedict TN (PC liaison), Christy Firestone (NE), Emily Austin (DUI), Micah Holmes (OK), Amanda Stroud (SC), Patty Allen (AFWA), Janine Massey (DUC), Valerie Fellows (FWS), Whitney Polich (DUI), Sarah Semmler (PHJV), Justyn Foth (FWS), Tamara Zeller (FWS), Dave Kostersky (DUC), Lisa Bidinosti (ECCC), Diane Eggeman (DUI), and Christian Roy (ECCC).

2 Tony Roberts (Chair), Josh Dooley, Alan Hanson, Kate Martin, Kyle Spragens, and Greg Yarris.

References

Christiano, Ann and Annie Neimand. 2018. "The science of what makes people care." *Stanford Social Innovation Review* 16 (4), 26–33. (https://doi.org/10.48558/GW2V-5279).

D.J. Case and Associates. 2023. *A Marketing Assets Inventory for the North American Waterfowl Management Plan.* 20pp. (https://nawmp.org/sites/default/files/2023-09/NAWMP%20Marketing%20Assets%20Inventory_Final%20%281%29.pdf).

Dayer, Ashley, Ashley Gramza, and Jessica Barnes. 2019. "Incorporating human dimensions into Joint Venture Implementation Plans." *NABCI Human Dimensions Subcommittee Report.* 12pp. (https://nabci-us.org/how-we-work/human-dimensions/).

Harshaw, Howie W. 2018a. *North American Birdwatching Survey: Canadian National Technical Report.* University of Alberta, Faculty of Kinesiology, Sport, and Recreation. Edmonton. (https://nawmp.org/sites/default/files/2018-11/18-11-26_CAN_NAWMP_Birdwatcher-NationalTechnical-Report.pdf).

Harshaw, Howie W. 2018b. *North American Waterfowl Hunting Survey: Canadian National Technical Report.* University of Alberta, Faculty of Kinesiology, Sport, and Recreation. Edmonton.

(https://nawmp.org/sites/default/files/2018-11/18-11-23_CAN-NAWMP_Hunter-National-Technical Report_0.pdf).

Holsman, Robert and Cynthia Longmire. 2023. *A 2023 Survey of Waterfowl Professionals: Insights in Support of the 2024 Update of the North American Waterfowl Management Plan.* Final Report to the Update Steering Committee. 61pp. (https://nawmp.org/sites/default/files/2023-09/2023%20NAWMP%20Professionals%20 Survey%20Reort%20FINAL.pdf).

Howerter, David, Michael Anderson, Diane Eggeman, Kathy Fleming, Dean Smith, Mark Vrtiska, Barry Wilson, Barbara Avers, Kristin Bianchini, Michael Brasher, Andrew Don Carlos, Annette Glick, David Gordon, Ashley Gramza, Heath Hagy, Alan Hanson, Howie Harshaw, Kristina Hick, Joe Lancaster, James Leafloor, Kurt Mazur, Jake Messerli, Shawn Meyer, Jena Moon, Eric Reed, Anthony Roberts, Katie Sainsbury, Emily Silverman, Jacey Scott, Scott Stephens, Josh Vest, and Corie White. 2024. *Technical Report in Support of the 2024 Update for the North American Waterfowl Management Plan.* 113pp. (https://nawmp.org/sites/default/files/2024-07/combined-2024-nawmp-technical-report-final.pdf).

McGraw Center for Conservation Leadership. 2018. *A White Paper for Sustaining and Improving Waterfowl Conservation in Canada.* McGraw Foundation, Dundee, IL. 136pp. (https://mcgraw.org/wp-content/uploads/2022/01/White-Paper.pdf).

Partners in Flight. 2021. *Avian Conservation Assessment Database, Version 2021.* (https://pif.birdconservancy.org/avian-conservation-assessment-database/) Accessed 26 September 2023.

Patton, Stephanie. 2021a. *National Survey of Birdwatchers: Nationwide and Flyway Comparisons.* Report to the National Flyway Council from the Minnesota Cooperative Fish and Wildlife Research Unit and University of Minnesota. St. Paul, MN. 141pp.

Patton, Stephanie. 2021b. "National Survey of Waterfowl Hunters: Nationwide and Flyway Comparisons." *Report to the National Flyway Council from the Minnesota Cooperative Fish and Wildlife Research Unit and University of Minnesota.* St. Paul, MN. 167pp.

Roberts, Tony, Joshua Dooley, Alan Hanson, Kate Martin, Kyle Spragens, and Greg Yarris. 2023. "North American Waterfowl management plan species prioritization – 2023 revision." *North American Waterfowl Management Plan Science Support Team Technical Report* 2023-01. 13pp.

Soulliere, Gregory J., Brad A. Andres, Mark J. Petrie, Anne E. Mini, William G. Vermillion, and Josh Vest. 2022a. "Assessing Joint Venture approaches for integrated planning and habitat delivery across bird taxa." *Unified Science Team and North American Waterfowl Management Plan Science Support Team, Technical Report* No. 2022-01. 25pp.

Soulliere, Gregory J., Mark J. Petrie, Diane R. Eggeman, Dale D. Humburg, Michael G. Brasher, Ashley R. Gramza, Jessica C. Barnes, Anne M. Bartuszevige, Barry C. Wilson, Kyle A. Spragens, and Barbara A. Avers. 2022b. "Status of integrating human dimensions into Joint Venture bird conservation planning and habitat delivery." *Unified Science Team and North American Waterfowl Management Plan Science Support Team, Technical Report* No. 2022-02. 39pp.

23 Expanding the Partnership – The 2024 Update

Like all updates, the 2024 edition was led by a Plan Committee (PC)-appointed Update Steering Committee (USC), this time cochaired by Tom Moorman (retired Chief Scientist for DU Inc.) and Tim Sopuck (former Executive Director of the Manitoba Habitat Conservancy).[1] They were supported by three technical working groups and a writing team[2]. As in 2004 and 2012 when a Plan Update involved substantive changes in technical guidance, the USC decided to produce both a policy-level update for a broad audience (NAWMP Committee 2024) and a more detailed companion document, the 2024 Technical Report (Howerter et al. 2024). The Update followed the form of the 2012 Revision goals with foci on populations, habitat, people, and integration among them.

The tag line for 2024, "Expanding the Partnership" should sound familiar. This notion has been attendant to the Plan from the beginning but emerged more explicitly in 1998 with objectives of working closely with other bird initiatives and at larger landscape scales. The idea was restated in 2012 and 2018, but more in the context of trumpeting other ecosystem benefits of conserving wetlands and associated habitats like grasslands or riparian forests. The theme came into focus more sharply this time for a few reasons. Despite expenditures of more than US $8 billion and conservation of more than 264 million acres (Chapter 24), in some places, habitat losses were still exceeding North American Waterfowl Management Plan (NAWMP) gains. Second, with changing social values and recreation preferences of North Americans, waterfowl alone appeared to be insufficient motivation to conserve enough wetland habitat to sustain Plan goals. Yet, it seemed imperative that the NAWMP footprint be expanded.

The 2024 Update tackled several other material issues, including a revision of NAWMP population objectives for many species, constructive probing of habitat accomplishments at the Joint Venture (JV) scale, surveying the incorporation of people objectives in NAWMP plans, assessing progress with targeting high-priority landscapes, and integrating multiple objectives in conservation planning. The heavy lifting in preparing the Update was done by three Work Groups. I will summarize the major findings and recommendations from each team.

Population Objectives

The 2018 Update recommended that Plan partners reconsider population objectives on about a ten-year cycle. The last time these were adjusted was in the 2014 Addendum, and with recent developments in surveys and assessment methods, the USC agreed it was time to do so in 2023. With analytical support from the USFWS and CWS, the Populations Work Group reviewed available data, commissioned some new analyses, and brought their recommendations to the table. Analytical details and rationalization for the recommendations are available in Howerter et al. (2024). More on the history of Pan population objectives can be found in Appendix H

DOI: 10.4324/9781003732310-23

online at (www.routledge.com/9781041224389). Largely quoting from the Update, the main recommendations for mid-continent and eastern duck population objectives were:

1 Adjust the period used for Traditional Survey Area objectives by making 1974–2023 the basis for long-term averages. An analysis of changing survey designs and protocols during the earliest years of the WBPHS led the Group to that conclusion. The 1974–2023 time series was from a period of consistent survey effort, better documentation of survey design, and was sufficiently long (50 years) to include a wide range of habitat conditions and waterfowl population levels. Using the 1974 start date resulted in only minor changes in long-term objectives for most species, and those species previously below the 2014 goal levels remained below the new goal levels.
2 For eastern mallards and black ducks, NAWMP objectives should include estimates from all consistently surveyed portions of Eastern North America. This is an expanded region beyond the historical Eastern Core Survey Area that includes state and provincial surveys, and an extended time period (1998–2023). For other duck species in the east, the Group also recommended that population objectives from the entire eastern survey area be used. This spatial expansion produced higher NAWMP population objectives than the 2014 Addendum but represented a more comprehensive estimate of true population sizes in the east. These revised objectives will help inform more habitat JV planning areas compared to previous coverage.

The USC also urged that before the next update, the NSST reviews how population objectives are formulated. The purpose would be to ensure that NAWMP population objectives remain relevant and useful for setting habitat objectives and assessing conservation progress. Specifically, the NSST should: (1) consider the utility of the current geographic scale of NAWMP objectives for conservation planning; (2) assess the capacity of current monitoring to provide what JVs need for effective objective setting and identify gaps that should be filled; (3) undertake analytical work, if needed, to derive alternative population objectives that are more useful; and (4) clarify the dual population objective concept put forward in the 2014 Addendum. Two specific questions are what data might best be used to inform habitat conservation planning for Western-breeding ducks, and the potential utility of incorporating State population estimates for conservation planning by mid-continent JVs. The USC also concurred with the recommendation from 2018 that a review of population objectives every ten years was appropriate. That frequency is compatible with the conservation planning cycles of many JVs.

From the beginning, the NAWMP has been distinguished by a commitment to evidence-based management. Monitoring breeding populations by the WBPHS and other surveys has provided valuable long-term information on annual population abundance, distribution, and variation. These surveys also provided vital benchmarks for indexing NAWMP success at continental and regional levels. Increasingly, waterfowl managers are using other data and analytical methods that provide additional information about populations. Band recoveries and harvest estimates have long been used to estimate survival or account for mortality in population models, and Bayesian estimation frameworks are now commonly used in integrated population models that combine these data streams. Some NAWMP population objectives are based on such model outputs already (e.g., Eastern survey area estimates, some goose population objectives). Of new concern is that results from recent banding analyses, especially increasing differences in male and female survival rates for some species, have challenged some of the assumptions of population surveys (e.g., do lone males represent breeding pairs?), leading to potentially divergent conclusions about waterfowl population trends (e.g., Alisauskas et al. 2014; Arnold et al. 2018; Riecke et al. 2022a, 2022b). Given the importance of population data for planning and evaluating success under NAWMP, the Update strongly recommended that the waterfowl management community undertake a critical assessment of waterfowl population estimation approaches, including an evaluation of the assumptions and potential biases of all estimation methods and data sources.

For reviewing the status of geese, swans, and other ducks, the population team engaged the Arctic Goose Joint Venture (AGJV), the Flyways, the Sea Duck Joint Venture (SDJV) Continental Technical Team (CTT), and other biologists/managers of species breeding mainly outside the mid-continent survey areas. As has been customary, this Update adopted the current Federal/Flyway and AGJV population objectives for goose and swan populations. The SDJV CTT has been working to improve estimates of sea duck populations, the geographic extent of those populations, and enhancing surveys as funds allow. This Update recommended several adjustments to the 2014 sea duck objectives based on their work. Among other ducks, notable progress had been made in estimating and monitoring the West Gulf Coast Mottled Duck population, and the Update endorsed a new objective and monitoring strategy for those birds.

Box 23.1 The Job is Far from Finished

> Despite the success of the NAWMP, ... waterfowl habitat losses continue at scales and rates that challenge the Plan's goal to sustain continental waterfowl populations.
>
> (NAWMP Plan Committee 2024, p. vii)

Habitat Objectives

Across the plains of Canada and the United States, historical grassland losses have approached 70%, with 34 million acres of grasslands lost, mostly to row crops, since 2012 (World Wildlife Fund 2024). In the Prairie Pothole/Prairie Habitat Joint Venture regions, the rate of loss has been estimated at 0.23% annually and that rate is predicted to increase (Fields and Barnes 2019). The latest data on the status and trends of wetlands in the United States estimated a 50% increase in the rate of loss from 2009 to 2019 compared to the previous ten-year period. Approximately 670,000 acres of palustrine wetlands were lost in that reporting decade, including prairie pothole wetlands (Lang et al. 2024). Thus, the broad challenge for Plan Partners is to deliver habitat conservation (whether directly or via policy solutions) at a scale that offsets habitat losses which, if unchecked, could lead to long-term declines of waterfowl, other bird populations, and other valuable ecological functions.

To do this, JVs have been developing specific goals and objectives, stepped down from continental population objectives (Fleming et al. 2019; Lancaster and Wilson 2024). For this Update, Howerter et al. (2024, pp. 97–99) surveyed JVs to assess their progress in establishing and fulfilling habitat goals. About half of JVs had quantified habitat objectives and adequate monitoring systems to enable such reporting of net change. Those JVs can also report on the PC's requested metric – the proportion of stepped-down population goals that are currently supported by the JV landscape. Some JVs reported significant net progress toward their habitat goals, but others could not. The PC should help ensure that all JVs develop the ability to assess net progress toward their habitat objectives and reiterate the expectation that JVs be able to do so. The Update further urged that the PC continue to promote information sharing and advancements among JVs so that the best methods and processes become widely adopted.

More than at any time in the four-decade history of NAWMP, it had become clear by 2024 that it will require mobilizing broader support of governments and citizens for the conservation of sufficient wetlands and grasslands, including those on working lands, to create the conditions for long-term sustainability of waterfowl populations.

Waterfowl and their habitats will always be the core focus of the Plan, but NAWMP partners need more than waterfowl aficionados to succeed. Framers of the 2024 Update firmly believed that opportunities exist to engage new supporters because of the multiple ecosystem benefits that accrue to people and other wildlife from Plan projects. Benefits include improved water quality and quantity, flood attenuation, nutrient sequestration, carbon sequestration, biodiversity conservation, and places for people to recreate and enjoy nature. Plan partners have contracted credible independent researchers to quantify many of these functions, but more studies are needed. Such scientific investments are essential for the credibility of Plan agencies in advocating these protections. Update authors hoped that this approach would be one effective strategy for substantially expanding conserved habitats on the North American landscape.

Some Plan partners have already developed strategies based on ecosystem services. Examples include wetland protection or restoration to reduce flooding, and nitrogen and phosphorus entering waterways in Iowa (Janke and Shannon 2023), publicizing wetland values in the Prairie Habitat JV region (Pattison-Williams et al. 2018) and delivering wetland and floodplain restoration projects along the Mississippi River (Herbert 2023). Restoring playa wetlands to recharge groundwater to support both agriculture and municipal drinking water supplies (Playa Lakes Joint Venture 2024) is another strategy, as well as working with ranchers to maintain or restore forage in flood-prone areas to benefit cattle production and wildlife in the Rainwater Basin (Rainwater Basin Joint Venture 2024). In rural Mexico provision of wastewater treatment technology to people sharing landscapes with waterfowl has improved habitat quality there too (http://dumac.org/en/dimensiones-humanas/).

Box 23.2 Broader Partnerships are Essential

> The future success of waterfowl conservation hinges on the NAWMP's ability to grow and diversity its partnership base, achieving conservation at a scale that reverses habitat loss.
>
> (NAWMP Plan Committee 2024, p. viii)

Also renewed in the Update was recognition that public policy actions have the potential to protect wetlands at a scale and pace that intensive conservation actions like purchase, engineered structures, or other interventions cannot match. Multiple reviewers told the USC that they believed that widespread success in wetland conservation could only happen with supportive public policies. Yet, such policies are vulnerable to shifting politics as experienced in the benefits and then the loss of protection for hydrologically isolated wetlands under the US Clean Water Act and its various interpretations by the courts. Wetland policies across Canada vary among provinces, currently from useful frameworks to non-existent, so policy protection there too demands constant attention. Large-scale grassland protection/restoration programs have at times impacted huge acreages of waterfowl habitat (e.g., the USDA Conservation Reserve Program) and similar initiatives are one goal of the new JV8 quest for a North American grassland initiative (JV8 2024).

Jim Patterson, one of the early visionaries of NAWMP, believed that sustainable success would only come when the benefits of direct conservation interventions by wildlife advocates were overtaken by policy changes or incentives provided by society broadly to sustain agriculture and the environment. Experience after 40 years shows that conservation partners were able to achieve more than the founders dreamed, but thus far not on the conservation policy front. The prescience of Patterson's vision is also supported by the continuing rate of habitat loss and degradation in many places without sufficient policy protection. Policy advancement is generally difficult to achieve without support from broad sectors in society. Thus, the focus of the 2024 Update was squarely on recruiting others to recognize and advocate for actions to realize the multiple benefits of conserving waterfowl habitat.

A major difference from Patterson's time is that Plan partners have greatly expanded scientific knowledge about wetland functions and values – hydrologically, biologically, chemically, and even psychologically. Wetlands like many other natural habitats provide a host of ecological functions that matter to people and many other creatures (Howerter et al. 2024, Table 5). Some benefits we have known about for decades, others are newly discovered, such as the importance of wetlands and their edges as habitat for pollinators valuable for some agriculture crops (e.g., canola). Across the continent, there are emerging numerous examples where conservation and working lands can co-exist with creative co-management of scarce water resources, or local governments can buffer cities and infrastructure from risk of floods while restoring or creating wetland habitats. There are great examples to emulate elsewhere.

Interesting new data on public subsidies for agriculture (OECD 2025) show that such support has declined from around 30% of gross farm receipts in the 1980s to 8.2% in Canada in 2024, perhaps mainly for fiscal reasons. The United States sits at 7% and the 54-nation average at 13.2%. A useful next step would be to see if rates of clearing of marginal lands have declined similarly or at least correlate with rates of subsidization. I suspect that other factors may be offsetting this declining incentive.

The 2024 Update working groups also reviewed the growing scientific literature on climate change, both experienced and predicted, for major waterfowl habitats and urged that every region continues to monitor and adapts their actions as warranted. Presently, climate change expectations are varied and uncertain in some places (Prairie Pothole Region [PPR]) but clearer in others (Gulf Coast sea-level rise). The Update urged everyone in the Plan community to monitor developments and adapt their long-range plans as evidence warrants. Seeking nature-based climate solutions where possible might also become a material source of conservation funding.

People Objectives

The first NAWMP people objectives were published in the 2014 Addendum. Those were centered on users and recreationists, whether hunters, birders, canoeists, campers, or others who appreciate time spent in marshes and grasslands. A decade later, Plan partners had come to think more broadly about stakeholders, recognizing that all people benefit in some way from wetland conservation because of wetlands' numerous ecological functions. There also remained the unresolved debate about whether the Plan's people goals were fundamental ones – things we want to accomplish "just because" – or means ones – things we want to accomplish because they help us achieve other ends like habitat conservation. Broader thinking about people made this an even more nuanced debate.

By 2024, this issue was not fully resolved. Many partners were inclined to think of people goals as both fundamental and means, but attitudes differed among stakeholders, by geography, and other factors. The ecology and human dimensions of the Central Valley of California are profoundly different from those in the Western Boreal Forest, or the Gulf Coast, or rural Mexico. Consequently, the advice in the 2024 Update was for JVs to consider their own unique mix of people (farmers, fishers, loggers, urban, rural, hunters, birders) and habitat systems and judge what set of people objectives makes the most sense in their jurisdictions. The Update reflected a belief that explicit consideration of people in each of these regions is vital for conservation success, but the diversity of settings seemed to prevent a simple universal approach for erecting and pursuing people goals.

A dozen specific people recommendations were offered in the Update (NAWMP 2024, p. 16) ranging from reviewing the experiences of JVs with people objectives, assessing barriers to the integration of people objectives, seeking to identify other societal benefits with obvious connections to NAWMP, adding to regional knowledge of ecosystem services of waterfowl habitat, and broadening the diversity of JV partners. The Update also endorsed continuing to provide support for the NAWPEP education initiative.

Status of Integrating Plan Objectives and Actions

As noted above, interviews with JVs revealed substantial progress with integrating waterfowl population and habitat objectives (Howerter et al. 2024). Of the 23 JVs interviewed, 15 indicated that they had quantified habitat objectives linked to population objectives. Only 2 of 23, however, had incorporated priorities for people into their spatial priorities for habitat, and none of them had devised linkages between waterfowl population and people objectives. The People working group did not think that these results indicated a lack of interest in human dimensions but rather how much uncertainty remained surrounding the process. The interviews revealed a near universal belief among JVs that people objectives comprise at minimum critical means objectives toward accomplishing waterfowl population and habitat objectives.

Increasing Conservation Efficiency

Many Plan partners expressed hope for improving conservation efficiency through more rigorous targeting of activities and landscapes. The science that relates waterfowl population change with habitat conditions continues to strengthen. Population models have been developed that can estimate the contribution to population growth of processes at many life-cycle stages (e.g., Hoekman et al. 2002; Flint et al. 2006; Coluccy et al. 2008; Johnson 2009; Wilson et al. 2012; Howerter et al. 2014; Koons et al. 2014; Arnold et al. 2018; Zhao et al. 2020). With achievable additional work, Plan partners also should be able to improve spatial targeting of resources to places and times that most affect population growth rates.

With nearly four decades of experience delivering NAWMP habitat programs, Plan partners also have, collectively, extensive knowledge of how habitat delivery costs vary by program and geography. Thus, for a finite set of resources available to invest in habitat, we should have considerable ability to optimize efficiency of habitat delivery (where to invest, what, and how) to maximize impacts on populations. Recognizing that there will always be political and operational constraints to achieving optimum efficiency, the USC still believed that we could do better with information already in hand. In that spirit, the 2024 Update authors offered three related recommendations:

1 The PC should encourage JVs to engage broader segments of society in the waterfowl enterprise, most likely via investments in science that quantify ecosystem benefits from actions targeted to improve conditions for waterfowl.
2 The PC should support the NSST, HDPET, and JV science and planning staff to build on existing prioritization tools and apply them to enhance biological and social integration and to allow examination of trade-offs of management alternatives. This should include

incorporating landscape-specific risks of habitat loss, contributions to waterfowl population growth, and relative costs of conservation delivery.

3 The PC should support and encourage approaches and tools that can be applied at various spatial scales to incorporate ecosystem services and econometrics in refining conservation plans.

In the closing paragraph of the 2024 Update (p. 23), the team offered this: "Done well, these efforts will ultimately lead to a broader, more diversified base of motivated stewards who value waterfowl habitat and the many benefits that improve quality of life in their communities. That is the opportunity".

Box 23.3 NAWMP on the Canadian Prairies

By Tim Sopuck, Manitoba Habitat Conservancy, Retired. 2024 Update Co-Chair

The Canadian Prairies before NAWMP

When first envisioned, a priority for NAWMP was to increase the effectiveness of waterfowl conservation in the PPR. The challenge was more than new resources; NAWMP needed to catalyze delivery of programs where the effects on waterfowl populations would be greatest. In the Canadian PPR, more than 90% of those landscapes were privately owned and managed for agriculture.

Interactions among groups that today seem like natural conservation partners were limited prior to NAWMP. Government agencies focused on regulatory matters and management of public lands, while NGOs mostly focused on conserving large wetlands. While these efforts were laudable, they could not match the scale of habitat loss on private lands.

While farming was, and continues to be, the main driver of habitat loss, relationships between conservation interests and farmers were mostly adversarial. Conservation groups spoke of precipitous waterfowl declines, but in the eyes of farm organizations, the "duck problem" was too many ducks at the wrong place and time, causing significant loss of harvest-ready crops. Such fundamental differences in values were part of the social landscape that NAWMP needed to address.

Arrival of NAWMP

Planning for NAWMP in the PPR required two new developments: (1) Institutions to support partnerships and enable effective planning and implementation, both within and among the provinces; and (2) Creation of incentive-based conservation programs targeting agricultural landowners. NAWMP was signed in 1986, but its arrival in the Canadian PPR was a sequenced affair (Chapter 4). Some initiatives in the 1980s laid the groundwork for more effective interactions among conservation organizations. An early-generation computer planning tool supported scenario planning for a new suite of incentive-based conservation approaches for agricultural lands. Such front-end work helped lay the foundation for NAWMP's partnership approach. Manitoba was the first province to develop these biologically based targeting tools between 1986 and 1988.

Prior to the arrival of NAWCA funds in 1991, a series of pilot activities, called First and Second Step Projects (Chapter 3), rolled out in the PPR. These included the "Prairie Pothole Project" and the "Habitat Enhancement Land-use Project (HELP)" that piloted "farmer-friendly" conservation options in Saskatchewan and Manitoba. By the time NAWCA funds were available, a regional oversight body – the Prairie Habitat JV – and provincial coordinating bodies were functioning with multiple partners and a first suite of conservation options. These delivery approaches would be subjected to biological evaluations beginning in 1991.

Evolution of NAWMP in the Prairie Provinces: The Manitoba Example

The scale of work needed and the diversity of agencies and organizations were major considerations in the decision to establish provincial coordination bodies in Alberta, Saskatchewan, and Manitoba. In Alberta and Saskatchewan, these bodies were eventually rolled into other institutions. Manitoba took a different path, with some interesting and enduring results.

The Manitoba Habitat Heritage Corporation (MHHC), a Crown Corporation (basically an operating agency working at arm's length from the provincial government) coordinated implementation in Manitoba. To support local development of the NAWMP partnership, its board of directors represented conservation, agriculture, and rural government groups. This provincial NAWMP meeting place functioned for more than 25 years, helping to break down barriers that had limited collaboration in areas of common interest.

During its first two decades, MHHC was led by two senior provincial staff, Richard Goulden (former Assistant Deputy Minister) and later, Lorne Colpitts (former Director of Wildlife). Each had previous regional, national, and continental experience. They not only came to their positions dedicated to achieving NAWMP objectives but also believed that NAWMP might leverage complementary conservation goals. Over time, that vision took MHHC down interesting paths.

MHHC's first Board Chair also brought different and valuable perspective. Ted Poyser, a soil scientist by training (and keen duck hunter when he wasn't pushing a soil probe), had a long and varied career applying his agricultural knowledge to questions of land use. When tasked to lead a major adjustment program on agricultural lands in Manitoba in the 1970s, he welcomed conservation considerations in planning and implementation. The most significant expansion of Wildlife Management Areas in southern Manitoba came from the agriculture-focused initiatives that he led.

For NAWMP, Poyser embraced two key principles. The first was a "farmer-friendly" approach. He had a major role in conceptualizing DU Canada's initiative, *Prairie CARE*, which stood for Conservation of Agriculture, Resources and the Environment. Rolled out across the PPR, Prairie CARE featured incentive payments for private-land habitat conservation. Farmers became engaged in conservation in new ways. The second was meaningful partnerships that extended across conservation, agriculture, and other rural interests. Organizations got to know each other around the MHHC board table and began to appreciate that they had areas of common interest, an achievement that served the NAWMP partnership well.

MHHC started with limited delivery capacity, which grew as additional revenue was brought in. A key factor in this funding expansion was that NAWMP projects came to be seen as beneficial for additional conservation priorities like watershed management and species at risk. That recognition drew other Canadian sources of funding that supported complementary conservation activities. MHHC became one of the largest holders of conservation easements in Canada, although it's only active in one province. Most of those easements are held in Manitoba's NAWMP delivery area, though many were funded from Canadian sources with objectives not tied to NAWMP. Lorne Colpitts played a major role in the development of the easement program, and I took over a well-oiled machine when I became CEO in 2007.

From 2008 to 2020, MHHC continued to expand programming that was both within and complementary to NAWMP. An example was a partnership with the Manitoba Beef Producers for species at risk programs on cattle ranches. Historically skeptical of conservation agencies, and especially wary of anything tied to species at risk, beef producers came to see that a cooperative, proactive approach was beneficial to their industry. For MHHC, having the backing of the Beef Producers increased its credibility with ranchers immensely. The gain for NAWMP was new funding for grasslands conservation in an NAWMP focus area. A history of participation by Manitoba Beef Producers on the MHHC board played a key role in building a positive relationship.

MHHC took a major leap forward when the Manitoba government developed a new unconventional funding stream for conservation. Beginning in 2018, the Province established three conservation endowment funds – the Conservation Trust, GROW Trust, and Wetlands GROW Trust – with total investments of $204 million. Those funds generate about $10 million annually for grants operated by MHHC, which are available to other, provincially based conservation groups. The original provincial contributions are held by a foundation outside of government, are irrevocable, and will deliver a stream of funding in perpetuity for conservation. Wetland conservation is a major priority for the Trusts, including dedicated funding (Wetlands GROW Trust) for the conservation of shallow at-risk temporary wetlands that are critical to waterfowl production, one of NAWMP's highest priorities.

To further protect against possible rollback by future governments, MHHC was allowed to become a private, registered charitable organization now called the Manitoba Habitat Conservancy (https://mbhabitat.ca/). Stephen Carlyle became CEO when I stepped down in 2021. I was gratified when the Board chose a stellar internal candidate who was already well-schooled in the workings of NAWCA, NAWMP, and other MHHC initiatives. He leads a "two-headed beast", with a major conservation delivery program and one of Canada's largest, provincially based conservation granting program.

The Manitoba example is but one that shows how NAWMP not only continues to focus on its founding objectives for waterfowl conservation but has also become a catalyst for related conservation activities that its founders could not have envisioned. I'd like to think that founders, like Harvey Nelson and Jim Patterson, would look upon this evolution and smile.

Box 23.4 Reflections on NAWMP and the 2024 Update

By Tom Moorman, Ducks Unlimited Chief Scientist, Retired. 2024 Update Co-Chair

The NAWMP was launched in 1986 by visionary waterfowl managers to address habitat conservation needs across North America. Those leaders had remarkable foresight that gave rise to NAWMP JVs, Plan implementation and evaluation, and vitally, periodic updates to keep the Plan relevant. These updates, mandated every five years, depended heavily upon working relationships among individuals representing the partners. Indeed, without a continental and JV business model, and the resulting trust and collaboration among the people representing the Partners, it is hard to envision how NAWMP would have persisted, much less achieved the success it has over four decades.

The history of Plan development and implementation is sketched in this book. Often, Plan leaders identified issues in need of resolution as landscapes and social circumstances changed. Experts were asked to come together to address such matters. Sometimes these were formal working groups, sometimes ad hoc, but they always sought to bring together the brightest minds to wrestle with and solve challenges. Such work strengthened the foundation and advanced Plan implementation. Sometimes the participants already had working relationships, but often the work forged new, valued relationships based on shared passions for waterfowl. Many of these trusting relationships gave rise to lasting professional and personal friendships.

The largest such endeavors were the mandated five-year updates. The strategic evolution of NAWMP can be surprisingly well summarized by simply considering the subtitles of the original Plan and each of the subsequent updates. These were:

1986 – *A Strategy for Cooperation* (It was far short of a detailed implementation plan).
1994 – *Expanding the Commitment* (Bringing focus to the big job ahead, and engaging Mexico).
1998 – *Expanding the Vision* (Conserving whole landscapes for waterfowl and many other species).
2004 – *Strengthening the Biological Foundation* (Evaluations, reducing uncertainties to better inform future actions).
2012 – *People Conserving Waterfowl and Wetlands* (Broader goals for people, populations, and habitats).
2018 – *Connecting people, waterfowl, and wetlands* (Continuing work on the expanded 2012 vision).
2024 – *Expanding the Partnership* (Seeking greater impact by engaging more people).

As we set about the 2024 Update, we had directions from the PC to address the scale and rate of implementation, adaptation planning with respect to climate change, and the state of diversity, equity, and inclusion among Plan partners. We also reviewed technical elements based on new information since the 2012 Revision and 2018 Update. We were challenged to complete the 2024 Update in a relatively short time compared to previous updates, mostly because the schedule had been delayed by the COVID pandemic. We engaged Plan partners through surveys, multiple presentations to partner groups (JVs, Flyway Councils, AFWA, etc.), and by creating work groups comprising individuals from many Plan partners. In all cases, the groups came together and did a remarkable amount of work rapidly to accomplish the Update.

Together, over the past four decades, Plan partners have achieved much worth celebrating, but it is also clear that our collective work will not end soon. Landscapes important to waterfowl have never been static – they change, sometimes in rapid, unexpected ways. New challenges arise due to issues unforeseen in 1986. Long-standing challenges of grassland and wetland loss continue, sometimes accelerated by new causes. Water scarcity and quality, sea-level rise, dynamic social change, and other factors exacerbate old challenges or give rise to new ones. Through dedication, perseverance, and commitments to learning and adaptation, Plan partners have repeatedly risen to new challenges.

Now, we reasoned, partners must strengthen and diversify those partnerships that are central to sustaining populations of waterfowl at desired levels. The 2024 Update called for increased rates and scales of habitat conservation. While continuing to earn the support of waterfowl hunters and other active conservationists, we also must strive to attract more and diverse supporters who can bring additional resources to help. The Update proposed that we do this by promoting the broad array of other ecological benefits of waterfowl habitat conservation that help sustain communities of people.

Opportunities to increase the rate and scale of habitat conservation rest in the ability of JVs to thoughtfully, strategically engage new partners whose values and goals align with, but do not necessarily include, sustaining populations of birds. This is not a new idea. Indeed, in the original Plan and many of the subsequent Updates, the notion of communicating the broad benefits of waterfowl habitat conservation to strengthen and diversify our supporter base can be found. However we engage new supporters, we hope they will help us achieve greater extents of valuable habitat conservation.

Given the challenges we face, we must be innovative and embrace opportunities embedded in change while maintaining essential focus on sustaining these magnificent birds. Each subsequent Update should probe deeply into emerging information to help ensure that the Plan will remain responsive to ever-changing landscapes, societies, and conservation challenges. Those tasked with future revisions of the Plan should strive to respect the Update process, especially to ensuring that all partners have time and opportunity for substantive input as drafts develop. Future Update committees should seek diverse participation from Plan partners to ensure consideration of a wide range of ideas to improve the Plan's focus and implementation. Work under the NAWMP will not end because of the dynamic nature of the landscapes where waterfowl live. It is incumbent upon Plan partners to continue learning and adapting to emerging challenges. The Plan Updates are the critical part of ensuring sustained waterfowl populations, habitats, and ecological benefits for generations of people to come.

Notes

1 **Update Steering Committee**: Tim Sopuck (Canada Co-Chair), Tom Moorman (US Co-Chair), Kathy Fleming, Michael Anderson, Mark Vrtiska, Dean Smith, Barry Wilson, Dave Howerter, Jacey Scott, Mike Brasher, Kyle Spragens, David Gordon, Diane Eggeman, Shaun Oldenburger, Vanessa Charlwood, Mark Gloutney, Jennyfer Tolley, Leonel Urbano, Miguel Flores, and Carlos Piedragil.

2 **Populations Work Group**: Michael Anderson, Kathy Fleming, Alan Hanson, Jim Leafloor, Eric Reed, Shawn Meyer, Mike Brasher, Joe Lancaster, Josh Vest, Emily Silverman

Habitat Work Group: Dave Howerter, Barry Wilson, Jake Messerli, Scott Stephens, Cary Hammel, Heath Hagy. Kurt Mazur, Jena Moon, Josh Vest, Kristin Bianchini

People Work Group: Dean Smith, Mark Vrtiska, Jacey Scot, Howie Harshaw, Katie Sansbury, Annette Glick, Ashley Gramza, Corie White, Andrew Don Carlos, Barb Avers

Writing Team: Tim Sopuck, Shaun Oldenburger, Mark Gloutney, Tom Moorman, Dave Howerter, Jacey Scott.

References

Alisauskas, Ray T., Todd W. Arnold, James O. Leafloor, David L. Otis, and James S. Sedinger. 2014. "Lincoln estimates of mallard (Anas platyrhynchos) abundance in North America." *Ecology and Evolution* 4, no. 2: 132–43.

Arnold, Todd W., Robert G. Clark, David N. Koons, and Michael Schaub. 2018. "Integrated population models facilitate ecological understanding and improved management decisions." *Journal of Wildlife Management* 82: 266–74.

Coluccy, John M., Tina Yerkes, Rebecca Simpson, John W. Simpson, Llwellyn A. Armstrong, and Jonas Davis. 2008. "Dynamics of breeding mallards in the Great Lakes States." *Journal of Wildlife Management* 72, no. 5: 1181–87.

Fields, Sean and Kevin Barnes. 2019. *Grassland Assessment of North American Great Plains Migratory Bird Joint Ventures.* (htps://ppjv.org/conservation/projects/#grasslands-birds). 70pp. Prairie Pothole Joint Venture. Great Falls, MT.

Fleming, Kathy K., Michael K. Mitchell, Michael G. Brasher, John M. Coluccy, J. Dale James, Mark J. Petrie, Dale D. Humburg, and Gregory J. Soulliere. 2019. *Derivation of Regional, Non-Breeding Duck Population Abundance Objectives to Inform Conservation Planning—2019 Revision.* NSST Technical Report 2019-01. 44pp.

Flint, Paul L., J. Barry Grand, Thomas F. Fondell, and Julie A. Morse. 2006. "Population dynamics of greater scaup breeding on the Yukon-Kuskokwim Delta, Alaska." *Wildlife Monographs* 162: 1–22.

Hassan, Rashid, Robert Scholes, and Neville Ash. (Editors). 2005. *Ecosystems and Human Well Being: Current State and Trends, Vol. 1. Millenium Ecosystem Assessment Project.* Island Press. London. 47pp.

Herbert, Ellen. 2023. *Building Data-Driven Partnerships to Accelerate Wetland Conservation to Benefit Wildlife and People*. (https://nawmp.org/content/nawmp-webinar-series).

Hoekman, Steven T., L. Scott Mills, David W. Howerter, James H. Devries, and I. Joe Ball. 2002. "Sensitivity analyses of the life cycle of midcontinent mallards." *Journal of Wildlife Management* 66, no. 3: 883–900.

Howerter, David, Michael Anderson, Diane Eggeman, Kathy Fleming, Dean Smith, Mark Vrtiska, Barry Wilson, Barbara Avers, Kristin Bianchini, Michael Brasher, Andrew Don Carlos, Annette Glick, David Gordon, Ashley Gramza, Heath Hagy, Alan Hanson, Howie Harshaw, Kristina Hick, Joe Lancaster, James Leafloor, Kurt Mazur, Jake Messerli, Shawn Meyer, Jena Moon, Eric Reed, Anthony Roberts, Katie Sainsbury, Emily Silverman, Jacey Scott, Scott Stephens, Josh Vest, and Corie White. 2024. *Technical Report in Support of the 2024 Update for the North American Waterfowl Management Plan.* 113pp. NAWMP Plan Committee. (https://nawmp.org/sites/default/files/2024-07/combined-2024-nawmp-technical-report-final.pdf).

Howerter, David W., Michael G. Anderson, James H. Devries, Brian L. Joynt, Llwellyn M. Armstrong, Robert B. Emery, and Todd W. Arnold. 2014. "Variation in mallard vital rates in Canadian aspen parklands: The Prairie Habitat Joint Venture assessment." *Wildlife Monographs* 188: 1–37.

Janke, Adam and Mike Shannon. 2023. *Exploring the Intersections between Wetlands, Water Quality, and Waterbird Conservation in Iowa's Agricultural Landscapes*. North American Waterfowl Management Plan Webinar Series. (https://nawmp.org/content/nawmp-webinar-series).

Johnson, Fred A. 2009. *Variation in Population Growth Rates of Mottled Ducks in Texas and Louisiana.* Report to the U.S. Fish and Wildlife Service, USGS Administrative Report. 26pp.

JV8. 2024. *A Business Plan to Recover North America's Central Grasslands: Strategies, Funding, and Partnerships Needed to Save One of the World's Most Endangered Ecosystems.* JV8 Central Grasslands Conservation Initiative. (https://jv8.org/wp-content/uploads/2025/12/JV8-PLAN-Digital-111025.pdf).

Koons, David N., Gunnar Gunnarsson, Joel A. Schmutz, and Jay J. Rotella. 2014. "Drivers of waterfowl population dynamics: From teal to swans." *Wildfowl, Special Issue* 4: 169–91.

Lancaster, Joseph D. and Barry C. Wilson. 2024. *Guidance for Selecting between Autumn and Winter Regional Duck Abundance Objectives for Conservation Planning.* North American Waterfowl Management Plan Science Support Team Technical Report 2024-02. 10pp.

Lang, Megan W., Jeff C. Ingebritsen, and Rusty K. Griffin. 2024. *Status and Trends of Wetlands in the Conterminous United States 2009–2019.* U.S. Department of Interior, U.S. Fish and Wildlife Service. Washington, DC. 43pp.

North American Waterfowl Management Plan Committee. 2024. *2024 North American Waterfowl Management Plan Update: Expanding the Partnership.* Environment Canada. Ottawa. 62pp. (https://nawmp.org/sites/default/files/2025-01/nawmp-update-2024-en-web.pdf).

OECD. 2025. *Agricultural Policy Monitoring and Evaluation 2025: Making the Most of the Trade and Environment Nexus in Agriculture.* OECD Publishing. Paris. 660pp. (https://doi.org/10.1787/a80ac398-en).

Pattison-Williams, John K., John W. Pomeroy, Pascal Badiou, and Shane Gabor. 2018. "Wetlands, flood control, and ecosystem services in the Smith Creek drainage basin: A case study in Saskatchewan, Canada." *Ecological Economics* 147: 36–47.

Playa Lakes Joint Venture. 2024. *Tomorrow's Water: Connecting People, Playas and the Ogallala Aquifer.* (https://pljv.org/playas/tomorrows-water/). Playa Lakes Joint Venture. Longmont, CO.

Rainwater Basin Joint Venture. 2024. *Wetland Management Tools for Landowners.* Rainwater Basin Joint Venture. (https://www.rwbjv.org/wetland-management/). Rainwater Basin Joint Venture, Grand Island, NE

Riecke, Thomas V., Benjamin S. Sedinger, Todd W. Arnold, Dan Gibson, David N. Koons, Madeleine G. Lohman, Michael Schaub, Perry J. Williams, and James S. Sedinger. 2022a. "A hierarchical model for jointly assessing ecological and anthropogenic impacts on animal demography." *Journal of Animal Ecology* 91, no. 8: 1612–26.

Riecke, Thomas V., Madeleine G. Lohman, Benjamin S. Sedinger, Todd W. Arnold, Cliff L. Feldheim, David N. Koons, Frank C. Rohwer, Michael Schaub, Perry J. Williams, James S. Sedinger. 2022b. "Density-dependence produces spurious relationships among demographic parameters in a harvested species." *Journal of Animal Ecology* 91, no. 11: 2261–72.

Wilson, Heather M., Paul L. Flint, Abby N. Powell, J. Barry Grand, and Christine L. Moran. 2012. "Population ecology of breeding Pacific common eiders on the Yukon-Kuskokwim Delta, Alaska." *Wildlife Monographs* 182: 1–28.

World Wildlife Fund. 2024. *Plowprint 2024.* (https://files.worldwildlife.org/wwfcmsprod/files/Publication/file/8mq6fdcmt4_PlowprintReport_2024_FINAL.pdf). World Wildlife Fund. Washington, D.C.

Zhao, Quing, Todd W. Arnold, James H. Devries, David W. Howerter, Robert G. Clark, and Mitch D. Weegman. 2020. "Using integrated population models to prioritize region-specific conservation strategies under global change." *Biological Conservation* 252: 108832.

24 Lessons Learned, Opportunities Ahead

A wise friend urged that this book be written as a "love letter" to the Plan Community in recognition of the unprecedented partnership and what it has accomplished for conservation over its first 40 years. There is, indeed, much to celebrate, much to build on for the future. It would be out of character for me, however, not to also offer a little advice as I reflect on the evolution of the enterprise over those four decades. So, this chapter will be a little of both that love letter and a few suggestions from one who has been involved in some manner since the beginning.

I will begin with a brief overview of accomplishments, some apparent keys to success, a few lessons learned, and the challenge of expanding the impact of the Plan. Lastly, I will look ahead at some ascending issues.

So Much to Celebrate

People Coming Together

Accomplishments of NAWMP took many forms. Stimulating formation of lasting partnerships of 23 habitat Joint Ventures (JVs), three species JVs plus two action groups, spanning all three countries may be the most important achievement of all. Traditionally, we think of dollars invested and acres conserved as indices of conservation accomplishment. I will get to those numbers, but the foundational and hopefully lasting thing NAWMP has done is bring people together, all over the continent, striving to accomplish common goals. This work included knowledge gained and perspectives changed. We learned through the journey. Partners worked in vital habitat regions and across borders. State Grant programs began and expanded. Conservation easement legislation and ecological gift legislation were passed in some jurisdictions. The development of native plants ecovars supported responsible grassland restoration. Repurposed pipe, rice rollers, and other clever initiatives helped build relationships with owners of lands important to the birds. Commitments to science-informed planning evolved into multiple-loop adaptive management. Critical thinking and five-year renewals kept the Plan relevant. Pride of accomplishment reinforced the partners and solid bi-partisan funding sustained the work. These were all outgrowths of the bold notion of NAWMP decades ago. Indeed, there is much to feel good about. Jim Patterson once told Stew Morrison it was amazing to him, "that everyone who was involved in the development of the NAWMP felt they had made an important contribution…and they did"!

The Birds

Recognizing that duck populations fluctuate for many reasons other than the work of Plan partners, it still seems reasonable to ask how are the ducks doing after all this work? Comparing

DOI: 10.4324/9781003732310-24

recent waterfowl populations with the initial 1986 objectives (NAWMP Committee 2024, Appendices) shows that several populations are close to those goal numbers (mallard, northern shoveler, canvasback), others somewhat higher (gadwall, green-winged teal, blue-winged teal, redhead), though concerns remain for species persistently below those objective levels (northern pintail, scaup, American wigeon). Goose populations are generally healthy, though there are a few exceptions. Many sea duck species still require better monitoring to track their status.

Funds Brought to Bear

Of course, funding mattered. The Plan was signed without a nickel of commitments from any government or nongovernment entity to support its ambitious ideas. This was at a time when US budget deficits were climbing steeply as the Cold War peaked. Interest rates for borrowers were sky high. Imagine the audacity of floating a 15-year plan, with an (under)estimated $1.5 billion price tag, with exactly $0 committed when the agreement was signed! Nevertheless, leadership commitments by a handful of states, then Ducks Unlimited (DU) and the US National Fish and Wildlife Foundation, launched proof-of-concept projects in each of the first habitat JVs. Well-positioned private citizens, duck hunters all, also invested political capital to encourage early progress. The notion of pooling resources to do work broadly acknowledged as vital for waterfowl populations carried over to the birth of the North American Wetlands Conservation Act (NAWCA) in 1989, a most consequential source of federal funding to support NAWMP. For a Plan that was first and ambitiously conceived to be a 15-year, $1.5 billion challenge, it has greatly exceeded those hopes (Table 24.1). In round numbers, an investment of $2.29 billion in US federal NAWCA funding has stimulated $6 billion in matching funding (from Canada, states, provinces, and NGOs) between 1991 and the first quarter of 2025, resulting in some 264 million acres of habitat conservation.

Acres Conserved

Tallying the habitat work under NAWMP is challenging. Some obvious pitfalls were solved early, like the risk of double counting when more than one partner worked on the same project. Added to NAWCA, there have been major investments by the US Refuge system, USDA's Wetland Reserve Program (WRP), and more. Under NAWCA, however, there are solid records of dollars invested, and acres of wetlands and associated habitats conserved for all three countries. Canada has also tracked investments and accomplishments of projects under NAWMP but not directly connected to NAWCA. The United States tried to do that too for a few years, but with so much work done under other federal directorates, like the Refuge system's acquisition program, comprehensive tracking seemed untenable and was dropped. Thus, US acreage accomplishments and investments in Table 24.1 are underestimated by some unknown amount. The accomplishment categories in each country also differ slightly, so numbers are additive in some sub-categories, but not all. For instance, "influenced" acres in Canada are not strictly like the US categories. Lastly, "projects" in the Canadian context are frequently multi-year "programs" consisting of many "projects" each in the parlance of individual US projects. Thus, I have not totaled project numbers below. Despite these limitations, the numbers are staggering compared to any other recent conservation initiatives.

The accomplishment tallies are constantly updating too so, please consult the websites for JVs of interest. The annual report of the Canadian partners, "Canadian Habitat Matters", is a great source for accomplishments in that country (https://nawmp.wetlandnetwork.ca/publications/).

Table 24.1 NAWCA and Related Expenditures and Accomplishments, 1991–2024

Nation	*Acres secured*	*Acres influenced*	*Acres Managed*	*Total Acres*	*NAWCA $$*	*US Non-Fed Match $$*	*Canadian Match $$*[a]	*Total $$ US*	*# Partners*	*# Projects*
Canada				16.68 M	$ 753.1 M	$ 754.1 M	$ 1,460 M	$ 2,967 M	172	609
The United States	6.46 M		5.27 M	11.72 M	$ 1,453.1 M	$ 3,602.5 M		$ 5,056 M	6,240	2,411
Mexico	2.68 M	0.20 M	1.31 M	4.19 M	$ 81.7 M	$ 176.4 M		$ 258 M	647	373
NAWCA Totals					$ 2,287.9 M	$ 4,533 M		$ 8,281M	7,059	3,393
Canada, including non-NAWCA	24.1 M	236.4 M	4.3 M	248.1 M						

[a] Canadian match $CA 1.825M converted to $US @.80 over 33 years. Note Canadian match on NAWCA project proposals limited by agreement with USFWS; actual amounts are usually higher.

Source: Dave Gordon, Pers. Comm, December 3, 2024. Data from: USFWS Division of Bird Habitat Conservation (2024). And for Canada from: https://nawmp.wetlandnetwork.ca/Media/Content/files/Habitat%20Matters%202024%20Annual%20Report.pdf. For US updates see https://www.fws.gov/program/north-american-wetlands-conservation/what-we-do.

Suffice to say that since 1987, the JV partnerships have accomplished millions of acres of habitat protection and enhancement, a few billion dollars in conservation investments, and a quantum leap in the scientific underpinnings of avian conservation from basic research to habitat data bases and decision-support models. They did this while engaging thousands of landowners, agencies from cities to federal governments, corporations, and nongovernment organizations in the pursuit of regional conservation visions. Collectively, they achieved a staggering amount of work and serve as a hopeful example for habitat conservation on a continental scale. It began with ducks but it's now much broader with benefits for all birds, resident wildlife, watershed health, biodiversity, and human well-being.

Thinking ahead, for planning and marketing purposes, it might be even more useful to have regular monitoring of net landscape change. On the prairies, trends for wetlands and perennial cover could index progress in maintaining sustainable conditions for breeding waterfowl. For coastal systems subject to rising sea levels, net change is also a vital metric. I will expand on this notion later in this chapter.

Keys to Success

Nations Working Together

A handful of early decisions by NAWMP planners were key in the Plan's success to date. Even before 1986, the waterfowl management enterprise was built on decades coordinated work between Canada and the United States. Despite sometimes feisty disagreements among the parties, the Migratory Bird Treaty of 1916 and the enabling legislation that followed led to the birth of the Flyway System in the late 1940s. Through a handful of early international planning forums, the government partners had established a tradition of consulting one another on major decisions about harvest matters. Waterfowl hunters had long been advocates for conservation, and nongovernment organizations, most conspicuously DU, had shone a light on the need for habitat conservation and cross-border cooperation to enable it since the 1930s. So, when the habitat and population crises of the 1980s came along, many members of the community already had the instinct to work together, but it took a handful of leaders with vision and determination to make things happen.

Flexibility

Part of the wisdom of the 1986 Plan, though many of us rued it at the time, was that it offered merely a "strategy for cooperation" with a primary focus on habitat. No one was "in charge". No funding was identified. It was not prescriptive, other than offering population objectives to strive toward, and it left open how to accomplish this. It identified landscapes most important to waterfowl populations across the annual cycle. The message was, if you want ducks, geese, and swans at these levels, you need these key breeding, migration, and wintering areas to remain substantially intact.

The JV Model

Critically, the Plan did not recommend a massive federal program that would have been difficult to deliver efficiently and might have generated resistance in many quarters. There also would have been the challenge of sustaining single-source funding over 40 years – something that only a few federal programs like the US defense budget, Social Security, and the National Science Foundation, could boast doing during that same period. Rather it conceived of something new in the conservation world – voluntary associations of willing partners in regional "joint ventures".

It was an inspired idea. This mechanism would only work if partners came together with mutual interest in addressing habitat loss – something most biologists had long recognized as essential despite the annual preoccupation with hunting regulations. The only thing "top-down" about this Plan was federal leadership in coordinating the effort and, in time, a little seed money, but it was the regional partners who had to make it work.

Given the passion and capability of private sector players in waterfowl conservation, encouraging the JVs to become public/private partnerships was also vital for success. The players varied from place to place – think California with partners like Audubon, California Waterfowl Association, and DU; the Atlantic Coast with 13 states initially; and the expansive Canadian PPR with multiple federal, provincial and NGO players. The combination of national/international framing with regional knowledge and innovation was proven brilliant. The relationship has been like a long dance as Jim McCuaig (CWS) described it, with both parties, JVs and national players, leading from time to time and issue to issue. It was not without tension but with experience and goodwill, the JVs made progress. Lessons from the JV experience continue to inform other conservation initiatives too (https://r2rbirds.org//assets/R2R-July-2021-Guidance-Document-v3.pdf).

Updates

Another part of the "secret sauce" was the commitment from the beginning to regular renewal of the Plan at approximately five-year intervals. I will expand on this experience later in this chapter but for now, I want to acknowledge that this provision has been vital to keeping the Plan current, relevant, and responsive to changes, both social and biological. The updates provided periodic motivation for Plan partners to think critically together about what they were doing and where they needed to head.

Evaluations

By the early 1990s, scientists engaged with the Plan realized that it was imperative to evaluate the efficacy of the work and thus enable programmatic and strategic changes informed by science. This was a funding challenge of unprecedented scope and scale given the breadth of the habitat work underway. Fortunately, it happened at a time when the conservation science community was making conceptual strides in adaptive management and the fit of our challenges to those concepts was obvious. Beginning with a few progressive JVs (LMVJV, PHJV, PPJV), these investments soon bore fruit and by the 2004 Update, Plan partners were focusing much attention on biological foundation issues. With the Continental Progress Assessment (Chapter 15) that followed, the entire Plan partnership took a critical constructive look at the state of planning, delivery, and evaluations throughout the enterprise and urged all to continue pursuing the science needed to help inform program decisions.

Other innovations followed but these provisions in the first couple of decades – a JV delivery structure, matching funding, regular updates, and a commitment to evaluation and adaptation – were vital in positioning the Plan for success.

Some Lessons Learned

Scale and Coordination

No one had tried to design and execute massive habitat conservation actions on a continental scale before NAWMP. With a diversity of ecosystems, cultures, political jurisdictions, and focal species and habitats, the prospects were daunting. The challenge was not only to lead the initiative at a conceptual and international level but also to encourage creativity and problem-solving by partners at smaller scales. Finding a working balance between coordination and innovation was difficult at the outset and remains so today. A material complication arose from 1999 onward when the JVs, one by one, adopted some degree of an all-bird mission and no longer looked only to the Plan Committee (PC) for guidance.

The JVs never fully "reported to" the PC in a typical hierarchical business sense, although at the onset of each JV, the PC was asked to endorse their conservation plans. The continuing persuasion of CWS or FWS leaders who provided coordination at both federal and regional levels also helped connect the evolving enterprise. Despite the loose structure, the partners found paths that worked with the PC and its ancillary bodies like the NAWMP Science Support Team (NSST) assuming the role of oversight for the biological foundations of the waterfowl work but including the JVs in each step of that process. NSST membership by JV science coordinators was one vital connection beginning in 2000.

Following the 2005–2007 Continental Progress Assessment (Chapter 15), the JVs began providing in-person, usually triennial reports to the PC. These were designed to be substantive discussions of progress and forward plans. Given 17 years of experience with these sessions, it may be worthwhile to revisit the form and schedule of this important routine connection to ensure that constructive, penetrating dialog is the norm.

Assessment and Adaptation

As the first Continental Assessment (2007) was completed, there was widespread appreciation by Plan leaders and the JVs that it had been a valuable exercise, more for stimulating fresh thinking about priorities than simply assessing progress. The Assessment Steering Committee recommended repeating a similar comprehensive assessment a decade later. Today (2026), 18+ years have passed without a second assessment. Importantly, we made fundamental changes in Plan goals with the 2012 Revision. In 2007, NAWMP was "simply" about waterfowl populations and habitat objectives. I strongly suggest that it would be valuable to undertake some version of a continental assessment again soon, recognizing the game-altering impact of the broadened NAWMP goals embraced in the 2012 Revision and nearly two decades of JV evolution since 2007.

Plan partners have learned that maximizing returns on investment (ROI) of conservation expenditures is complicated. Some JVs have done an impressive job of assessing program performance and adjusting their project delivery approaches accordingly. More of this appears to be happening every year, but the perennial challenge is to find sufficient funds for the applied research to inform these choices. An even greater challenge lies in potentially adjusting allocation of conservation efforts among regions and among species. The PC recommended such actions as early as the 1998 Update, but at least two problems persist: First, only a few habitat funds are truly movable, i.e., the organization with funding can shuttle resources from one area to another or one project to another. National organizations like the USFWS, CWS, or DU can sometimes do that, but even with them matching funding, typically attached to nearly every project, may not be as portable. Nevertheless, with time, matching funds for higher priority geographic areas should be attainable. Another interesting wrinkle involves federal allocation to the various JVs. When JVs were single-line items in the FWS budget, each partnership lobbied Congress for their dollars. Once the JV line items were combined, the proportional allocation to each JV was fixed at that point in time. Changing geographic allocations now would require agreement on what factors are most important for distribution of conservation dollars (Tony Roberts, pers. comm., May 2025).

Cultural elements also can limit maximizing ROI. Two previous senior NAWMP managers have told me, they preferred to do whatever they could, whenever they could with willing partners, as long as the projects made sense biologically, trusting that over time the highest-priority work would be accomplished. I believe that's a risky proposition. Perhaps we could do better? The recent Krainyk et al. (2022) analyses provided new advice for scoring NAWCA proposals, and Roberts et al. (2023) revision of species' prioritizations are steps that could help move the partnership in the direction of increased average biological value per dollar invested for any suite of species. Other questions obtain, of course, such as on what basis would such prioritization decisions be made, how might it work mechanistically, and what authority could mandate those decisions?

Population Matters

It has been decades since Mark Koneff and the Continental Evaluation Team first used harvest data to step down continental population objectives to JV jurisdictions as a basis for habitat planning. Fortunately, led by the NSST, there have been many improvements on this process in recent years (Fleming et al. 2019; Lancaster and Wilson 2024). By utilizing banding and harvest data, now including Canadian and Mexican data, these analyses have achieved the most technically sound means to date of connecting wintering, migration, and breeding area populations. Some decisions remain, for instance about incorporating breeding source areas beyond the TSA, especially in Western North America, but these were framed in the 2024 Update and should be reviewed soon. A related challenge is to reconcile those regional population objectives with habitat objectives sufficient to support populations at goal levels. Programs within JVs have been variably successful in figuring out what to deliver where, and how much. The scientific evidence required to do that is improving everywhere but remains incomplete, and the ability to report on the status of those habitats is uneven. Prioritizing work within JVs to accomplish these stepped-down population and habitat objectives remains important for every JV habitat partnership.

Looking ahead and thinking about other informative metrics that the JVs and the PC could track one stands out: improved monitoring of landscape and land use changes that matter to waterfowl populations. For breeding areas, net trends in wetland abundance and grasslands in target landscapes would be metrics of basic importance to judging JV success. One such tool presently available is the "Plowprint" monitoring program of World Wildlife Fund (https://files.worldwildlife.org/wwfcmsprod/files/Publication/file/8mq6fdcmt4_PlowprintReport_2024_FINAL.pdf) that tracks rates of change in cultivated land in the mid-continent grasslands. Metrics like this account for background loss to agriculture and other developments, gains through conservation actions and improved agricultural policies, and report on net changes. If an essential goal is to maintain the productive capacity of a JV landscape, then data such as these could be vital. Strategically, knowing the net effect of public policies, private landowner actions, and conservation programs seems essential for judging progress toward long-term sustainability goals. Wetland and land-use tracking in migration and wintering JVs might be useful as well.

Research is also needed to better understand population status and trends of species that breed and/or winter in less accessible locations and are only sketchily enumerated by current surveys. The Sea Duck JV's Continental Technical Team has been working on this problem but needs more resources to accelerate its work. The statuses of some of these species/populations remain poorly known and they may not be secure. Some of them are important resources for First-Nations communities and, of course, have value beyond human use as well. Many southern breeding species like fulvous and black-bellied whistling ducks, mottled ducks, cinnamon teal, and several resident species in Mexico and Central America are not well surveyed either.

An urgent population issue that partners must resolve soon is the disparate conclusions about the trends of duck populations from two sources: (a) the annual WPBHS survey estimates and (b) trajectories predicted from banding analyses using Lincoln Estimators that have revealed changing sex-specific survival rates for some species. Changing sex ratios could affect estimations of breeding pairs from long-standing surveys. Howerter et al. (2024) reviewed this conundrum in more detail. Presently, these two sources of information suggest quite different statuses and trends for some mid-continent duck populations. Both methods rely on critical assumptions that if violated could affect the veracity of their conclusions. The 2024 Update highlighted this dilemma and urged that federal and university scientists work together to resolve these uncertainties. Until it is resolved, maintaining the integrity of both the annual surveys and the banding operations should remain a high priority. If there emerge opportunities to monitor populations in a more cost-effective way, those should be explored.

The decision of the US and Canadian partners to suspend the five-year continental population survey for Trumpeter Swans has left us with a formerly endangered species without an agreed-upon means of indexing their populations. Population segments appear to be expanding in some regions (Midwest) but struggling in others (Yellowstone Tri-State Area). Meanwhile the NAWMP goal for the species is in abeyance. I urge the Federal/Flyway authorities to find a solution for this gap. Vrtiska et al. (2024) discussed some options for this that should be considered further. There are many other species/populations where both compelling objectives and monitoring programs are lacking (NAWP Committee 2024, Appendices). Where there is cause for concern, it would be well to address those gaps as soon as possible.

Integration of NAWMP Goals

One of the themes emerging from the 2008 Summit and subsequently the 2012 Revision was a desire to see closer integration of NAWMP's population, habitat, and people goals. This included harvest management, at least in the United States where a formal system of adaptive harvest management is in place (Nichols et al. 2007). The "dream" ca. 2008 was for an integrated and adaptive decision process wherein objectives for harvest, users, and habitat were set simultaneously with knowledge of the interacting effects on each of them. Policies might then be constructed in a way that enabled something approaching an optimal decision solution. Monitoring system responses over time would then allow future adjustments in this linked system.

From the beginning of consultations in 2009, it was apparent that there were substantive obstacles in play. First, integration between habitat and harvest management in the simple way envisioned by the Joint Task Group (JTG) (Chapter 16) was challenged by how to translate "continental K" (carrying capacity) into tangible habitat objectives. That would become even more complicated if applying to multiple species simultaneously. Incorporating people objectives in the model was even more daunting – technically possible, perhaps, with lots of assumptions, but complex. Most fraught, however, was the matter of who would oversee this decision process and ultimately set direction for each management cycle. A deeply entrenched structure and process is in place for establishing US harvest regulations, but it is different in Canada and Mexico. A common system has been adopted for black duck harvest between Eastern Canada and the United States but that is the sole such example. For habitat decisions, an entirely different structure is in place – the PC/JV complex in each country. People objectives are still evolving at both national and regional scales. Thus, a truly integrated system would require investing in a great deal of technical work and devising a new, over-arching administrative system to decide on pursuing or adjusting the integrated objectives, ideally for all three countries. And lastly, managers lack robust system control to do this.

I was a naïve advocate for the JTG's elegant, simplistic recommendations in 2007, but I now think this is impractical, at least for the integration of all three goals in all three countries. I still believe that there is potential for useful integration at smaller scales in several ways: (1) It is probably most important for NAWMP partners to work out an accommodation between people objectives and waterfowl needs for habitat. The NSST and a few JVs have piloted some linkage ideas; for example, with planning to achieve both user and waterfowl objectives by proximity to urban areas in the upper Midwest, and with Playa Lake restorations for municipal water benefits and waterfowl habitat. These innovations seem promising for expanding the footprint of conservation if we can engage the public in funding such work. (2) On the harvest side, there may be opportunities for adjustments based on better information about what hunters most desire. Surveys to support the AFWA's R3 initiative, and some parallel surveys in Canada, are providing actionable information for harvest managers. Simpler ties between NAWMP habitat and harvest management may need to suffice, but it seems prudent that there remain cross-engagement of at least some technical experts between the HMWG and the NSST. At minimum, both management streams need to know what the other is doing and why.

People as Fundamental or Means?

The explicit commitment to people as fundamental objectives flowing from the 2012 Revision has not born much fruit yet, other than accelerating support in some jurisdictions for hunter recruitment (NAWMP Committee 2024). Uncertainty about how best to regard people objectives did stimulate research to provide a better understanding of current attitudes of hunters, birders, landowners, and the general public about conservation, which should help us engage more people. Bringing people, in all their diversity, into formal decision models has proven even more difficult for harvest management, but we have some promising new tools from the NSST's Priority Landscape Committee to help spatially target habitat conservation for birds and some user groups. The Public Trust Doctrine (https://www.law.cornell.edu/wex/public_trust_doctrine) of our societies puts long-term benefits for people at the apex of resource management policy. Practically, however, I think the Plan's approach to people goals can be regarded usefully as both fundamental and means.

There are some people, hunters most prominently, who largely got us where we are today with waterfowl conservation. Serving them seems to me like a fundamental goal given the long synergistic history of hunting and habitat conservation (Anderson and Padding 2015). Landowner engagement seems to fit the notion of fundamental goals too for different reasons. People who own the land on which the birds depend – mainly farmers and ranchers – are the most important people whose interests must be accommodated for sustainable habitat solutions. Their competing interests with NAWMP vary greatly from place to place and can change over time with evolving market forces and agronomic practices.

For most of the public, however, it may make sense today to regard them as means to an end of more habitat conservation. Perhaps the best view is dynamic, thinking about "moving" more groups of people from the category of means to the category of fundamental as the views of more people align with the NAWMP conservation ethic and they come to engage more actively with us as partners co-benefiting from NAWMP work.

The Many NAWMP Committees

Given the current state of the NAWMP enterprise it seems evident that the NSST, the Human Dimensions/Public Engagement Team (HDPET), the Integration Steering Committee (ISC), and the new education team (NAWPEP) all have important work to do as components of the broadened 2012 NAWMP vision. Their connections with each other, with the PC and with the JVs, are essential, and it is evident that this complexity comes with challenges of time management and funding. I expect that these institutions will continue to evolve and adapt and find ways to become more efficient. The recent merger of the HDWG and the PET was one progressive change.

The Pintail Action Group, and especially the Scaup Action Team have become inactive. Yet for these species, important questions remain unanswered, and the populations have only partially recovered from the low levels that motivated the creation of the groups more than two decades ago. I think it would be valuable to assemble again scientists and managers currently involved in scaup and pintail work to look critically at the status of the birds, key uncertainties that remain, and recommend a path forward to both Plan partners and harvest managers. As I finish this book there are plans brewing for a new comprehensive northern pintail demographic study as a first project for a new "North American Center for Waterfowl Science".

The PC, in addition to coordinating all its subordinate parts, would doubtless benefit from occasional formal interactions with at least the US and Canadian NAWCA Councils, the NABCI Council, and perhaps the active Road to Recovery Initiative as well. These bodies have important, overlapping strategic interests with the PC and all may benefit from regular personal interactions. I also think it would be valuable for the PC to expand its tracking of landscape change in the JVs vital to waterfowl populations. A high-level tracking of such changes would provide important insights and a sense of dynamic priorities for attainment of Plan objectives.

Box 24.1 Reflecting on the NAWMP "Fire"

When I contemplate NAWMP today I feel the warm embers from a well-built fire. Still valued, still warming the community gathered around, still powering useful progress, but also a little smothered by a complex, intertwined pile of sticks that are demanding much of its energy. I yearn for people to gather 'round and blow hard, to rekindle the flames, then to heap on some new wood and build the fire higher again!

NAWMP needs to grow fresh flames from these warm embers. Newness, simplicity, crises and passion powered our metaphorical fire through much of the '90s and '00s. The Revision in 2012 made our quest broader, but also more complex. Now we can envision a path to new growth, but how to feed a newer, bigger fire isn't entirely clear. Might we need to simplify our approach again before we can magnify? Or maybe the 2024 Update's emphasis on broad ecological values of waterfowl habitats can finally guide us to the new fuel we need?

Expanding the Impact of the Plan

"Expanding the Partnership" was chosen as the main theme for the 2024 Update with good reason. Comparing the current "footprint" of Plan accomplishments with best estimates of what is needed to secure the future for waterfowl, the USC concluded (Howerter et al. 2024) that we must grow our spatial impact on the land and provide stronger backstops of regulations and public policies to slow losses of habitat that continue to offset gains by Plan partners. It is, of course, net landscape change that matters most. A broadening partnership was seen as key to achieving that larger footprint.

Policy initiatives have great potential but are also unreliable. On one hand it is possible to provide protection of large areas with words on paper, but on the other, such protection is vulnerable to being removed just as quickly. Witness the highs and lows of the US Clean Water Act and its come and gone protection of isolated wetlands (Ryan 2020; EPA 2024).

Plan partners should continue to seek and support bird-friendly policy initiatives. Assessing those that emerge will hopefully position conservationists to be effective advocates for durable policies in the future. Waterfowl interests did this with the WRP in the Mississippi Valley and USDA's Conservation Reserve Program (CRP) on the Prairies. Many of those CRP contracts expired nevertheless, and much of that land returned to annual grain production as the prices of corn, soybeans, and land jumped with incremental effects of biofuel markets.

As much as we might wish it otherwise, the policy game requires vigilance and renewal too (e.g., Leopold et al. 2018). Habitat acquisition remains valuable. Incentive programs help too. But regulatory backstops remain vital (e.g., Clean Water Act, Sodbuster in the US Farm Bill, Provincial wetland policies in Canada) for achieving habitat protection at large scales. USDA programs in my lifetime have included Soil Bank in the 1950s and 1960s, CRP in the 1990s and 2000s, and WRP currently. Smaller scale but similar initiatives have had brief tenures in Canada. We need to stay vigilant and be prepared to push hard at the nexus between NAWMP habitat needs and water, soil, market, or landcover conservation policies when they arise. Working with agricultural interests to devise such policies, as Larry Jahn and others did in advance of the 1985 Farm Bill in the United States, seems like wise time investments.

The 2024 Update concluded that presently, the greatest opportunities for Plan partners may be in building connections between waterfowl habitat needs and other ecological benefits of conserving wetlands and grasslands. Though encouraging examples of this exist (Chapters 19 and 22), we need to develop more ecosystem services value propositions to grow partnerships and capacity and inform public policies. We need to communicate about these among the JVs, evaluate such efforts, and adapt to what we learn. Those recommendations from the 2024 Update seem compelling to me.

Perhaps part of the solution to the sense of plateauing support for waterfowl conservation is to tell a bigger story about crises with water (scarcity, flooding, quality), pollinators, biodiversity broadly, Gulf hypoxia, sequestration of greenhouse gases, soil conservation, and more. There are plenty of ecological crises to build cases upon. What Plan partners can offer is delivery experience, a track record second-to-none of getting things done on the land. We need NOT evolve into the North American Everything Management Plan, but we can lead by example, we can be a partner that can make a difference. This will need to occur on at least two levels – the JVs and nationally. Plan partners are poised to do more of this in several JVs (e.g., PLJV, IMJW, UMGLJV) and we can do more elsewhere. How we do it nationally/continentally is less apparent but it implies a stronger role for the PC or maybe an evolved or augmented PC. Networking at the national level, team building, and marketing are ways the PC might lead. Advocating for public policies is where affiliated nongovernment leadership could add experience and clout.

A strong example of preparations for marketing NAWMP conservation to encourage broader societal support has been developing in Canada's Prairie Habitat JV. Starting in the early 1990s, DU's Institute for Wetland and Waterfowl Research (IWWR) conducted one of the largest breeding waterfowl studies ever undertaken to assess the efficacy of a suite of JV habitat interventions (Emery et al. 2005; Arnold et al. 2007; Howerter et al. 2014). In time, these results along with landcover information led to a series of spatial targeting tools that now assist conservation delivery teams in prioritizing investments based on predicted impacts of various actions on waterfowl population growth (Devries et al. 2023). Building on this, a team of other IWWR researchers and their collaborators have created a new tool, classifying 10,800, 16-mi2 pixels across the Canadian Prairie Pothole Region for their estimated values in water retention, phosphorus and nitrogen retention, soil carbon storage, and biodiversity conservation. These data on ecological values can be overlapped with the importance of waterfowl habitat in each area (Bortolotti 2024). Coupled with the waterfowl decision-support tool, cost of land, and an index to risk of loss, this product for the first time offers a prediction of ROI for conservation actions in each of these small areas across the entire PHJV landscape. ROI for waterfowl conservation and ROI for any combination of other ecological values can be merged. Thanks to 30 years of applied research, analyses, and modeling, the PHJV is now in position to not just claim broader ecological benefits of waterfowl conservation but also to demonstrate those benefits and target any 16-mi2 segment of the pothole country with knowledge of the synergies and trade-offs involved in incremental conservation actions.

Each JV should be able to do something conceptually similar, although each will be limited initially by available scientific knowledge about landscape influences on waterfowl vital rates, and ecological functions of wetlands and grasslands/woodlands in their jurisdictions. Different ecological parameters might be of greater interest in different places (e.g., aquifer recharge in Playa Lakes). Related tools are under development in Canada's Eastern Habitat JV and the US Prairie Pothole JV but are more limited by available data than is PHJV. This first comprehensive ecosystem services targeting tool reinforces the importance of Plan partners' finding and allocating resources for technical work in broader ecological and social sciences as well as quantifying waterfowl values of NAWMP projects. Ecosystem service studies as well as combining elements like human recreation vs. waterfowl values in choosing project locations will likely become increasingly valued for project planning and JV marketing. Marketing proven ecosystem services should help Plan partners broaden NAWMP supporters because these values accrue to every person in a community.

Investing in Innovation

The ecosystem services progress described above sprang from forward-looking initiatives by a few thought leaders, people like Richard Bonnycastle a Board member of IWWR and DU Canada (DUC), whose visions in the 1990s stretched beyond the current view of how conservation might be supported. His mantra was simply, "Everyone cares about water, but only a few of us care about ducks", and he thought that if we were smart, we might someday get "ducks for free".

Learning and adjusting has been a hallmark of effective JVs. Learning in response to local/regional conditions is inherently a JV activity, not a national or continental one. JVs have been creative, entrepreneurial, and willing to try new things for the benefits of broader partnerships and more conserved acres. NAWMP partners should seek to reinforce this, perhaps with an innovation fund, or an innovation recognition award, but at minimum through active encouragement. Innovation must be celebrated in some manner. Most importantly, experiences with new projects and new ideas must be shared across JVs. JV reporting to the PC is important, but perhaps even more vital is sharing ideas among the JVs. By their nature – their size, their diverse boards, and their connections with landowners and local communities, JVs will forever be the most frequent source of innovation for NAWMP, and we must enable them.

A related notion is exploring the value of creating time-limited innovation teams within or among JVs. We need to find ways to innovate more often and more effectively while minimizing the burden of extra meetings. In my view, NAWMP should become an even more effective "learning organization" wherein when one learns everyone learns (Senge 2006; Ortenblad 2020). Every JV likely has at least one unique, creative idea from which others could benefit. Joint problem-solving has been greatly enhanced at the NSST by combining the perspectives and talents of JV science people and those from national organizations, and we need to ensure that such cross-member teams thrive.

John Eadie in his essay (Chapter 21) offered a bold vision for better engaging the academic community in NAWMP and its JVs. From long experience with a few university professors working on the Plan, I want to strongly endorse his ideas. The independent, critical, and innovative thoughts those people brought to our enterprise have been valuable and it's vital for such connections to grow while also acquainting aspiring professionals with the NAWMP enterprise. NAWMP partners should aspire to be a tantalizing career choice for new graduates.

Another source of ideas and influence might be found by looking back to the future. The nongovernment advisory board created by Harvy Nelson, Larry Jahn, Matt Connolly, and others in the late 1980s, shortly before NAWCA was envisioned, was set up to bring the talents, perspectives, and the operating flexibility of the private sector to bear in support of the Plan. This group was an adjunct to the governing NAWMP PC then comprising exclusively government agency representatives. This nongovernment "implementation board" (Chapter 3) was abandoned with the coming of NAWCA and the NAWCA Councils about 1990. But it may be useful to consider if a 2026 version of that model, assembling periodically, might help Plan partners tackle some of the current challenges before it and add additional momentum to the work. Perhaps using the occasion of NAWMP's 40th Anniversary could bring a group of NGO leaders together again with the PC to acknowledge their collective achievements, but more importantly to think hard about the next couple of decades. What should we set ourselves to achieve? What might the path to success look like? What can each organization do to help?

Countering the Notion That Enough Has Been Done for Waterfowl

In interviews for this book, several colleagues observed that NAWMP is aging. Some committees are finding it more difficult to attract top agency personnel to their tables. Fewer State or Provincial Directors are personally engaged than in the 1990s. Technical staff or mid-level managers are now more likely to populate some committees. Their expertise is valuable, but this does signal that the "new shine" of NAWMP has faded in some minds. There are so many conservation priorities in North America that despite its attraction as a successful program, NAWMP has in some minds become "old business". The perception of crisis that brought leaders to the table in 1986 has passed, leaving NAWMP with a heightened challenge to capture the attention of agency leaders and politicians.

Outside the waterfowl profession, the mantra "we've done enough for waterfowl" has been heard ever since the first NABCI "State of the Birds" reports emerged in the United States in 2009 and Canada in 2012. That narrative needs to change. We need to add nuance to this mainly positive story. Recall that NABCI has as a main strategic plank the objective of keeping common species common while aiding the recovery of those in more dire straits.

There are at least three counters for this naïve pronouncement of "done enough" for waterfowl. First, while the waterfowl guild collectively has fared well (Figure 24.1), not all duck species have increased in numbers. Several species are still of concern. Much of the uptick in Figure 24.1 was the result of some very abundant goose populations and increases of common ducks during the wet years on the prairies in the early 2000s. This uptick was not universal for waterfowl species. Second, the waterfowl family, because of our collective experience, infrastructure, and diverse funding, has become a "poster child" for keeping common species common while helping to lift the status of other species who rely on the same habitats that support waterfowl. Note, for instance, the positive trend for other wetland birds combined (Figure 24.1). Several JVs have or are now collecting data to build multi-species decision-support tools that

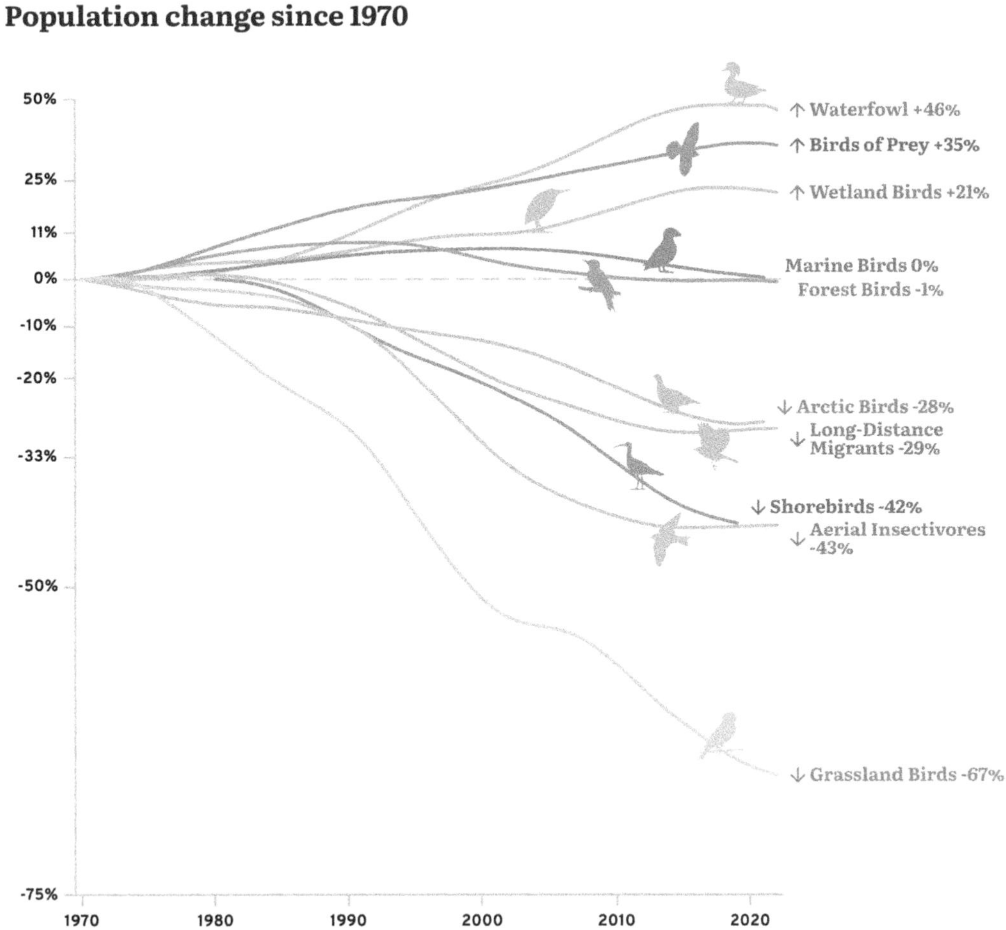

Figure 24.1 Breeding population trends for major groups of North American birds since 1970. Waterfowl populations combined have fared better than most. Data from Birds Canada and ECCC (2024). The parallel US report (North American Bird Conservation Initiative 2025) split the waterfowl group differently so the results need more parsing.

can help direct contributed dollars to programs and places that will benefit ducks and other birds. No enterprises other than the JV partnerships are presently in position to do this on a large scale. None. We should be celebrating this and touting the potential of the Plan partnership to do even more to assist other bird conservation actions. Third and finally, there is no reason today to assume that all the partners who have championed and funded NAWMP to date would be willing to simply abandon their waterfowl focus and direct those same efforts to aerial insectivores or some other group of birds. We must build together and not fixate on reallocating resources from the waterfowl Plan.

Time, Money, and the Essential NAWMP Updates

NAWMP partners have two kinds of assets, mainly: funds, and the time and passion of their people. If we want to accomplish more, we have two ways to do that: (1) find more money and/or more people; or (2) use those resources more effectively. While we strive to attract more

Table 24.2 Estimated Fate of Explicit Recommendations Made in Each of the 1994–2018 Plan Updates. Recommendation Statuses Were Scored as Completed, Partially Completed, or Not Addressed

	1994		*1998*		*2004*		*2012*		*2018*	
Number of recommendations	22		11		12		7		8	
Complete	6	27%	3	27%	3	25%	2	28%	2	25%
Partial	12	54%	8	73%	9	75%	3	43%	5	62%
Not done	4	18%	0	0%	0	0%	2	28%	1	13%

partners and funding, we should always seek ways to increase our efficiencies. Before going further, I want to declare with a loud voice that regular updates of the Plan have been absolutely essential for its endurance over four decades of operations. Regularly engaging the community in reflection, introspection, and gazing ahead has been vital for team building and keeping the Plan fresh and relevant under changing conditions – socially, biologically, and institutionally. I have personally been engaged at various levels in five of the six updates since 1998, and I believe strongly in their value. But I want to share a couple of ideas.

Every update offered recommendations for immediate attention by Plan partners. I thought it might be interesting to ask at a high level, what has been the fate of those recommendations with each iteration. None of the recommendations were directly related to achieving the Plan's numeric objectives (e.g., these were not comparing current vs. desired canvasback population levels); rather they were strategic or programmatic steps to move us along the path to achieving NAWMP population and habitat objectives. Above (Table 24.2) I report the fate of the 60 specific recommendations made by Updates from 1994 to 2018. The 2024 Update was published just recently, so any report card for it seems premature. The prescribed five-year updates have occurred at intervals from four to eight years, averaging 6.3 years. Given the exigencies involved with each iteration, with time for partner reviews, and with getting ministerial signatures, it is a minor miracle that the average has been that close to five. In any case, I looked over the 60 recommendations and estimated to what extent each of them had been dealt with before the next update was concluded. Someone else might judge completeness differently, of course. The general pattern, however, is for about one-quarter of the recommendations to be enacted within that 6.3-year average timespan. About two-thirds of them were partially addressed within the same window. Only a few were not advanced at all, usually because of changing circumstances between updates that obviated the need to pursue them.

This led me to wonder if the five-year cycle is still most appropriate. While updates are of enormous value in building consensus and momentum within the Plan Community, it seems that devoting more time to execution, analytics, and marketing of the ideas solidified in each update might be worth contemplating. The 2018 and 2024 Updates have recommended a ten-year cycle for comprehensive renewal of population objectives. Surveys of the JVs during the 2024 Update revealed that strategic planning by the JVs seldom occurs on less than a ten-year interval, although some strive for five. Thus, it might be worth debating lengthening the NAWMP update cycle from five years to something longer, perhaps with some sort of mid-point progress check. A corollary may be that any benefits of a longer cycle would only accrue if the human and financial resources invested in frequent Updates would be invested in executing recommendations between updates.

Working with People Who Own the Land

This topic would be worthy of book treatment alone. Agriculture in myriad forms dominates land use in North America, and often our interests in wetland conservation and management of perennial cover have clashed with production-focused farmers. There have been encouraging exceptions where NAWMP managers and local people have found common ground whether with a farmer in Mexico, a rancher in Oregon, a duck club owner in Mississippi, a beef producer in Alberta, or a small town in Texas in need of reliable water. Connecting with people whose interests overlap with NAWMP is a proven model for progress and will likely be even more essential in the decades ahead.

Planning increasingly needs to be both top-down (what is needed for the birds) and bottom up (what is socially achievable) to enable impactful things to happen on the land. How might we do that better? There is not enough money, nor social license, for duck interests to do it all. Jim Patterson told us this in 1984 and Bob Bailey reminded us in 1995. The experience with the USDA CRP years in the Dakotas reinforced this in 1998. We must work with agriculture, cities, watershed districts, legislatures, and more. Public policies and public attitudes are key. Plan partners might continue to use their dollars in the most effective ways and to secure the most valuable places for waterfowl, but broadly relevant public policies and attitudes seem essential for sustainability.

Engaging additional stakeholders, especially landowners, remains challenging, but again there are encouraging stories to share. Working lands must be part of NAWMP's "bigger tent" as several JVs have shown us. Examples include the Intermountain West JV, the Northern Great Plains, the Rainwater Basin, the Playa Lakes, and the prairie JVs. Partnerships with rice farmers in California's Central Valley have been in place almost since the beginning of NAWMP. The challenge of accommodation with agriculture remains difficult, however, particularly in waterfowl breeding areas. At the onset of the prairie JVs, Plan partners experimented with subsidizing zero-tillage cereal crops, rotational grazing systems, delayed hay cutting, chemical vs. mechanical fallow treatments, flushing bars on hay-cutting equipment, and several other techniques. These were ideas worth testing and were found to be variably acceptable to farmers and ranchers, but of only middling value to nesting ducks so the programs were reduced or discontinued.

Canadian partners have lately been investing more in understanding what farmers think about conservation practices and what might be necessary to spark their interest in participating (Howie Harshaw, University of Alberta, Canadian Agri-Food Production Institute) *(https://capi-icpa.ca/)*. Previous results have not been encouraging (e.g., van Kooten and Schmitz 1992). Much has changed in agriculture in the last 40 years. Individual land holdings have grown. Equipment is larger. Fewer farmers live on the land surrounded by nature that most appreciate at some level. World markets fluctuate frequently changing demand for various crops. Operating costs have soared with fuel prices and chemical input costs. We need to understand better what conservation measures are possible in today's world from the producer's perspective. Is common ground with industrial agriculture possible? If not, what options are left? More effort to sustain waterfowl habitats in the forested systems to the north?

Public policy is one element that might help. Canadian governments have regulatory, national, and international obligations to reconcile land use with social values additional to agricultural production (Nudds 2025). Among them is biodiversity conservation using best available evidence from science and traditional knowledge systems. We know a lot about what waterfowl need to be successful. We are learning more about what landscape characteristics are associated with protection of other ecosystem services. Working with agriculture interests to understand the trade-offs of these values with their production needs might help reach accommodations that could provide the basis for public policy incentives or prohibitions. It seems vital that agricultural stakeholders, from individual farmers to industry associations, be engaged in these processes with NAWP partners.

Looking Over the Hill

Leaders of any major business have responsibility for thinking ahead and trying to anticipate problems and opportunities that may be on the horizon. In trying to peek over such a hill, as the JTG was fond of saying, I see a few signs of change worth pondering.

The first is gnawing uncertainty about the status of duck populations in the Prairie Pothole Region. Signals of possible changing sex ratios, especially among ground-nesting ducks, are concerning for what that might mean to assumptions used to generate breeding population estimates from annual surveys, and more fundamentally, what that might imply about breeding female survival and the long-term viability of those populations. It is vital that waterfowl scientists determine with greater confidence what these mixed signals mean for various species and geographic areas. If this becomes a new crisis for harvest and habitat management, it soon will be for duck hunters too, and we will have to do what the waterfowl community has done repeatedly – confront troubling results with determined, cooperative efforts to adjust whatever harvest and habitat management approaches may be needed to turn the tide for the birds. As concerning as this news might be, it might also comprise an opportunity to rekindle agency interests in NAWMP or at least help motivate funding the crucial research needed to resolve this uncertainty.

Government funding via NAWCA, generous State Grant programs, and Provincial and Federal funding in Canada and Mexico has been unprecedented. The wallets and voices of waterfowlers in States and nongovernment organizations have provided generous matching commitments to keep these vital government funds flowing for more than 35 years. AFWA's Fall Flight program continues to impress as a way of amplifying cross-border resources with 48 of 50 states contributing at least some money annually for matching NAWCA grants to Canada (in 2024, $5.55 million of a $10 million stretch goal) that are matched by Ducks Unlimited, Inc. and matched again by Canadian partners. Despite this remarkable record, we must acknowledge that government funding is always at risk. As public debt continues to rise, and new priorities compete for waterfowl habitat funding, the threat of reduced revenue grows. The chorus of hunter voices is not as loud as it was in the 1980s either. It seems incumbent on us to do everything we can to invest wisely the dollars we receive and continue to tell the stories about the multiple-benefit ROI that NAWMP offers. But should setbacks come, are we sufficiently prepared in terms of continuing management funds, to keep those projects productive? Cautious managers like Rod Fowler and Len Warren at DUC created mechanisms to help ensure this in Canada decades ago. I don't know how widespread such provisions are across all the JVs. As much as the quest for public policy gains is essential, so too are steps to safeguard NAWMP's hard won habitat assets.

We should seek creative ways to buffer our programs from possible short-term challenges by any government of the day. These are human "potholes" we hope to avoid, but for any long-term project some are inevitable. NAWMP's diversity of partners should allow it to weather some such bumps on the road. Through 2024, bi-partisan Congressional support for NAWCA, as one example, has been hugely helpful thus far in buffering appropriations from the worst of partisan politics. However, as 2025 has shown us, politics can enter deep into conservation affairs and like for anyone's investment portfolio, diversification of support seems prudent.

With the COVID-19 pandemic still fresh in our minds, **the risk of waterfowl being involved in some devastating disease has caught our attention**. Highly pathogenic avian influenza seems to be a quickly evolving threat to wild and domestic birds with the potential that this or some other pathogen could become an epizootic affecting birds and people. Threats like this are being monitored by responsible government agencies, and the waterfowl community needs to do whatever we can to assist and keep pressure on governments to sustain their vigilance.

Climate change remains a concern for waterfowl and wetland conservation, but I concur with the 2024 Update assessment that much uncertainty remains about its probable impacts. Some things are clear. Sea-level rise will steal productive waterfowl habitat from many North American coastal zones. Coastal JVs are already striving to mitigate those coming losses by accreting sediment, acquiring property at higher elevations where possible, coordinating with adjacent JVs and more. For the Prairie Pothole Region, it remains unclear how precipitation patterns will change and what agricultural impacts will result.

One of the less known threats of climate change is that long-term directional change in the environment of any species might fundamentally affect their population dynamics (Conroy et al. 2011). Many ducks are examples of species, populations of which naturally fluctuate in large part as wetness fluctuates in their environments. Over the long-term, however, those populations tend to return to average or equilibrium levels. If those environmental trends change directionally, however, those equilibrium points may change too or simply fail to retain their former stable properties. This "non-stationarity" in system dynamics could undermine waterfowl harvest models, for example, and pose challenges to the management systems that have been in place for decades to allocate harvest opportunities.

Another insidious risk is "shifting baselines" in perceptions of conservation opportunities. Papworth et al. (2008) argued that there is some inevitability that people alive today view the circumstances of their youth as a baseline against which future losses are reckoned. Where I grew up in ND, a 15-year-old child now might be excited to see a Bobolink a time or two each summer or occasionally thrill to the call of a meadowlark. Sixty years ago, when I was that kid, these were unremarkable daily occurrences. My Dad, in turn, somewhat ruefully remembered abundant Greater Prairie Chickens feasting on his stooks of drying wheat, but I saw none of them. A challenge for conservation communication is to counter the risk of diminishing expectations, of setting our sights too low. In contrast, as the result of our collective work on waterfowl, we can point with satisfaction to the current abundance of wood ducks, or the return of trumpeter swans to the Midwest. People need to know about such conservation victories; these can encourage an eco-weary heart.

Human populations continue to grow and concentrate, straining the planet and perceptually disconnecting people from what supports their lives. It is stunning to recognize that with a global population in 2025 of about 8.2 billion people, that is 65% more than when NAWMP was signed in 1986. We add net +85million human mouths each year and consume the annual productive capacity of 1.6 planet earths. How will we feed, clothe, house, and provide clean water for 9–10 billion people and still have room for wildlife? How will we power that society? And how will we achieve all this with a manageable level of social strife? The scale of the challenges can feel overwhelming, so listen to the words of Jane Goodall…do what you can, locally, every single day and collectively it will matter.

Box 24.2 We All Need to Act

> It's important to take action and realize we can make a difference, and that will encourage others to take action, then we realize we are not alone and our cumulative actions truly make an even greater difference. This is how we spread the light, and this of course, makes us all ever more hopeful.
>
> (Goodall and Abrams 2021)

As Jane Goodall articulated to millions, hope needs: realistic goals, a realistic pathway, confidence, and support along the way (Goodall and Abrams 2021). This is what she called the "hope cycle". The 1986 NAWMP had these things once the first-step projects were accomplished. The pathway was mostly based on hope until NAWCA came along, but it did. Others came on board, so NAWMP followed this path of hope too. In the late 1980s that's how NAWMP was born. We had an idea, we had hope and we had each other.

A final observation – **partnerships are built on trust**, on relationships, and these don't flourish on Zoom or Teams calls. Travel is expensive and time consuming. I know because we did a lot of it for NAWMP in the 1980s–2010s. But relationships are best built face-to-face, and I worry about the future of Plan partnerships with so many of us working from home and mostly connecting digitally. For those with established relationships, these connections seem to work fine. But for building new relationships, they simply do not. You need time together in meetings, over coffee, in the hallways, over dinner, while watching birds together on a knoll overlooking a wetland, or wading in a boot-sucking coastal marsh. Humans are social critters. That's how we bond. That's where trust is born.

In a recent essay, Diane Eggeman (2024) pointed out that it's mostly the people and the relationships that endure and shine bright across the arc of a career. I couldn't agree more. Diane said, "The best rewards are the friends, colleagues, mentors, allies and confidantes collected along the way. Cultivate them, serve them and invest in high-trust relationships". John Tomke, a long-serving DU board member, famously declared, "I came for the cause; I stayed for the people".

Box 24.3 Work that Inspires

We are at our best when working cooperatively with others in pursuit of shared goals.
(Routledge 2023)

Mostly, despite these worries, it's the NAWMP partnerships that sustain my hopes for tomorrow. Waterfowl advocates and others came together to start the crusade for the birds and the people who valued them leading to the Migratory Bird Treaty in 1916. In the 1930s and 1940s, citizens formed DU and Delta Waterfowl that have worked for the birds for more than 80 years, and governments came together to establish the Flyway System to manage waterfowl harvest responsibly. Banding and population monitoring commitments by federal agencies began soon after and still serve us today. Then 40 years ago, at another time of gloom and angst, NAWMP was created with a conviction and a passion to sustain the wetlands, grasslands, and forested habitats essential for the birds to flourish. Waterfowl conservation has been a winning wager for over a century, and as I "look over the hill", although I cannot claim to see all the coming trials, nor the coming campaigns to overcome them, I am confident that those who follow us into the marshes will find ways to sustain these marvelous birds and stir hearts yet unborn.

References

Anderson, Michael G. and Paul I. Padding. 2015. "The North American approach to waterfowl management: synergy of hunting and habitat conservation." *International Journal of Environmental Studies* 72, no. 5: 810–29.

Arnold, Todd W., Lea M. Craig-Moore, Llwellyn M. Armstrong, David W. Howerter, James H. Devries, Brian L. Joynt, Robert B. Emery, and Michael G. Anderson. 2007. "Waterfowl use of dense nesting cover in the Canadian prairie parklands." *Journal of Wildlife Management* 71, no. 8: 2542–49.

Birds Canada and Environment and Climate Change Canada. 2024. *The State of Canada's Birds.* (https://naturecounts.ca/nc/socb-epoc/report/2024/en/). Birds Canada. Port Rowan, ON.

Bortolotti, Lauren. 2024. *Maximizing Co-Benefits of Conservation: A Wetland Ecosystem Services Tool for Prairie Canada.* (https://nawmp.org/content/nawmp webinar series).

Conroy, Michael J., Michael C. Runge, James D. Nichols, Kirk W. Stodola, and Robert J. Cooper. 2011. "Conservation in the face of climate change: The roles of alternative models, monitoring, and adaptation in confronting and reducing uncertainty." *Biological Conservation* 144, no. 4: 1204–13.

Devries, James H., Llwellyn M. Armstrong, David W. Howerter, and Robert B. Emery. 2023. "Waterfowl distribution and productivity in the Prairie Pothole Region of Canada: Tools for conservation planning." *Wildlife Monographs* 211: 1–49.

Eggeman, Diane. 2024. "Leading with your head and heart, integrating people into waterfowl conservation." *The Wildlife Professional* 10 (12): 48–52.

Emery, Robert E., David W. Howerter, Llwellyn M. Armstrong, Michael G. Anderson, James A. Devries, and Brian L. Joynt. 2005. "Seasonal variation in waterfowl nesting success and its relation to cover management on the Canadian prairies." *Journal of Wildlife Management* 69: 1181–93.

Environmental Protection Agency. 2024. *Definition of "Waters of the United States" under the Clean Water Act.* (https://www.epa.gov/cwa-404/definition-waters-united-states-under-clean-water-act.). Environmental Protection Agency. Washington, D.C.

Fleming, Kathy K., Michael K. Mitchell, Michael G. Brasher, John M. Coluccy, J. Dale James, Mark J. Petrie, Dale D. Humburg, and Gregory J. Soulliere. 2019. "Derivation of regional, non-breeding duck population abundance objectives to inform conservation planning—2019 Revision*." NSST Technical Report* 2019-01. 44pp.

Goodall, Jane and Douglas Abrams. 2021. *The Book of Hope*. Celadon Books. MacMillan, NY. 272pp.

Howerter, David, Michael Anderson, Diane Eggeman, Kathy Fleming, Dean Smith, Mark Vrtiska, Barry Wilson, Barbara Avers, Kristin Bianchini, Michael Brasher, Andrew Don Carlos, Annette Glick, David Gordon, Ashley Gramza, Heath Hagy, Alan Hanson, Howie Harshaw, Kristina Hick, Joe Lancaster, James Leafloor, Kurt Mazur, Jake Messerli, Shawn Meyer, Jena Moon, Eric Reed, Anthony Roberts, Katie Sainsbury, Emily Silverman, Jacey Scott, Scott Stephens, Josh Vest, and Corie White. 2024. *Technical Report in Support of the 2024 Update for the North American Waterfowl Management Plan*. 113pp. (https://nawmp.org/sites/default/files/2024-07/combined-2024-nawmp-technical-report-final.pdf).

Howerter, David W., Michael G. Anderson, James H. Devries, Brian L. Joynt, Llwellyn M. Armstrong, Robert B. Emery, and Todd W. Arnold. 2014. "Variation in mallard vital rates in Canadian Aspen Parklands: The Prairie Habitat Joint Venture Assessment." *Wildlife Monographs* 188: 1–37.

Krainyk, Anastasia, James E. Lyons, Gregory J. Soulliere, John M. Coluccy, Barry C. Wilson, Michael G. Brasher, Mohammed A. Al-Saffar, and Dale D. Humburg. 2022. "Structured decision-making to rank North American Wetlands Conservation Act proposals within joint venture regions." *Journal of Fish and Wildlife Management* 13, no. 2: 375–95. (https://doi.org/10.3996/JFWM-21-089).

Lancaster, Joseph D. and Barry C. Wilson. 2024. "Guidance for selecting between autumn and winter regional duck abundance objectives for conservation planning." *NSST Technical Report* 2024-02. 10pp.

Leopold, Bruce D., Winnifred B. Kessler, and James L. Cummins (Editors). 2018. *North American Wildlife Policy and Law*. Boone and Crockett Club. Missoula, MT. 648pp.

Nichols, James D., Michael C. Runge, Fred A. Johnson, and Byron K. Williams. 2007. "Adaptive harvest management of North American waterfowl populations - recent successes and future prospects." *Journal of Ornithology* 148: 343–49.

North American Waterfowl Management Plan Committee. 2024. *2024 North American Waterfowl Management Plan Update: Expanding the Partnership.* Environment Canada. Ottawa. 62pp. (https://nawmp.org/sites/default/files/2025-01/nawmp-update-2024-en-web.pdf).

North American Bird Conservation Initiative. 2025. *The State of the Birds, United States of America, 2025.* StateoftheBirds.org. (https://www.stateofthebirds.org/2025/wp-content/uploads/2025/03/state-of-the-birds-sotb-2025-spreads.pdf). North American Birds Conservation Initiative. Washington, D.C.

Nudds, Thomas D. 2025. *Science, Policy, and Governance for the Conservation of Biodiversity and Species at Risk on Canadian Agricultural Landscapes*. The Canadian Agri-Food Policy Institute. Ottawa. (https://capi-icpa.ca/wp-content/uploads/2025/05/2025-05-23-Tom-Nudds-Revised-Species-at-Risk-and-Agricultural-Lands-EN.pdf).

Ortenblad, Anders (Editor). 2020. *The Oxford Handbook of the Learning Organization.* Oxford University Press. Oxford. 514pp.

Papworth, Sarah K., J. Rist, Lauren Coad, and Eleanor J. Milner-Gulland. 2008. "Evidence for shifting baseline syndrome in conservation." *Conservation Letters* 2, no. 2: 93–100.

Roberts, Tony, Joshua Dooley, Alan Hanson, Kate Martin, Kyle Spragens, and Greg Yarris. 2023. "North American Waterfowl Management Plan Species Prioritization — 2023 Revision." *NSST Technical Report* 2023-01. 13pp.

Routledge, Clay. 2023. *Past Forward. How Nostalgia Can Help You Live a More Meaningful Life.* Sounds True. Boulder, CO. 105pp.

Ryan, Mark A. (Editor). 2020. *The Clean Water Act Handbook, Fourth Edition.* American Bar Association. 438pp. Chicago, IL.

Senge, Peter M. 2006. *The Fifth Discipline: The Art and Practice of the Learning Organization (Revised edition).* Crown Currency/Doubleday. New York. 464pp.

Van Kooten, Gerrit C. and Andrew Schmitz. 1992. "Preserving waterfowl habitat on the Canadian Prairies: economic incentives versus moral suasion." *American Journal of Agricultural Economics* 74, no. 1: 79–89.

Vrtiska, Mark P., James A. Dubovsky, and Michael G. Anderson. 2024. "The North American trumpeter swan survey: Retain or find something new?" *Wildfowl Special Issue* 7: 57–75.

Epilogue

As I wrote this book through 2024–2025, the political and environmental agendas in the United States, Mexico, and Canada were evolving. The economies of the three countries had been functioning in a highly integrated manner. Goals for free trade, political alliances, and environmental conservation were similar across North America. NAWMP had long operated within this collaborative context. Plan partners expected this united approach to continue when the 2024 Update was signed by all three countries.

The political changes that began in 2025 were not entirely unexpected; however, the speed and extent of the changes surprised the conservation community. While these changes continue to unfold and remain unsettled, they have highlighted vulnerabilities that Plan partners should strive to minimize in the years ahead.

A first lesson is that public policies that safeguard conservation of wetlands, grasslands, forests, and tundra – habitats vital for waterfowl and other wildlife – are always at risk. Constant vigilance, timely communication, and broad alliances of stakeholders are essential to preserve such policy protections. Millions of acres can be protected, damaged, or lost forever at the stroke of a pen and Congressional/Parliamentary oversight, public pressure, and long-term vision are vital to counterbalance pro-development agendas to recklessly exploit natural resources.

The partnerships that have been essential for the success of NAWMP and other continental conservation initiatives also require nurturing. NAWMP as a continental strategy inspired Congress to create a cost-sharing program (NAWCA) that supported conservation of waterfowl habitat. Investments by states, provinces, and the private sector matched and helped stimulate the federal funds, which were intentionally allocated among the three nations based on biological needs. Cost-sharing programs among nations for migratory birds were never "foreign aid". They are cooperative agreements wherein all parties benefit from working together. For example, many ducks hatched in Canada are harvested on migration by US hunters, while others fly on to winter in Mexico and return north in spring. The loss of US federal matching funds for work in Canada and Mexico in 2025 was debilitating. There are hopeful signals now that these will be restored in time, though likely with new constraints. After 37 years of NAWCA functioning in this galvanizing manner, with broad bi-partisan support, the rationale for continuing this partnership seems overwhelming.

A related lesson is that the NAWMP partnership needs broad and diverse support to obviate overreliance on any single source of funds. Participation of state fish and wildlife agencies and NGOs, through initiatives such as AFWA's Fall Flights program, has provided critical matching funds for work outside the United States, but also reinforced the value of matching federal funding. The focus of the 2024 Update on broadening Plan partnerships seems now even more prescient.

DOI: 10.4324/9781003732310-25

NAWMP has progressed with a strong science base and vital monitoring programs (banding, bird surveys, JV assessments) that have allowed Plan partners to learn and adapt their actions over time and enhance returns on investments. Any diminishment of that capacity would be a threat to the foundations of NAWMP, NABCI, and the other bird conservation initiatives. Rapid and widespread turnover of experienced staff is another concern. It can take decades to replace knowledgeable program leaders, so here too, partners need to rally arguments and political support and help envision how the necessary human capital can be organized and funded to continue the work of NAWMP.

The NAWMP enterprise from the outset has been an attempt to change the landscape at a continental scale. There are few continental-level actions that can be taken, however. Most actions will be at the national, state/province, or JV/local level. Improving conditions for wildlife and people on such a large scale means addressing or reacting to opportunities and threats across a wide spectrum of human society – from agricultural policy to food systems and human health; from international relations and trade to climate change. Pursuing such ambitious conservation goals requires vigilance, patience, and persistence, i.e., playing the long game.

NAWMP has weathered crises before. It was "born" with no committed funding at the height of the Cold War when federal debt was soaring. Nevertheless, with determined, clever work by both non-federal partners and federal leaders, NAWCA was established in 1989. Across the last 40 years, there are examples, at all scales, of the Plan community responding to issues out of their direct control. Changes will continue to test the partnership. Periodic droughts, policy fluctuations, changing administrations, and public attitudes can shift quickly or slowly. The partnership needs a long-term vision but be nimble enough to respond assertively when circumstances are favorable for progress.

The Plan's strategic foundations, based on ecological and social science, should help keep the partners poised to make progress whenever they can. They also need leaders open to novel ways to take advantage of opportunities when they arise, much like the "First Step" coalitions did in the 1980s.

In the 40 years of the Plan's existence, we have never seen the scale and pace of change that is underway today. Let us learn from this, and as Harvey Nelson said in the first NAWMP Newsletter in 1988, "To meet our goals we will have to make the program strong enough to weather shifts in politics in the U.S. and Canada." As usual, Harvey was right.

February 25, 2026

Appendix A

Acronyms Used in this Book Arranged by Major Categories

Federal Agencies United States/Canada/Mexico

BLM	US Bureau of Land Management
BSFW	Former name for USFWS
CONABIO	National Commission for Knowledge and Use of Biodiversity (MX)
CONANP	National Commission of Natural Protected Areas (MX)
CWS	Canadian Wildlife Service
DBHC	USFWS Division of Bird Habitat Conservation
DOI	US Department of the Interior
ECCC	Environment and Climate Change Canada (includes CWS)
FWS	US Fish and Wildlife Service (also USFWS)
SEMARNAT	Secretaría de Medio Ambiente y Recursos Naturales (MX)
USDA	US Department of Agriculture
USFS	US Forest Service
USFWS	US Fish and Wildlife Service

NAWMP Institutions

ASC	Assessment Steering Committee
CET	Continental Evaluation Team
HDPET	The combined HDWG and PET committee
HDWG	Human Dimensions Working Group
IIC	Interim Integration Committee
ISC	Integration Steering Committee
JTG	Joint Task Group on Harvest and Habitat Management
JV or JVs	NAWMP Joint Ventures (also Migratory Bird Joint Ventures)
NAWMP	The North American Waterfowl Management Plan
NSST	NAWMP Science Support Team
PC	The international NAWMP Committee
PET	Public Engagement Team
TriST	Tri-Initiative Science Team
UST	Unified Science Team

NAMWP Joint Ventures (a.k.a. Migratory Bird Joint Ventures)

ACJV	Atlantic Coast JV
AGJV	Arctic Goose JV
AMJV	Appalachian Mountains JV
BDJV	Black Duck JV

C3JV	California Central Coast JV
CHJV	Central Hardwoods JV
CIJV	Canadian Intermountain JV
CVJV	Central Valley Joint Venture (formerly CVHJV)
EGCPJV	East Gulf Coastal Plain JV
EHJV	Eastern Habitat JV
GCJV	Gulf Coast JV
IWJV	Intermountain West JV
LMVJV	Lower Mississippi Valley JV
NGPJV	Northern Great Plains JV
OPJV	Oaks and Prairies JV
PAG	Northern Pintail Action Group
PCJV	Pacific Coast JV became the Pacific Birds Habitat JV (PBHJV)
PHJV	Prairie Habitat Joint Venture (Canada)
PLJV	Playa Lakes JV
PPJV	Prairie Pothole Joint Venture (US)
RGJV	Rio Grande JV
RWBJV	Rainwater Basin JV
SAT	Scaup Action Team
SDJV	Sea Duck JV
SFBJV	San Francisco Bay JV
SJV	Sonoran JV
UMGLJV	Upper Mississippi River/Great Lakes JV

Agency Offices

HAPET	Habitat and Population Evaluation Team USFWS
NCTC	National Conservation Training Center USFWS (WV)
NPWRC	Northern Prairie Wildlife Research Center USGS (ND)
NWI	National Wetlands Inventory
PMBRC	Prairie Migratory Bird Research Centre ECCC (SK)

Acts and Associated Institutions

CRP	USDA Conservation Reserve Program
CWA	Clean Water Act (United States)
MBCA	Migratory Bird Convention Act (Canada enabling legislation)
MBCC	Migratory Bird Conservation Commission of the US Congress
MBT	The 1918 Migratory Bird Treaty
MBTA	Migratory Bird Treaty Act (US enabling legislation)
NAWC	North American Wetlands Council (United States)
NAWCA	North American Wetlands Conservation Act
NAWCC	North American Wetlands Council Canada
WRP	USDA Wetland Reserve Program

Miscellaneous

AHM	Adaptive Harvest Management.
ARD	Assistant Regional Directors, USFWS
ARM	Adaptive Resource Management
ESA	Eastern Survey Area

FOW1	Future of Waterfowl Management Summit, 2008
FOW2	Future of Waterfowl Management Summit, 2017
GIS	Geographic Information Systems
PPR	Glaciated Prairie Pothole Region of southern Canada and northern United States
RAMSAR	Ramsar Convention on Wetlands of International Importance
TSA	Traditional Survey Area of the WBPHS
WBPHS	Waterfowl Breeding Pair and Habitat Survey
WWT	Waterfowl and Wetlands Trust, United Kingdom

Other Bird Conservation Plans

BCR	Bird Conservation Region (from NABCI)
CSCP	Canadian Shorebird Conservation Plan
NABCI	North American Bird Conservation Initiative
NAWCP	North American Waterbird Conservation Plan
PIF	Partners in Flight, a landbird plan
USSCP	US Shorebird Conservation Partnership

State/Provincial Organizations

AFC	Atlantic Flyway Council
AFWA	Association of Fish and Wildlife Agencies (formerly IAFWA)
CFC	Central Flyway Council
DNR	Department of Natural Resources for various states
IAFWA	International Association of Fish & Wildlife Agencies (became AFWA)
MFC	Mississippi Flyway Council
MHC	Manitoba Habitat Conservancy (formerly MHHC)
PFC	Pacific Flyway Council
SWCC	Saskatchewan Wetlands Conservation Corporation

Non-Government Organizations (NGOs)

ABC	American Bird Conservancy
CWA	California Waterfowl Association
DUC	Ducks Unlimited Canada
DUI	Ducks Unlimited Inc. (United States)
DUMAC	Ducks Unlimited de México
DWF	Delta Waterfowl Foundation (was NAWF – North American Wildlife Foundation)
IWWR	Institute for Wetland and Waterfowl Research (DUC)
NAS	National Audubon Society
NCC	Nature Conservancy of Canada
NFWF	National Fish and Wildlife Foundation (United States)
NRA	National Rifle Association
NWF	National Wildlife Federation (United States)
TNC	The Nature Conservancy (United States)
TWS	The Wildlife Society
WHC	Wildlife Habitat Canada
WMI	Wildlife Management Institute (sponsors the North American Wildlife and Natural Resources Conference (NAWNRC)

Appendix B

Glossary of People Quoted in the Book

Amos Eno	Director of Conservation, then Executive Director of the National Fish and Wildlife Foundation.
Andy Raedeke	MO Department of Conservation. Waterfowl biologist and human dimensions expert.
Barry Verbewski	MB Department of Natural Resources. Held several positions including Provincial waterfowl biologist
Basile van Havre	Canadian Wildlife Service. Director of Population Conservation and Management, 2007–2016. Former PC Co-Chair.
Bob Andrews	Director of Wildlife Management, AB Fish and Wildlife Service. Member of NAWWP negotiating team.
Bob Bailey	CWS member of the NAWMP negotiating team and a principal author of the Plan. Later worked with Delta Waterfowl and as a private consultant.
Bob Clark	Research Scientist. CWS and Adjunct Professor University of SK.
Bob Davison	Primary author of NAWCA as staff member on the Senate Committee on Environment and Public Works. Later Deputy Assistant Secretary for Fish and Wildlife and Parks in Clinton administration. Then with Wildlife Management Institute and Defenders of Wildlife. See Box 3.6.
Bob Ford	Presently, executive director of the Hatchie River Conservancy. Formerly with Nature Conservancy of TN, US NABCI coordinator, and PIF Coordinator 2015–2022 with USFWS.
Cal Dubrock	Pennsylvania State University. Former Wildlife Director, PA Game Commission. First Chair of HDWG.
Carey Smith	Long career with the USFWS, mainly as Pacific Coast Joint Venture Coordinator, renamed Pacific Birds Habitat Joint Venture.
Carl Madsen	Wetland and Habitat biologist, 37 years mainly with USFWS. Fergus Falls, MN. Then worked with Harvey Nelson in the early days of NAWMP.
Charles Baxter	First Coordinator of LMVJV. An intellectual leader among the early JV coordinators. Early supporter of evaluations and all-bird conservation mission. See Box 13.3.
Chip Collins	Early executive director of NFWF. Played an important role in first-step funding of NAWMP projects.
Chuck Hunter	Thirty-seven years with USFWS, mainly in the southeast region and the Refuge System. Important technical work with PIF and NABCI.
Dale Humburg	MO Department of Conservation and later Ducks Unlimited Inc. Involved in 2012 NAWMP Revision; then first Chair of the IIC. See Box 19.2.
Dave Case	President of D.J. Case and Associates. Communication consultant. Long-term and extensive engagement with NAWMP, AHM, and more. See Boxes 3.2 and 22.2.
Dave Gordon	Research biologist with Delta, then DU Inc. Last position Chief, Branch of Science and Planning, DBHC. Extensive engagement with NAWMP and the Joint Ventures.
Dave Paullin	USFWS. Long-serving CVJV Coordinator. Later Chair of the Continental Assessment Committee. See Boxes 3.3 and 5.1.

Dave Sharp	Biologist USFWS. Multiple roles including technical assistant to Harvey Nelson in the early days of NAWP. Later Central Flyway Biologist.
Dave W. Smith	USFWS. Coordinator IWJV, 2007 to present, Innovator especially of private-lands conservation programs.
David Pashley	Leader in PIF, NABCI and more. See Box 9.1.
David A. Smith	Former Executive Director US NAWMP office and PC co-chair.
Dean Smith	Formerly with Agriculture Canada and PHJV, Canadian representative at AFWA. See Box 22.1.
Deanna Dixon	CWS. longest serving JV coordinator, both AGJV and PHJV. See Box 7.1.
Diane Eggeman	FL Game Commission, multiple NAWMP committee roles, then Chair Integration Steering Committee and NAWPEP. See Box 21.1.
Eduardo Carrera	CEO DUMAC, formerly chief biologist. See Box 11.1.
Fred Johnson	Research scientist USFWS and later USGS. A leader in adaptive harvest management and other evaluation applications to NAWMP
Gary Myers	TN Director of wildlife. Former President IAFWA. Early advocate for NAWMP funding. See Box 3.5.
George Finney	CWS, Eastern Region Director. NAWMP negotiating team. EHJV and SDJV leader, then Bird Studies Canada
George Mitchell	Senator from Maine; primary sponsor of NAWCA
Gerald McKeating	Former CWS Regional Director, Edmonton. Former PHJV Chair.
Gray Anderson	Wildlife Division Chief, VA. Active PC member for many years.
Greg Soulliere	Upper Mississippi/Great Lakes Joint Venture Science Coordinator. Active NSST contributor.
Harvey Nelson	USFWS. NAWMP Founder and first Executive Director. See Box 2.2
Hazard Campbell	Former President of DU Inc., early supporter of NAWMP
Jack Capp	USDA Forest Service; a leader in supporting migratory bird conservation in Mexico and Central America.
Jack Grieb	Biologist with CO Division of Wildlife. Member of the NAWMP negotiating team and a primary author of the Plan.
James D. McCuaig	Director of Water and Habitat Conservation Environment Canada, Co Chair of PC for several years. Active liaison with Canadian JVs.
Jerry Johnson	GCJV coordinator at the outset through 1993.
Jeff Nelson	Biologist with DUC at the outset of PHJV planning. Then DUI Chief Biologist, then the Director of DU's Great Plains office in Bismarck. Later CEO, DUC, and Regional Director WWF
Jerome Ford	Long-time USFWS manager. Current US Cochair of the Plan Committee.
Jim Patterson	Founder of NAWMP; led Canada's negotiating team. Former president of IAFWA and IWRB. Later, Government Affairs lead for DU Canada. See Box 2.1.
Jim Ringelman	Waterfowl biologist with CO Division of Wildlife, then DU Inc. Deeply engaged in NAWMP for many years. Led consultations prior to 2012 Revision. See Box 18.3.
John Eadie	Professor and Raveling Chair, University of California, Davis. Heavily engaged in NAWMP for two decades. See Box 21.2.
Keith McAloney	NS biologist. Led much of early NAWMP work in Atlantic Canada. See Box 4.4.
Larry Jahn	President of Wildlife Management Institute. Deeply involved in organizing and funding NAWMP in the early days. Vital work on conservation provisions in the 1985 USDA farm bill.
Mark Koneff	USFWS biologist. Vital member of the CET and then the NSST. Presently Chief, Branch of Migratory Bird Surveys.
Matt Connolly	Formerly CEO of Ducks Unlimited Inc. Active supporter of NAWMP in the early days. First Chair of the NAWCA Council.
Mike Brasher	GCJV, later DUI. Long-serving member of NSST. See Box 20.1.
Mike Carter	Former NABCI biologist, then long serving and creative Playa Lakes JV Coordinator.
Mike Tome	USFWS biologist with North American Waterfowl and Wetlands Office. Early Chair of the CET. Later a supervisor for the Cooperative Research Units Program with USGS.

Paul Schmidt	USFWS, Director of Migratory Birds. US Cochair of the PC through 2012 Revision. Later Chief Conservation Officer for DU Inc.
Ray Alisauskas	Research Scientist CWS and Adjunct Professor at the University of SK. Long active in the AGJV and Arctic Goose research.
Rex Johnson	USFWS biologist for 27 years. Led Service in conceptual development of Strategic Habitat Conservation. Active in 2004 Update.
Richard Goulden	Early NAWMP leader in MB and Canada. See Box 4.3.
Robert Blohm	USFWS, long career in the Office of Migratory Bird Management; many years as Chief of the Division.
Rod Fowler	DUC leader on PHJV and beyond. Influential in building partnerships. Supported broadening NAWMP. See Box 12.2.
Rollie Sparrowe	US Cochair of NAWMP negotiating team. Former head of Coop Wildlife Research Units. Chief of Migratory Bird Management for FWS.
Ross Melinchuk	Former SK biologist. Later worked with AFWA, DU Inc. Texas Parks and Wildlife. A leader in building state support for conservation in Canada.
Sandy Macaulay	DUC biologist delegated to help Patterson set up NAWMP in Canada. First PHJV Coordinator. See Box 4.2
Scott Sutherland	Led government affairs office for DU in Washington, D.C. Vital role in stewarding support for NAWCA.
Seth Mott	Held many NAWMP positions with USFWS, regionally and nationally. My closest NAWMP colleague for decades. See Box 14.2.
Stephen Wendt	CWS population ecologist. Cochair of the AGJV and the PC. Active in Tri-national affairs, PIF and NABCI. A leader in movement to all-bird JVs.
Stew Morrison	Executive Vice-President of DU Canada for 27 years. Early supporter of NAWMP. Led DUC through dramatic changes to adapt and lead the JVs.
Terry Rich	PIF National Coordinator 2000–2014; previously non-game bird program coordinator for BLM.
Tim Sopuck	Long-serving leader with MB Habitat Conservancy, Cochair of 2024 NAWMP Update. See Box 23.3.
Tom Nudds	University of Guelph. Post-doctoral associate with Patterson. Eclectic interests. Joint Task Group member.
Tom Moorman	Held regional and national leadership positions in DUI. Cochair of 2024 NAWMP Update. See Box 23.4.
Tom Rothe	AK State Waterfowl biologist. Engaged with SDJV and AGJV.
Trevor Swerdfager	Held several regional positions in CWS before Director General in Ottawa. Then several ADM appointments. Now at U. of Waterloo. Broad perspective on NAWMP's unusual structure and governance.

Appendix C

Common and Scientific Names of North American Waterfowl

Common Name	*Species*
Black-bellied Whistling-Duck	Dendrocygna autumnalis
Fulvous Whistling-Duck	Dendrocygna bicolor
Emperor Goose	Anser canagicus
Snow Goose	Anser caerulescens
Ross's Goose	Anser rossii
Greater White-fronted Goose	Anser albifrons
Lesser White-fronted Goose	Anser erythropus
Brant	Branta bernicla
Cackling Goose	Branta hutchinsii
Canada Goose	Branta canadensis
Hawaiian Goose	Branta sandvicensis
Mute Swan	Cygnus olor
Trumpeter Swan	Cygnus buccinator
Tundra Swan	Cygnus columbianus
Muscovy Duck	Cairina moschata
Wood Duck	Aix sponsa
Blue-winged Teal	Spatula discors
Cinnamon Teal	Spatula cyanoptera
Northern Shoveler	Spatula clypeata
Gadwall	Mareca strepera
American Wigeon	Mareca americana
Laysan Duck	Anas laysanensis
Hawaiian Duck	Anas wyvilliana
Mallard	Anas platyrhynchos
Mexican Duck	Anas diazi
American Black Duck	Anas rubripes
Mottled Duck	Anas fulvigula
White-cheeked Pintail	Anas bahamensis
Northern Pintail	Anas acuta
Green-winged Teal	Anas crecca
Canvasback	Aythya valisineria
Redhead	Aythya americana
Ring-necked Duck	Aythya collaris
Greater Scaup	Aythya marila
Lesser Scaup	Aythya affinis
Steller's Eider	Polysticta stelleri
Spectacled Eider	Somateria fischeri
King Eider	Somateria spectabilis
Common Eider	Somateria mollissima
Harlequin Duck	Histrionicus histrionicus
Surf Scoter	Melanitta perspicillata

Common Name	*Species*
White-winged Scoter	Melanitta deglandi
Common Scoter	Melanitta nigra
Black Scoter	Melanitta americana
Long-tailed Duck	Clangula hyemalis
Bufflehead	Bucephala albeola
Common Goldeneye	Bucephala clangula
Barrow's Goldeneye	Bucephala islandica
Common Merganser	Mergus merganser
Red-breasted Merganser	Mergus serrator
Masked Duck	Nomonyx dominicus
Ruddy Duck	Oxyura jamaicensis

Source: Chesser, R. T., S. M. Billerman, K. J. Burns, C. Cicero, J. L. Dunn, B. E. Hernández-Baños, R. A. Jiménez, Oscar Johnson, N. A. Mason, and P. C. Rasmussen. 2025. Check-list of North American Birds (online). American Ornithological Society. (https://checklist.americanornithology.org/taxa/).

Appendix D
People of NAWMP

Throughout this book, we have identified people who made extraordinary contributions to the evolution of the North American Waterfowl Management Plan. They and many others have been recognized with NAWMP's formal awards (Appendix I (www.routledge.com/9781041224389)). In some cases, we heard their own voices reflecting on pivotal periods in the Plan's history. Plan accomplishments, however, are the sum of efforts by field staff, office staff, scientists, volunteers, and landowners across the continent. There are literally thousands of people who have contributed to the success of the NAWMP partnership.

Among these are people who knocked on doors and convinced a rancher to set up a rotational grazing system to improve beef production while improving nesting cover. Others sold the idea of a permanent easement on a family's marginal lands to enhance farm profitability while improving watershed management and conserving biodiversity. Some developed efficient project tracking systems or streamlined grant agreements. Others made landowners aware of opportunities under WRP funding in the Mississippi River flood plain. Below, representing the many, are ten people nominated by their peers who made NAWMP happen. I hope that by celebrating their varied contributions, I might in some small way recognize the hundreds more whose names do not appear in this book.

Danielle Thibeault

It was a great day for NAWMP in 1989 when Danielle came over to the Canadian Wildlife Service from Agriculture Canada. She quickly became an integral member of the CWS Headquarters team, learning how to implement the fledgling NAWMP at all levels. She filled many roles including communications, national liaison, coordinator of meetings, project approvals for partners, and administration of international funding. Vitally, she also led development of a national tracking and reporting system (NRS-NTS) to enable reporting on all partners' contributions and the habitats secured.

Danielle worked closely with Joint Ventures and at all levels: federal, provincial, regional, and local and with NGO partners. Her liaison and communications roles involved people from field operators to Provincial Wildlife Directors, USFWS senior staff, and communications officers. She was well liked and respected for her gentle, effective and professional way of performing her roles. She exemplified the partnerships and accomplished much by quietly bringing together disparate groups of organizations, individuals, and interests. People (me included!) found her friendly and efficient to work with.

One of her legacies, the Tracking System, demonstrated early successes of the Plan, much to the encouragement of all those who had dreamed of such conservation accomplishments. That documentation did much to ensure continuation of the program.

(Nominated by Jim McCuaig and David Smith)

Lucian A Newton

Lucian worked for the USFWS at the Yazoo NWR Complex. He played a vital role in the implementation of the Mississippi Partners Project, an effort by the Lower Mississippi Valley Joint Venture to provide wintering waterfowl habitat on private lands in the Mississippi Delta. The heart of the program was repurposing steel pipe donated by gas transmission companies to fabricate and then distribute as water control structures to landowners in return for ten-year agreements to winter-flood croplands or wetlands. Using his local knowledge and contacts, Lucian was able to enlist many landowners and enrolled tens of thousands of acres in the program. Additional achievements were realized when many of the project landowners subsequently enrolled some of their acres in permanent Wetland Reserve Program easements and restored natural wetland vegetation. Lucian's stewardship of these partners made a difference.

(Nominated by Seth Mott)

Doug McGeoghegan

A Central Valley rice grower, Doug McGeoghegan and his father were instrumental in solving a farming problem that also benefited countless wintering waterfowl over the following three decades. Rice farmers in the Central Valley need to manage their copious stubble after harvest. Stubble has to be burned, extracted or flooded and mashed before the next growing season. In the 1980s, many growers used to drain off their water and burn the residual vegetation leading to grim air quality problems in the Valley.

Figure D.1 Doug McGeoghegan and the first rice roller. (Photo courtesy of Doug McGeoghegan).

Doug was then Chairman of the California Rice Industry, whose members went to Sacramento and helped negotiate the "Rice Straw Burning Reduction Act" in 1991. Their goal was to affect the speed of "phase-down" to give farmers some time to work on alternatives. McGeoghegan and his father, working with Fritz Reid and Jack Payne of Ducks Unlimited, constructed an experimental roller to try on the harvested fields as they sought an alternative to burning straw.

The rice roller worked well for acreages where fall water was available, thereby enlisting Pacific Flyway waterfowl to help with seed removal and decomposition. DU helped obtain a grant from The Wildlife Conservation Board of California to finance construction of additional rice rollers. Construction was accomplished at the McGeoghegan ranch shop. Units were then loaned to the five counties in the Sacramento Valley with substantial rice acreages. Rolling the rice straw eliminated the smoke and allowed water to remain on the land providing foraging habitat for waterbirds of all kinds. DU's Mark Petrie and others later estimated that roughly half of waterfowl food resources in the Central Valley came from harvested rice fields.

These prototype rollers provided a mechanical means to crush the stubble. Techniques have evolved. Farmers used to pull the rollers with old tracklayers. Others have now designed rollers to replace the big tires on articulating tractors. Those, along with a tow-along roller for the area between the wheels, cover the ground quicker and do an even better job of incorporating crop residue.

(Nominated by Ruth Ostroff)

Wendy Eliot

Working with the San Francisco Bay Joint Venture, Wendy Eliot of the Sonoma Land Trust took the lead in securing private lands that came up for sale and adding them to the San Pablo Bay National Wildlife Refuge or the Sonoma County Parks and Open Space District. Some of these were the two Dickson Ranches that became the Sears Point Restoration Project, the Haire Ranch Unit of Skaggs Island, Tolay Creek and Tolay Ranch, Camp 4, and other wetland parcels. Wendy built trust with landowners so that when they were ready to sell, they worked directly with the Sonoma Land Trust rather than selling to developers.

(Nominated by Beth Huning)

Senator Diane Feinstein

Also working with the San Francisco Bay Joint Venture, Senator Feinstein facilitated brokering a deal to acquire the Cargill Salt ponds, over 4,000 acres in the South Bay to be added to San Francisco Bay National Wildlife Refuge and 1,400 acres of salt ponds in San Pablo Bay, which became the Green Island Unit of the Napa Sonoma Marshes State Wildlife Area. She then secured an annual $5 million congressional appropriation for wetland restoration in the area.

(Nominated by Beth Huning)

Ms. Margaret Willis

As the USFWS Wetlands Office was transitioning from Minnesota to Washington, Margaret Willis became a key administrative assistant, moving over from the Duck Stamp Office in 1991. Her focus was helping the staff implement NAWCA. She inherited the complex task of planning, coordinating, arranging, and reimbursing expenses for the travel of North American Wetlands Council members, state agency directors, NGO executives, and their staffs. Margaret quickly learned the travel preferences, requirements, and idiosyncrasies of these members who were usually consumed with the demands of their own organizations. Largely due to Margaret's efforts during the first ten years, Council members arrived to enjoy their meeting facilities, lodging, and support hassle free, allowing them to concentrate on their project selection work. Council members were effusive in their compliments for Margaret's efforts. Margaret, ever helpful, similarly arranged logistics for the NAWMP Plan Committee and their meetings as well often coordinating those arrangements with her counterparts in Canada or Mexico.

Bob Streeter recalls that Margaret was a person who came to work to work. She set that example for others, mentoring younger people about standards in the office. In an oral history interview, she recalled really liking the team in the office and what she was doing...and she learned a lot about wetlands too! Margaret retired after more than 21 years with the USFWS.

(Nominated by Bob Streeter and David Smith)

Dr. David K. Weaver

David, a respected manager from a park district in Minnesota, was recruited in the first year of NAWCA implementation to establish and manage the grants to states, NGOs, and Canadian and Mexican partners for projects approved by the Wetlands Council. Dave had to learn rapidly and work within the US government contracting rules to develop innovative and efficient processes to expedite the grants, especially for international projects. He also worked cooperatively with grantees to develop a monitoring and reporting system that met the Council's needs but was not overly burdensome for grantees. Dave developed positive rapport among the Council, the NAWMP Office, and grantees that assured successful outcomes for wetlands conservation.

(Nominated by Bob Streeter and David Smith)

Brent Wark

In the 1980s, Brent Wark was a DU Canada habitat biologist in Manitoba. When NAWMP pilot programs ensued, DUC began securing and enhancing complexes of small productive potholes in farmed landscapes. Brent was a proponent of these changes and with a team of agrologists, researchers and programmers helped develop computer planning models to guide habitat delivery. Planting nesting cover was Brent's forte. Building on experience with cover plantings in the upper Midwest by the USFWS, he worked with colleagues to establish upland cover and manage it across Prairie Canada. This included sourcing specialized machinery, developing native ecovar seed mixes, and implementing methods such as sculptured seeding and burn management to achieve the best results. Thanks to Brent's foresight, reliable sources of locally adapted affordable native seed were established. Brent co-authored influential technique manuals and delivered workshops that aided in program delivery across Prairie Canada and beyond.

Brent's expertise also resulted in the establishment of Native Plant Solutions in 2001. NPS is a for-profit business arm of DUC that establishes native cover and water management projects for businesses, institutions, cities and agencies. Although many projects take place in urban environments, they have allowed DUC staff to further improve restoration methods. NPS profits are channeled back into NAWMP programs, and projects serve to educate the public about the economic and environmental values of small wetlands and naturalized uplands.

These initiatives were new at the time, and Brent and other trailblazers faced skepticism. However, their determination, forthrightness, and measured approach produced results and won the day. We owe Brent Wark thanks for his innovation, energy, and leadership in the practical delivery of early NAWMP programs.

(Nominated by Rick Wishart and Wayne Cowan)

Bruce Taylor

Bruce Taylor was an effective, innovative, and long-serving Oregon Coordinator for the Pacific Coast Joint Venture. Working with Carey Smith and his team, Bruce helped develop innovative land trust partnerships that contributed materially to meeting PCJV goals. The trusts are local grassroots organizations that get things done on the ground. Bruce appreciated Cary's approach of carefully selecting state coordinators and then giving them the latitude to do whatever was needed to help the JV partners succeed in conserving habitat.

Taylor contributed in a very different way as well. He formerly worked as a congressional aide in Washington and helped lead as the NAWMP JVs begin their annual "lobbying" trips to Capitol Hill. Their quest for earmarked financial support resulted in millions of dollars designated for JV work.

(Nominated by Ian Barnett and Carey Smith)

Kathy Dickson

Kathy Dickson began her career with the Canadian Wildlife Service in Yellowknife. Later, she moved to Ottawa to work with Steve Wendt, and the two became a formidable pair that significantly influenced the evolution and implementation of NAWMP, and particularly its species Joint Ventures. While Steve was the public face of NAWMP in Canada, Kathy was the tireless professional in the background making things happen. She served on one or more of the Species Joint Ventures continuously from 1989 until her retirement in 2014, first serving as the Black Duck Joint Venture coordinator (1989–1995), later the Canadian Sea Duck Joint Venture

Figure D.2 Kathy Dickson banding birds on Baffin Island. (Photo courtesy of Jim Leafloor CWS).

coordinator (1999–2014) and as a key technical committee leader of the Arctic Goose JV from 1996 onwards.

A self-described introvert, Kathy was highly respected by her colleagues across North America. Her soft-tempered nature belied a highly determined personality who had waterfowl conservation and management as her top priority. She had a unique ability to work effectively with everyone, from field technicians to high-ranking executives. She harnessed these qualities to contribute immensely to the light goose overabundance issue and implementation of Special Conservation Measures in Canada. She also played a crucial role in managing Canada's response to the Atlantic Population Canada goose decline and subsequent harvest restrictions. Her influence with the Sea Duck Joint Venture's strategic planning was also notable among many other contributions. But what she loved most was spending time in the field. She was part of the goose banding crew at the Great Plains of the Koukdjuak, on southern Baffin Island, for over two decades, marking thousands of snow geese, brant and some King eiders – including an eider that was recovered in western Russia!

In nominating Kathy for a NAWMP award years ago the cochairs of the AGJV wrote that they "had relied on her deep knowledge, expertise, and understanding of the legalities of waterfowl conservation, of NAWMP, and of waterfowl science and policy, for many years. Her commitment and dedication to this international partnership has been outstanding". Few people have had such an influence on NAWMP in Canada.

(Nominated by Eric Reed, Jim Leafloor, and Scott Gilliland)

Index

Note: **Bold** page numbers indicate tables; ***bold italic*** page numbers indicate text boxes; *italic* page numbers indicate figures; and page numbers followed by 'n' indicate notes.

For Product Safety Concerns and Information please contact our EU
representative GPSR@taylorandfrancis.com
Taylor & Francis Verlag GmbH, Kaufingerstraße 24, 80331 München, Germany

www.ingramcontent.com/pod-product-compliance
Ingram Content Group UK Ltd.
Pitfield, Milton Keynes, MK11 3LW, UK
UKHW050230200926
475702UK00020B/2951
* 9 7 8 1 0 4 1 2 2 4 3 8 9 *